AF411453

New Traits of Agriculture/Food Quality Interface

New Traits of Agriculture/Food Quality Interface

Editor

Alessandra Durazzo

MDPI • Basel • Beijing • Wuhan • Barcelona • Belgrade • Manchester • Tokyo • Cluj • Tianjin

Editor
Alessandra Durazzo
CREA-Research Centre for
Food and Nutrition
Italy

Editorial Office
MDPI
St. Alban-Anlage 66
4052 Basel, Switzerland

This is a reprint of articles from the Special Issue published online in the open access journal *Agriculture* (ISSN 2077-0472) (available at: https://www.mdpi.com/journal/agriculture/special_issues/Agriculture_Food_Quality_Interface).

For citation purposes, cite each article independently as indicated on the article page online and as indicated below:

LastName, A.A.; LastName, B.B.; LastName, C.C. Article Title. *Journal Name* **Year**, *Volume Number*, Page Range.

ISBN 978-3-0365-4053-5 (Hbk)
ISBN 978-3-0365-4054-2 (PDF)

Contents

About the Editor

Alessandra Durazzo was awarded a Master's degree in Chemistry and Pharmaceutical Technology cum laude in 2003, and a PhD in Horticulture in 2010. Since 2005, she has been a Researcher at the CREA-Research Centre for Food and Nutrition. The core of her research is the study of the chemical, nutritional and bioactive components of food, with particular regard to the wide spectrum of substances classes and their nutraceutical features. For several years, she was involved in national and international research projects evaluating several factors (agronomic practices, processing, etc.) that affect food quality, the levels of bioactive molecules and their total antioxidant properties, as well as their possible impact on the biological role played by bioactive components in human physiology. Her research activities address the development, management and updating of bioactive compounds, nutraceuticals, and dietary supplements databases; particular attention is paid to the harmonization of analytical procedures and classification and codification of dietary supplements.

agriculture

Editorial

New Traits of Agriculture/Food Quality Interface

Alessandra Durazzo

CREA—Research Centre for Food and Nutrition, Via Ardeatina 546, 00178 Rome, Italy; alessandra.durazzo@crea.gov.it

Citation: Durazzo, A. New Traits of Agriculture/Food Quality Interface. *Agriculture* **2021**, *11*, 1182. https://doi.org/10.3390/agriculture11121182

Received: 20 October 2021
Accepted: 10 November 2021
Published: 23 November 2021

Publisher's Note: MDPI stays neutral with regard to jurisdictional claims in published maps and institutional affiliations.

1. Introduction

There is a close link between food and territory. The current challenges can be found in precision agriculture and food metrology from the perspective of monitoring and improving food quality and addressing the promotion of diversity of agroecosystems and diets. The study of Durazzo et al. [1] describes the metrology, agriculture, and food relationship by quantitative literature research analysis as a useful tool for identifying emerging research directions, collaboration networks, and suggestions for more in-depth literature searches.

Research studies describing factors affecting food quality—such as agronomic conditions, post-harvest elicitors, cultivar selection, harvest date, or environmental influences—are described. Sustainable environmental and innovative practices should be promoted.

Hu et al. [2] have compared the grain quality and starch physicochemical properties between Japonica rice cultivars with different contents of amylose, as affected by nitrogen fertilization: the selection of a low amylose content japonica rice cultivar grown without nitrogen fertilizer can reduce the amylose and protein contents, as well as improving the pasting properties, starch retrogradation properties and eating quality of the cooked rice.

Feledyn-Szewczyk et al. [3], by studying the milling and baking quality of spring wheat (*Triticum aestivum* L.) from organic farming, concluded on the basis of the 3-year study results that the most useful variety for organic production is Arabella, followed by Brawura, Izera, Kandela, Katoda, KWS Torridon, Waluta, and Zadra.

Boussahel et al. [4] described phenolic compounds and antioxidant properties of monocultivar olive oils from Northeast Algeria.

Ma et al. [5] showed how reasonable nitrogen fertilizer management improves rice yield and quality under a rapeseed/wheat–rice rotation system.

Jahangirlou et al. [6] showed a case study from the semiarid conditions of Iran on the study of grain quality of maize cultivars as a function of planting dates, irrigation, and nitrogen stress.

Botella et al. [7] studied bioactive compounds of tomato fruit in response to salinity, heat, and their combination: the results show the viability of exploiting abiotic stresses and their combination to obtain tomatoes with increased levels of health-promoting compounds.

Advanced techniques, such as mass spectrometry, infrared, and Raman spectroscopy, in the monitoring and control of foodstuffs for modeling the agrofood system should be considered. Innovative green technologies should be taken into account. Targeting food approaches should be promoted. Chemometrics applications are welcome.

For instance, Kędzierska-Matysek et al. [8] proposed a chemometric approach for studying the relationships linking the color and elemental concentrations of blossom honeys with their antioxidant activity.

Guilherme et al. [9] showed a comparative study on phenolics and antioxidant activity of green and red sweet peppers from organic and conventional agriculture: the results confirmed that the production system, as well as the maturation stage affect phenolic content and antioxidant activity, allowing their possible use as maturation–production biomarkers. Moreover, the authors mark how the chemometric approach reveals how phenolic composition together with the antioxidant capacities could be used to differentiate the production system and the maturation stage of sweet peppers.

Da Silva et al. [10] studied the effects of camu-camu (*Myrciaria dubia*) powder on the physicochemical and kinetic parameters of deteriorating microorganisms and *Salmonella enterica* subsp. *enterica Serovar Typhimurium* in refrigerated vacuum-packed ground beef: camu-camu powder addition decreases the lipid oxidation of vacuum-packed ground beef, although it influences color features leading to a decreased red hue in the camu-camu powder-containing meat. Regarding antimicrobial activity, camu-camu powder does not interfere in *S. enterica ser. Typhimurium* behavior and does not extend the shelf-life of vacuum-packed ground beef based on the concentration of certain spoilage microorganisms, acting only on the kinetic bacterial behavior parameters [10].

Souza et al. [11] showed the physicochemical properties and antioxidant activity of spouted bed dried *Rosmarinus officinalis* extract.

The perspective of Durazzo et al. [12] is focused on antioxidant properties of bee products derived from medicinal plants as beekeeping sources.

It is worth mentioning that the preliminary study of Das et al. [13] evidenced how microbiome analysis of the rhizosphere from wilt-infected pomegranate reveals complex adaptations in *Fusarium* by demonstrating the capabilities of the whole metagenome sequencing approach for rapid identification of potential key players of wilt disease pathogenesis wherein the symptomatology is complex.

The papers included in this Special Issue cover a broad range of interdisciplinary aspects from agriculture and biology, chemistry, and nutrition. It is hoped that this Special Issue will stimulate further research in these areas.

Funding: This research received no external funding.

Conflicts of Interest: The authors declare no conflict of interest.

References

1. Durazzo, A.; Souto, E.B.; Lombardi-Boccia, G.; Santini, A.; Lucarini, M. Metrology, Agriculture and Food: Literature Quantitative Analysis. *Agriculture* **2021**, *11*, 889. [CrossRef]
2. Hu, Y.; Cong, S.; Zhang, H. Comparison of the Grain Quality and Starch Physicochemical Properties between *Japonica* Rice Cultivars with Different Contents of Amylose, as Affected by Nitrogen Fertilization. *Agriculture* **2021**, *11*, 616. [CrossRef]
3. Feledyn-Szewczyk, B.; Cacak-Pietrzak, G.; Lenc, L.; Gromadzka, K.; Dziki, D. Milling and Baking Quality of Spring Wheat (*Triticum aestivum* L.) from Organic Farming. *Agriculture* **2021**, *11*, 765. [CrossRef]
4. Boussahel, S.; Di Stefano, V.; Muscarà, C.; Cristani, M.; Melilli, M. Phenolic Compounds Characterization and Antioxidant Properties of Monocultivar Olive Oils from Northeast Algeria. *Agriculture* **2020**, *10*, 494. [CrossRef]
5. Ma, P.; Lan, Y.; Lv, X.; Fan, P.; Yang, Z.; Sun, Y.; Zhang, R.; Ma, J. Reasonable Nitrogen Fertilizer Management Improves Rice Yield and Quality under a Rapeseed/Wheat–Rice Rotation System. *Agriculture* **2021**, *11*, 490. [CrossRef]
6. Jahangirlou, M.; Akbari, G.; Alahdadi, I.; Soufizadeh, S.; Parsons, D. Grain Quality of Maize Cultivars as a Function of Planting Dates, Irrigation and Nitrogen Stress: A Case Study from Semiarid Conditions of Iran. *Agriculture* **2020**, *11*, 11. [CrossRef]
7. Botella, M.; Hernández, V.; Mestre, T.; Hellín, P.; García-Legaz, M.; Rivero, R.; Martínez, V.; Fenoll, J.; Flores, P. Bioactive Compounds of Tomato Fruit in Response to Salinity, Heat and Their Combination. *Agriculture* **2021**, *11*, 534. [CrossRef]
8. Kędzierska-Matysek, M.; Teter, A.; Stryjecka, M.; Skałecki, P.; Domaradzki, P.; Rudaś, M.; Florek, M. Relationships Linking the Colour and Elemental Concentrations of Blossom Honeys with Their Antioxidant Activity: A Chemometric Approach. *Agriculture* **2021**, *11*, 702. [CrossRef]
9. Guilherme, R.; Aires, A.; Rodrigues, N.; Peres, A.; Pereira, J. Phenolics and Antioxidant Activity of Green and Red Sweet Peppers from Organic and Conventional Agriculture: A Comparative Study. *Agriculture* **2020**, *10*, 652. [CrossRef]
10. Da Silva, J.L.; Cadavez, V.; Lorenzo, J.M.; Figueiredo, E.E.d.S.; Gonzales-Barron, U. Effects of Camu-Camu (Myrciaria dubia) Powder on the Physicochemical and Kinetic Parameters of Deteriorating Microorganisms and Salmonella enter-ica Subsp. enterica Serovar Typhimurium in Refrigerated Vacuum-Packed Ground Beef. *Agriculture* **2021**, *11*, 252. [CrossRef]
11. Souza, C.R.F.; Baldim, I.; Bankole, V.O.; Da Ana, R.; Durazzo, A.; Lucarini, M.; Cicero, N.; Santini, A.; Souto, E.B.; Oliveira, W.P. Spouted Bed Dried *Rosmarinus officinalis* Extract: A Novel Approach for Physicochemical Properties and Antioxidant Activity. *Agriculture* **2020**, *10*, 349. [CrossRef]

12. Durazzo, A.; Lucarini, M.; Plutino, M.; Pignatti, G.; Karabagias, I.K.; Martinelli, E.; Souto, E.B.; Santini, A.; Lucini, L. Antioxidant Properties of Bee Products Derived from Medicinal Plants as Beekeeping Sources. *Agriculture* **2021**, *11*, 1136. [CrossRef]

13. Das, A.J.; Ravinath, R.; Usha, T.; Rohith, B.S.; Ekambaram, H.; Prasannakumar, M.K.; Ramesh, N.; Middha, S.K. Microbiome Analysis of the Rhizosphere from Wilt Infected Pomegranate Reveals Complex Adaptations in Fusarium—A Preliminary Study. *Agriculture* **2021**, *11*, 831. [CrossRef]

Article

Spouted Bed Dried *Rosmarinus officinalis* Extract: A Novel Approach for Physicochemical Properties and Antioxidant Activity

Claudia R. F. Souza [1], Iara Baldim [1,2], Victor O. Bankole [1], Raquel da Ana [3], Alessandra Durazzo [4], Massimo Lucarini [4], Nicola Cicero [5,6], Antonello Santini [7,*], Eliana B. Souto [2,3,*] and Wanderley P. Oliveira [1,*]

[1]   Faculty of Pharmaceutical Sciences of Ribeirão Preto, University of São Paulo, Av. do Café s/n, Ribeirão Preto 14040-903, SP, Brazil; souzacrf@gmail.com (C.R.F.S.); iara.baldim@usp.br (I.B.); vobankole@usp.br (V.O.B.)

[2]   CEB—Centre of Biological Engineering, University of Minho, Campus de Gualtar, 4710-057 Braga, Portugal

[3]   Department of Pharmaceutical Technology, Faculty of Pharmacy, University of Coimbra, Pólo das Ciências da Saúde, Azinhaga de Santa Comba, 3000-548 Coimbra, Portugal; quele.ana@gmail.com

[4]   CREA-Research Centre for Food and Nutrition, Via Ardeatina 546, 00178 Rome, Italy; alessandra.durazzo@crea.gov.it (A.D.); massimo.lucarini@crea.gov.it (M.L.)

[5]   Department of Biomedical and Dental Sciences and Morphofunctional Imaging, University of Messina, Polo Universitario Annunziata, 98168 Messina, Italy; ncicero@unime.it

[6]   Science4Life, a University of Messina Spin-Off Company, V.le Leonardo Sciascia Coop Fede Pal. B, 98168 Messina, Italy

[7]   Department of Pharmacy, University of Napoli Federico II, Via D. Montesano 49, 80131 Napoli, Italy

*   Correspondence: asantini@unina.it (A.S.); ebsouto@ff.uc.pt (E.B.S.); wpoliv@fcfrp.usp.br (W.P.O.)

Received: 15 July 2020; Accepted: 10 August 2020; Published: 11 August 2020

Abstract: In this study, a conical-cylindrical spouted bed dryer with Teflon® beads as spouting material was used for producing powdered rosemary (*Rosmarinus officinalis* L.) extract. The influence of the inlet drying gas temperature (T_{gi}) and the percentage ratio between the feed rate of concentrated liquid extract by the maximum evaporation capacity of the spouted bed (W_s/W_{max}) on selected physicochemical properties of the finished products were investigated. Antioxidant properties of the concentrated liquid extract and dried extracts were also evaluated by the 2.2-diphenyl-1-picrylhydrazyl radical scavenging (DPPH•) and lipid peroxidation induced by Fe^{2+}/citrate (LPO) methods; and compared with the values obtained for a lyophilized extract (used as a control). Colloidal silicon dioxide (Tixosil® 333) and maltodextrin (DE 14) at a 2:1 ratio was added to the concentrated extract before drying (4.4% w/w) to improve the drying performance. The drying variables W_s/W_{max} and T_{gi} have statistically significant influence on total polyphenols and total flavonoid contents of the dried powders. The concentrated extract (on dry basis—being absolute solid content) showed superior antioxidant activity (AA) compared to both the spouted bed dried and the lyophilized extracts; exhibiting IC50 values of 0.96 ± 0.02, 2.16 ± 0.04 and 3.79 ± 0.05 µg mL^{-1} (DPPH• method) and 0.22 ± 0.01, 1.31± 0.01 and 2.54 ± 0.02 µg mL^{-1} (LPO method), respectively. These results of AA are comparable to values obtained for quercetin, a flavonoid compound often used as a reference standard due to its potent antioxidant activity; with IC_{50} of 1.17 µg mL^{-1} (DPPH•) and 0.22 µg mL^{-1} (LPO). However, the dried rosemary extracts are about 13.5 times more concentrated than the initial concentrated extract (dry weight), with a concentration of total flavonoids and polyphenols compounds ranging from 4.3 to 12.3 and from 1.2 to 4.7 times higher than the concentrated extract values (wet basis). The AA per dry product mass was thus significantly higher than the values measured for concentrated extractive solution, irrespective of some losses of AA apparently due to the drying process.

Keywords: herbal medicinal product; dried extract; spouted bed drying; antioxidant activity; *Rosmarinus officinalis*; powder properties

1. Introduction

The use of herbal materials as medicines and food is as old as mankind. Plant-derived natural products are widely acclaimed to be cheap and safe, currently making them a preferred source of medicines and nutritional supplements [1–5]. These products are necessarily processed from their crude natural form to an acceptable finished product for reasons of convenience, standardization, stability enhancement, and improvement in physicochemical characteristics among other considerations. However, several factors affect the quality of herbal products, such as the climate, harvest period, and post-harvest treatments (e.g., drying and storage conditions) [6–11]. Therefore, a strict control of the steps involved in the production of a herbal product is needed to achieve consistent and appropriate level of bioactive substances, ensuring quality, efficacy and safety. The production control can be very difficult, since herbs and their preparations are a complex mixture of substances with varied physicochemical properties [8,9,12–16].

Herbal preparations are commonly marketed as a liquid, viscous preparations, or powders resulted from dried and comminuted plant materials (e.g., leaves, flowers, roots or the whole plant), or from the drying of an extractive solution. Compared to conventional liquid forms, the dried extracts show several advantages compared to liquid preparations such as lower costs of transport and storage, high concentration and stability of active substances. However, the processing conditions used during the manufacture of the dried extract affect the physicochemical properties of the product and might cause varying degrees of loss of active compounds [17].

Drying techniques, including spray drying (SD), freeze-drying, and fluidized beds, have been commonly used in the production of dry powders including herbal extracts [17–19], with spray drying being particularly used in herbal processing industries [20]. Various studies reported in literature focus on the search of alternative and innovative drying methods for herbal extracts drying, such as the spouted bed technology (SB). Spouted bed drying is considered a cheap, efficient, and reproducible method for drying herbal extracts, capable to produce high-quality powdered products suitable for pharmaceutical/nutraceutical applications [17,21,22]. A description of the fundaments of the spouted beds and their application in drying can be found elsewhere [23,24].

Polyphenols are an important class of plant metabolites, which attract high interest linked to their postulated health protecting properties, in particular their antioxidative activity, attributed to the ability to scavenge free radicals and/or to prevent oxidation of low-density-lipoprotein [8,9,25–27]. However, adequate intakes and absorption rate of polyphenols would be recommended to ensure any beneficial outcome [28].

Rosmarinus officinalis L. (rosemary) is a common aromatic plant grown in many parts of the world, being native of the Mediterranean region. It can be used as ornamental plant, spice in cooking, preservative in food, and medicinal plant. The profile of rosemary bioactive compounds includes principally phenolic compounds such as caffeic acid, chlorogenic acid, carnosic acid, carnosol, rosmarinic acid, and ursolic acid found in the extract; and constituents identified in the essential oil such as α-pinene, oleanolic acid, camphor, eucalyptol, rosmadial, rosmanol, rosmaquinones A and B, secohinokio, and derivatives of eugenol and luteolin [29,30].

Rosemary extracts are commercially available in Europe and USA for use as natural antioxidants in the food industry. They have received Generally Recognized as Safe (GRAS) status from the US Food and Drug Administration. The antioxidant activity of rosemary is linked to the high content of polyphenols, such as carnosic acid, carnosol, rosmanol, epirosmanol, and methyl carnosate, but over 90% of the antioxidant profile is attributed to carnosol and carnosic acid. The development of high quality standardized dried extracts of rosemary is a highly noteworthy study.

A comparison between SD and SB for drying of rosemary extract has been previously reported [31], in which the feasibility of the process at very strict processing conditions has been demonstrated.

The aim of the present work was thus to determine the effects of the SB operating variables, namely, the inlet drying gas temperature (T_{gi}) and the percentage ratio between the feed rate of concentrated liquid extract by the maximum evaporation capacity of the spouted bed (W_s/W_{max}), on concentration of some antioxidant compounds (flavonoids and polyphenols) and on product moisture, using statistical methods. The antioxidant activities of the concentrated liquid extract and SB dried extract were also evaluated by both DPPH$^\bullet$ and lipid peroxidation induced by Fe^{2+}/citrate (LPO) methods; and compared with the values obtained from a lyophilized extract (used as a reference). To the best of our knowledge this is the first time that antioxidant activity of SB dried rosemary extract was determined.

2. Materials and Methods

2.1. Herbal Material and Reagents

Dried leaves of *Rosmarinus officinalis* were purchased from Oficina de Ervas Pharmacy (Ribeirão Preto, São Paulo, Brazil). The dried vegetable material was milled in a knife mill (model MA 680, Marconi, Piracicaba, São Paulo, Brazil) until all particles passed through an 800 µm sieve. The reagents, standard materials, and drying carriers utilized in this study include ethyl acetate, methanol, isobutanol, acetone, aluminum chloride, ethanol, chloridric acid, *o*-phosphoric acid, sodium hydroxide, potassium chloride, dimethyl sulfoxide (DMSO) and sucrose obtained from Labsynth (Vinhedo, Brazil), hexamethylenetetramine, sodium tungstate, phosphomolibidic acid obtained from Vetec, (Duque de Caxias, Brazil), colloidal silicon dioxide (Tixosil® 333, Rhodia, São Paulo, Brazil), maltodextrin (Mor Rex® 1914, Corn Products, São Paulo, Brazil), dehydrated quercetin, gallic acid, 2.2-diphenyl-1-picrylhydrazyl (DPPH$^\bullet$), ammonium ferrous sulfate hexahydrate, thiobarbituric acid (TBA) and MDA (malondialdehyde tetrabutylammonium salt, purity >98%) from Sigma-Aldrich (Steinheim, Germany).

2.2. Experimental Procedure

2.2.1. Extraction of the Bioactive Compounds from Rosemary Leaves

The extractive solution was obtained by dynamic maceration in a bench top extraction system comprising a jacketed vessel coupled to a mechanical stirring unit. The vessel was connected to a thermostatic water bath (Marconi, MA-184), which circulate water at constant temperature; thereby maintaining desired extraction temperature. As described by Souza et al. [31], the extractive solution was obtained using a hydroalcoholic solvent (70% *v/v* ethanol) at the following conditions: extraction time of 1 h; extraction temperature of 50 °C; and plant to solvent mass ratio of 0.2. The crude extractive solution was vacuum filtered at 650 mm Hg through filter paper (grade 80G) and concentrated three times in a rotary evaporator at a temperature of 50 °C under a vacuum pressure of 650 mm Hg.

2.2.2. Physicochemical Characterization of the Powdered Raw Herbal Material and Extractive Solution

The powdered herbal material was characterized by determination of the mean particle diameter, moisture content, total extractable matter and total flavonoids and polyphenols content. The concentrated extract was characterized for its density (pycnometry), solids concentration, and total flavonoids and polyphenols contents.

The moisture content (X_p) of the powdered herbal material was carried out by gravimetric analysis (using the oven method). About 2 g of herbal material weighed into a dry Petri dish was dried to constant weight in an open tray oven operated at 102 ± 1 °C. X_p was determined as the percentage ratio of the difference between the wet and dried herbal material by the dried mas of herbal material (dry basis).

The total flavonoids and polyphenols contents were determined by UV-Vis spectrophotometry, using a HP 8453 spectrophotometer running the software HP Chem-Station® (Agilent Life Sciences and Chemical Analysis, Santa Clara, CA, USA). The procedure for flavonoid quantification involves the hydrolysis of the glycosides, the extraction of the flavonoids with ethyl acetate and the color development with the addition of a solution of aluminum chloride [32,33]. The reaction between aluminum chloride ($AlCl_3$) and flavonoids results in a bathochromic shift of absorption wavelength of flavonoids. In this manner, concentration of flavones and flavonols in the extractive materials can be determined spectrophotometrically, measuring the absorbance at λmax = 425 nm, after 30 min of the addition the $AlCl_3$ solution. Other phenolic compounds can equally complex with $AlCl_3$, but show λmax at 434 nm. Quercetin, a well-recognized flavonol, was used as reference standard; and the results expressed as quercetin equivalent (QE). The quantification of the total polyphenols content was determined using the Folin-Denis method [34]. The procedure is based on the reduction of the phosphomolybdic-phosphotungstic acid by the polyphenols in a basic medium producing a dark blue color measured spectrophotometrically at 750 nm [31]. Gallic acid, a standard polyphenolic acid, was used as the reference substance, being the results expressed as gallic acid equivalent (GAE).

The solids content (Cs) of the concentrated extract was carried out by gravimetric analysis (using the oven method). About 2 g of liquid extract weighed into a dry Petri dish was evaporated to constant weight in an open tray oven operated at 102 ± 1 °C and the solid content determined by the ratio between the dry mass by the corresponding mass of the liquid extract. Results are expressed as average of three determinations (±standard deviation).

2.2.3. Spouted Bed Drying

Experimental Apparatus

Powdered extract of rosemary was obtained from the extractive solution following incorporation of adjuvants denominated as drying carrier. Previous experiments by our group evaluating different proportions of the drying carriers showed that the best ratio for spouting bed drying (under the operating conditions studied in this work) was 2:1 (40:20 Tixosil®:maltodextrin) [31]. The drying carrier was added to the concentrated extractive solution at a proportion of 4.4% *w/w*. The concentrated extract plus carrier were standardized to a solids content of 11.3%. The drying runs were carried out in a stainless steel, conical-cylindrical spouted bed, constituted by a conical base, with internal angle of 40° and inlet orifice diameter of 33 mm, connected to a cylindrical column with diameter of 150 mm and height of 400 mm. The upper part of the equipment is constituted by another cone and a powder collecting system (cyclone). The drying gas, heated by an electric heater (total power of 5000 W), was supplied into the spouting chamber through a 7.5 hp blower. Teflon® beads having a mean diameter of 5.45 mm, density of 2160 kg m^{-3}, surface area of 5.27 cm^2 g^{-1} and shape factor of 0.96 were used as spouting particles. For spouted bed drying of pharmaceutical products, Teflon beads are an excellent choice, due to its inert nature, thermal stability, low coefficient of friction, insolubility and lack of toxicological effects [35]. The extract feed system consists of a 0.8 mm double fluid atomizer with internal mixing (installed at the top of the fountain), a peristaltic pump and an air compressor. A temperature control system, thermocouples, and rotameter were employed in the equipment instrumentation.

Drying Procedure

Drying operation started with the feed (through the inlet entrance orifice), of heated drying air to the spouted bed previously loaded with Teflon® beads (static bed height of 14 cm). At the desired drying temperature, the concentrated extract incorporated with the respective drying carrier was fed at the top of the fountain of the spouting bed through the double fluid atomizer, together with the atomizing air, at a preset flow rate. Measurements of the outlet gas temperature, T_{go}, were taken at regular intervals in order to detect the moment when the dryer reach the steady state (±15 min). Samples of the dried extract were withdrawn and used for determination of its physicochemical properties and AA. The studied process parameters were the inlet drying gas temperature, T_{gi} (80 and 150 °C); and the percentage ratio of the mass feed flow rate of the concentrated extract to the evaporation capacity of the dryer, W_s/W_{max} (15, 45 and 75%) (Table 1). Study on the evaporation capacity for this dryer have been reported elsewhere [31]. The mass flowrate of the drying gas was maintained at 0.0340 kg s^{-1}, corresponding to 1.4 times the minimum spouting mass flowrate (Q_{ms}). The atomizing air feed flow rate was fixed at 20 L min^{-1} at a pressure of 196.1 kPa.

Table 1. Spouted bed drying parameters used in manufacture of the dried extract.

T_{gi} (°C)	W_s/W_{max} (%)	W_s (g min^{-1})
80	15	6.0
80	45	18.0
80	75	30.0
150	15	10.0
150	45	33.0
150	75	49.0

T_{gi}, inlet drying gas temperature; W_s/W_{max}, the percentage ratio of the mass feed flow rate of the concentrated extract to the evaporation capacity of the dryer; W_s, mass feed flow rate of the concentrated extract (g min^{-1}).

A portion of the crude concentrated extract was lyophilized and the product used as a reference sample, since dehydration occurred at a low temperature. The lyophilization was carried out in a Thermo Fisher Scientific freeze dryer model SNL 108 (Waltham, MA, USA) containing a Micromodulyo 1.5 L freeze-drying unit (305 × 330 × 432 mm), stainless steel condenser; 1/4 hp compressor and 0.30 kw power; ultra-vacuum pump VLP 195 FD-115 and freeze-drying vials with independent valves. No adjuvant was added to the concentrated extract prior to dehydration.

2.2.4. Physicochemical Characterization of Dried Rosemary Extract

The physicochemical characterization of the spouted bed dried rosemary extract was carried-out by measurement of product moisture content, total flavonoids, total polyphenols contents, extract bulk density (ρ_b); tapped density ($\rho_{,1250}$), and the flow and compressibility parameters; Hausner ratio ($I_{Hausner} = \rho_{,1250}/\rho_b$) and Carr index ($I_{Carr} = (\rho_{,1250} - \rho_b)/\rho_b$), and product morphology.

Moisture content, total flavonoids and polyphenols contents were determined by the methods described in Section 2.2.2.

The bulk density (ρ_b) was obtained as the ratio of the mass of a powder sample to the volume occupied by the same sample when placed freely in a measuring cylinder without any compaction; while the tapped density ($\rho_{,1250}$) was determined by a similar ratio using the volume occupied by the powder after the cylinder was tapped 1250 times through a distance of 3 cm using a tapped density tester mod. TDT 22 (Caleva, Frankfurt, Germany). Determinations were carried out in triplicates [36].

The product morphology was assessed from photomicrographs of the spouted bed dried extracts acquired in a scanning electronic microscope (SEM) with magnification of 1000×. Samples were placed on an aluminum foil placed on a carbon tube. An auto fine coater was used to coat the samples by sputtering with platinum and analyzed on a Quanta 200-ESEM system (FEI, Eindhoven, The Netherlands) using gaseous secondary electron detector holding the chamber at 27 °C and imaging at 20 kV and 600 Pa.

2.2.5. Antioxidant Activity of the Rosemary Extracts

The DPPH$^\bullet$ scavenging and the lipid peroxidation assays were the methods selected to estimate the antioxidant activity of the concentrated extract, lyophilized extract and the spouted bed dried rosemary extract.

DPPH$^\bullet$ Radical Scavenging Assay

Preliminary tests were carried out to determine concentration ranges of the concentrated extract and the dehydrated test samples required for the DPPH$^\bullet$ assay. Ten μL of the concentrated extract at various dilutions (1:10; 1:20; 1:50; 1:100 and 1:200) and of the dried rosemary extract (0.375–6.25 mg mL^{-1}) in ethanol were added to the reaction mixture containing 1 mL of 0.1 M acetate buffer pH 5.5, 1 mL of ethanol, and 0.5 mL of 250 µM of 2,2-diphenyl-1-picrylhydrazyl (DPPH$^\bullet$) in ethanolic solution. The concentrations of the concentrated extract and dried rosemary extracts in the reaction medium ranged from 0.15 to 2.93 µg mL^{-1} and from 0.94 to 15.6 µg mL^{-1} (dry basis), respectively. The change in absorbance was measured after 10 min at room temperature. The hydrogen donating ability of the dried extracts of rosemary to DPPH$^\bullet$ was determined from the change in absorbance at 517 nm using the spectrophotometer HP 8453 running the HP Chem-Station$^\circledR$ software [37]. All measurements were made in triplicate. The same test was also performed for the flavonoid quercetin (positive control), being a reference antioxidant substance, which present a dose-dependent response [38–40].

Lipid Peroxidation Method

The lipid peroxidation (microsomal lipid peroxidation induced by Fe^{+2}/citrate) was assayed by malondialdehyde generation [41] in the presence of different concentrations (determined in a preliminary experiment) of rosemary extracts. Mitochondria was prepared by standard differential centrifugation techniques as described by Rodrigues et al. [41]. The protein content was determined by the biuret reaction [42]. 10 µL of the concentrated extract at various dilutions (1:10 to 1:200) and of the dried rosemary extract at different concentrations (12.5 to 300 µg mL^{-1} in DMSO) were added to 1.0 mL of a reaction mixture (125 mmol L^{-1} sucrose, 65 mmol L^{-1} KCl and 10 mmol L^{-1} Tris-HCl, pH 7.4—medium I). Mitochondria was added to yield a final concentration of 1 mg of protein plus 50 µM (NH$_4$)$_2$Fe(SO)$_4$ and 2 mM sodium citrate for 30 min at 37 °C. The solids concentrations of the extractive solution and of dried extract in the reaction medium ranged from 0.15 to 6.25 µg mL^{-1} and from 0.38 to 6.25 µg mL^{-1}, respectively. 1 mL of 1% thiobarbituric acid (TBA) (prepared in 50 mM NaOH), 0.1 mL of 10 M NaOH and 0.5 mL of 20% H$_3$PO$_4$ were added to the reaction medium, followed by incubation for 20 min at 85 °C. Malondialdehyde-thiobarbituric acid (MDA-TBA) complex was extracted with 2 mL of isobutanol. The samples were then centrifuged at 1660× g for 10 min. The measurement was performed on the supernatant at 535 nm using a UV-Vis spectrophotometer. Two controls were used for this test, a positive control (without the samples) and a negative control (without iron). The blank was prepared in the same way as the reaction mixture, but without the mitochondria. All measurements were made in triplicate.

2.3. Experimental Design

The drying runs were conducted according to two-factors and three-levels (2 × 3) experimental design, aiming to determine the effects of the processing drying variables (factors) W$_s$/W$_{max}$ and T$_{gi}$ on total flavonoids and polyphenols contents, and on product moisture (Xp). The experimental data was submitted to an analysis of variance (ANOVA) test to detect the factors effects with statistical significance on the experimental responses (α = 0.05). Response surfaces were plotted using the Statistica 13.5 software (TIBCOTM, Palo Alto, CA, USA), to show the impact of the studied factors on the desired outcomes.

3. Results and Discussion

3.1. Physicochemical Characterization of the Powdered Herbal Material and of Extractive Solution

The vegetable material was milled to increase the surface area for greater solvent contact, thereby enhancing the extraction efficiency. The milling process resulted in powdered leaves with mean particle diameter of 0.3 mm (sieve), considered as moderately fine. The starting vegetable material showed moisture content of 10.4 ± 0.2 (% *w/w*), extractable matter of 22.2 ± 0.01 (% *w/w*), and total flavonoids and polyphenols contents of 3.32 ± 0.02 (mg.QE g^{-1}—dry basis), and 31.2 ± 0.31 (mg.GAE g^{-1}—dry basis).

The extraction of the bioactive compounds from the powdered leaves of *Rosmarinus officinalis* was carried in an extraction system at a controlled temperature of 50 °C, adequate for extraction but not to cause substantial degradation of the extracted bioactive compounds. Hydroalcoholic solvent was used for the extraction process to facilitate the extraction of the polar and non-polar substances with ethanol preferred due to its relatively low toxicity profile. The extractive solution obtained containing about 2.8% (*w/w*) of solids, was concentrated approximately 2.6 times by rotary evaporation of the solvent, thus increasing the solids content prior to spouted bed drying. The concentrated extract had a density of 0.98 ± 0.01 (g cm^{-3}), a solids content of 7.32 ± 0.06 (% *w/w*), total flavonoid contents of 11.87 ± 0.02 (mg.QE g^{-1}—dry basis) and a total polyphenols concentration of 146.07 ± 0.23 (mg.GAE g^{-1}—dry basis).

3.2. Spouted Bed Drying and Physicochemical Characterization of the Dried Rosemary Extracts

Although spouted bed driers have been previously employed to process vegetable drug materials and phytopharmaceuticals [31], the initial attempts at drying the concentrated rosemary extracts were challenging. High amounts of the atomized feed material adhered to the surfaces of the inert material and the dryer wall, resulting in dryer collapse within a short processing time. In an attempt to solve this problem, drying carriers were added to the concentrated extract before the drying procedure [31]. A mixture of colloidal silicon dioxide (Tixosil 333®) and maltodextrin DE 14 at proportion of 40:20 (2:1) relative to the total solids content improved considerably the spouted bed dryer performance; with respect to atomization of the feed, adherence of extract on the inert material and dryer wall.

Table 2 shows the experimental results of total flavonoids and total polyphenols contents of the spouted bed dried extracts obtained at various processing conditions, together with the moisture content (Xp).

Table 2. Total flavonoids and total polyphenols contents of the spouted bed dried extracts.

T_{gi} (°C)	W_s/W_{max} (%)	T_F (mg.QE g^{-1}) *	T_p (mg.GAE g^{-1}) *	Xp (% *w/w*)
80	15	6.4 ± 0.05	13.1 ± 0.98	13.7 ± 1.12
80	45	5.6 ± 0.03	25.0 ± 1.87	8.7 ± 0.87
80	75	4.7 ± 0.04	36.6 ± 2.11	13.9 ± 0.95
150	15	5.0 ± 0.04	32.4 ± 1.13	2.0 ± 0.05
150	45	3.8 ± 0.05	48.6 ± 2.05	7.9 ± 0.55
150	75	3.7 ± 0.01	49.9 ± 1.29	11.1 ± 1.03

T_{gi}, inlet drying gas temperature; W_s/W_{max}, the ratio of the mass feed flow rate of the concentrated extract to the evaporation capacity of the dryer; * dry basis.

The experimental data of T_F, T_p, and X_p were subjected to analysis of variance (ANOVA) to detect any significant effects of the processing variables on the responses. The sum of the interaction effects between the variables were used as an estimate of the experimental error. ANOVA results reveal that the linear effects of the factors T_{gi} and W_s/W_{max} exerts statistically significant influence on both T_F (*p*-values of 0.026 and 0.033) and T_p (*p*-values of 0.026 and 0.033), with a F_{calc} at least 1.52 times higher the critical value for an $\alpha = 0.05$ ($F_{1,2-0.05} = 18.513$). On the other hand, the ANOVA results for Xp did not show statistical significance of the investigated factors, even using an $\alpha = 0.10$, with the maximum F_{calc} value (obtained for the linear effect of T_{gi}), at least 3.7 times lower than the critical tabled value ($F_{1,2-0.10} = 8.526$). Figure 1a,b present the response surface plots showing the effects of drying variables W_s/W_{max} and T_{gi} on total flavonoids and total polyphenols contents in the dried rosemary extracts, respectively. It can be seen from Figure 1a that increases in W_s/W_{max} and T_{gi} exert negative effects on T_F. The decrease observed for T_F might be linked to the thermal degradation of flavonoids, for example due to the occurrence of oxidative reactions or decomposition of thermolabile compounds, increasing concentration of other phenolic substances. The extent of degradation varies from compound to compound and is dependent on the dehydration method [43]. Flavones (e.g., luteolin) and flavonols (e.g., quercetin) concentration have been shown to reduce by up to 50% within 15 min of exposure to heat above 130 °C [44]. This behaviour could explain the corresponding increase of T_p as values of study variables increase (Figure 1b). For example, thermal degradation of quercetin resulted in accumulation of protocatechuic acid, a phenolic compound [45]. Such compounds might be detected by the quantification reaction of the Folin-Denis method [34], which measures total phenolics substances.

The phenolic compounds comprise a large number of organic molecules with heterogeneous structure, including the flavonoids. These substances present high reactivity to O_2 and have an important role as antioxidants [46].

The moisture content of biological dried materials and the water activity are important quality parameters, since they are linked to product stability. The physicochemical and microbiological stability tended to increase conversely with the water content (linked to water activity) of the dried products. The decrease in product moisture content implies a reduction on water activity; although there is a minimum value given by the monolayer moisture content. Below this value, the water is strongly bonded to material structure; reducing the rate of degradation reactions and hindering microbial growth. The monolayer moisture content is usually considered the lowest value to which food and agricultural material can be dried to ensure product stability [47,48]. Although ANOVA did not evidence any statistically significant effects of the drying variables on Xp ($F_{calc} < F_{1,2\text{-}0.10}$), the data presented in Table 2, give evidences that the increase of the inlet gas temperature and/or reduction in the feed flow rate of the concentrated extract tends to reduces the product moisture content; as can be seen qualitatively on Figure 2. For dried extracts of medicinal plants, the Brazilian Pharmacopoeia recommends maximum moisture content of 4%. From Table 2, it can be seen that only the dried extracts produced at $T_{gi} = 150\,°C$ and $W_s/W_{max} = 15\%$ falls in the recommended range. The powder production at this drying condition provided the higher powder production rate (yield $\approx 65\%$), perhaps due to the decrease of the adhesion tendency of the powder product to equipment wall; a positive effect of its low moisture content. Since the biological activities of rosemary extract is attributable to several phenolic constituents rather than a singular compound, it is appropriate that simultaneous retention of these bioactive principles is put in perspective in order to set the best processing conditions.

The characterization of the flow properties of the spouted bed dried extracts were carried out through the determination of the loosely packed bulk density, ρ_b, tapped bulk density, ρ_t, the real density, ρ_r, Carr's index, and the Hausner ratio. These properties have been shown to have direct relationship with the behavior of the product during storage, manipulation and industrial processing. The spouted bed dried rosemary extracts has an Hausner ratio ≈ 1.81 and compressibility index $\approx 45\%$; mean loosely packed bulk density around 430 ± 10 kg m^{-3} and tapped bulk density $\approx 780 \pm 20$, being the true density, measured with a helium pycnometer, equal to 1493.7 ± 2.9 kg m^{-3} [31]. These results are indicative of powders with poor flow and compression properties.

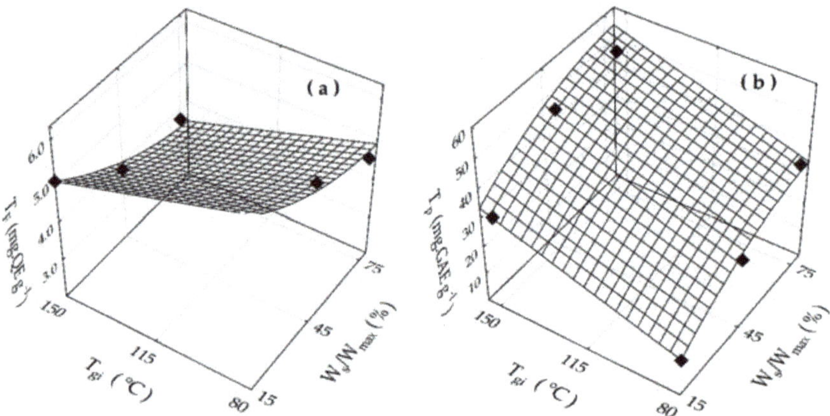

Figure 1. Response surface plots showing the effects of drying variables W_s/W_{max} and T_{gi} on total flavonoids (**a**), and total polyphenols content in the dried rosemary extract (**b**).

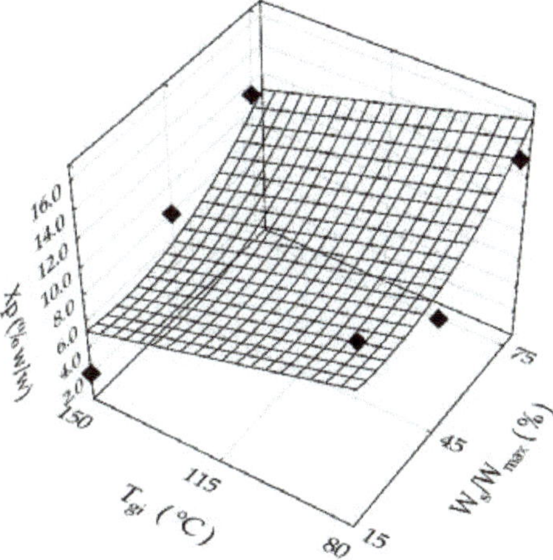

Figure 2. Qualitative surface plot showing the effects of drying variables W_s/W_{max} and T_{gi} on product moisture content (Xp) of the dried rosemary extract.

Figure 3 shows photomicrographs obtained by SEM at two distinct processing conditions; showing the presence of irregular to rounded powder particles, with polydisperse characteristics. Figure 3b corresponds to the product obtained at high W_s/W_{max} (45%) and low T_{gi} (80 °C), which shows Xp almost 4.4 times superior to the value measured for the product of Figure 3a (8.7 vs. 2.0% *w/w*—see Table 2). The bigger particles in Figure 3b might be caused by the increase in atomized droplets size and on Xp by the high feed flowrate of rosemary extract to the spouted bed [49], and lower drying temperature.

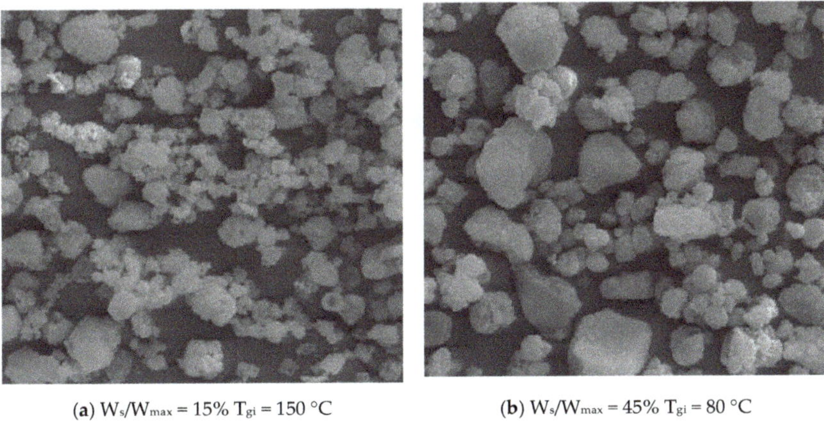

(**a**) W_s/W_{max} = 15% T_{gi} = 150 °C   (**b**) W_s/W_{max} = 45% T_{gi} = 80 °C

Figure 3. Typical SEM photomicrographs (magnification of 1000×) of spouted bed dried rosemary extracts produced at distinct operating conditions: (**a**) at W_s/W_{max} (15%) and T_{gi} (150 °C), (**b**) at W_s/W_{max} (45%) and T_{gi} (80 °C) (see Table 2).

3.3. Antioxidant Activity of the Concentrated and Dried Rosemary Extracts

Deleterious free-radical reactions have been linked to a variety of degenerative pathological conditions including cancer, autoimmune disorders, cataract, rheumatoid arthritis, cardiovascular and neurodegenerative diseases [50–53]. This effect has been described as a consequence of the involvement of reactive oxygen species in initiating autoxidation of cellular lipid membranes, which are mainly composed of polyunsaturated fatty acids, leading to cellular necrosis [54,55]. There is no single method that alone can evaluate the antioxidant activity accurately and quantitatively, since oxidative processes involve multiple active species, reaction characteristics, and mechanisms [56]. The literature recommends the simultaneous use of several methods based on different mechanisms of the inhibition of oxidation to examine the antioxidant capacity of a given sample. In this study, the evaluation of the antioxidant activity was carried out for the concentrated extract, SB dried extract obtained at W_s/W_{max} of 15% and T_{ig} of 150 °C (which presents good T_F and T_P values, and has moisture content below the maximum value recommended by the Brazilian pharmacopoeia), and for a sample of lyophilized extract (used as a control). The antioxidant activity was determined by two distinct methodologies, namely: DPPH$^\bullet$ scavenging (DPPH$^\bullet$) and the inhibition of the lipid peroxidation (LPO) methods. Figure 4a–c shows the results of DPPH$^\bullet$ inhibition produced by the concentrated extractive solution, spouted bed dried, and lyophilized extracts, respectively (dry basis). It can be seen from the Figures that DPPH$^\bullet$ inhibition exhibits a dose dependent behavior regardless of the sample analysed. The concentrated extract exhibited an IC50 = 0.96 ± 0.02 µg mL^{-1} (dry-basis), with a maximum DPPH$^\bullet$ inhibition of approximately 78 ± 1.6% at concentration of 2.93 ± µg mL^{-1}, being the steady state obtained thereafter (Figure 4a). Figure 4b,c show a maximum DPPH$^\bullet$ inhibitions of 79 ± 1.1 and 81 ± 1.3%, respectively for the spouted bed dried and the lyophilized extracts. These results were reached at concentrations of 15.6 ± 0.2 µg mL^{-1} of the dried extract sample; with IC50 values of 2.16 ± 0.04 and 3.79 ± 0.03 µg mL^{-1}, respectively.

The AA of the spouted bed dried extract determined by this method was lower than the one measured for the concentrated extract (in dry basis), but higher than the value determined for the lyophilized extract, used as a control. Nevertheless, considering that the spouted bed dried extract has in its composition 60% of drying carriers, the concentration of active constituents of the extract is about 62.5% diluted in relation to those present in the concentrated extract. Performing this correction, it can be seen that the loss of AA due to drying is less than 30%. Tests (control) performed with synthetic quercetin, a flavonoid with potent antioxidant properties, resulted in an IC_{50} of 1.17 µg mL^{-1} [57]. The rosemary extracts caused instantaneous decrease in the absorbance of DPPH$^\bullet$, in a manner similar to that presented by quercetin, which supports its AA.

The effects of the rosemary extracts on lipid peroxidation induced by Fe^{+2}/citrate were accessed by the production of MDA-TBA complex (absorbance reading at 535 nm). Figure 5a–c show, respectively, the experimental results of the inhibition of lipid peroxidation obtained for the concentrated extract, spouted bed dried, and lyophilized rosemary extracts, respectively (dry basis). It can be seen that the inhibition of the LPO activity produced by the rosemary extracts samples were dose dependent, the same behavior observed with the DPPH* method. The maximum inhibition falls between 82% and 86%.The IC_{50} values obtained for the concentrated extract (a), spouted bed dried rosemary extract (b), and for the lyophilized sample (c) were, respectively, 0.22 ± 0.01, 1.31 ± 0.02 (0.82 ± 0.01 with dilution correction), and 2.54 ± 0.02 µg mL^{-1} in the reaction medium; while the flavonoid quercetin showed an IC_{50} of 0.22 µg mL^{-1} [58]. The significant AA exhibited by the rosemary extracts samples by DPPH* and LPO assays are in agreement with studies reported in the literature [59]. IC_{50} values reported for concentrated extract and SB extracts of *Bauhinia forficata* L. were 15.2 and 12.2 µg/mL (DPPH$^\bullet$ assay) [60], and 22.5 and 25.9 µg/mL for the LPO method; values significantly higher than the values here reported. The maximum inhibition of DPPH exhibited by the actual spouted bed dried rosemary extract was also superior to those reported for SB dried *Bauhinia forficata* extract (79% × 69%), and (86.3% × 80.3%). Notwithstanding, these values were reached at significantly lower concentrations for the rosemary SB dried extract compared to *Bauhinia forficata*. Altogether, these results evidenced a higher AA for the SB dried rosemary extract compared to *B. forficata* dried extract [60].

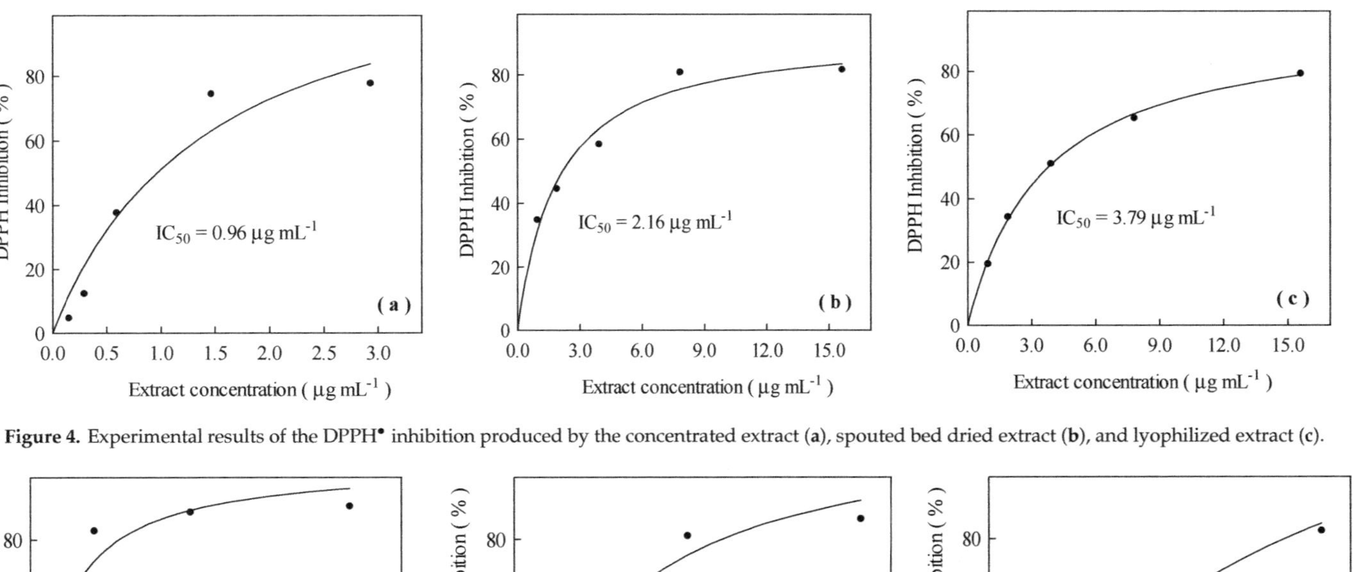

Figure 4. Experimental results of the DPPH• inhibition produced by the concentrated extract (**a**), spouted bed dried extract (**b**), and lyophilized extract (**c**).

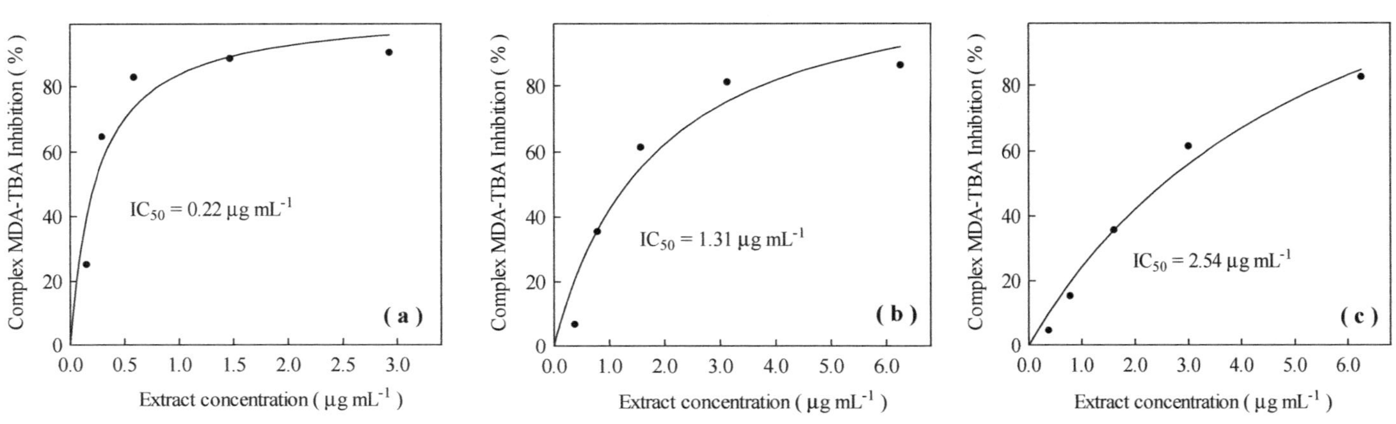

Figure 5. Experimental results of the inhibition of the lipid peroxidation induced by Fe^{+2}/citrate promoted by the concentrated extract (**a**), spouted bed dried extract (**b**), and lyophilized extract (**c**).

The results of AA of the rosemary extracts assessed by DPPH$^{\bullet}$* and LPO assays evidenced a slight to moderate decrease in the AA of the rosemary extracts caused by the SB drying operation, apparently due to degradation of bioactive compounds on exposure to the processing conditions used in this study; mainly heated air. It is known that polyphenolic compounds are thermolabile and their efficacy may be compromised on exposure to heat [61]. However, the SB dried rosemary extracts here produced are near 13.5 times more concentrated than the concentrated extract (in dry weight); and have high concentration of bioactive compounds by extract mass (in wet basis). A simple calculation with the experimental results of T_F and T_p presented in Table 2, shows that the concentration of total flavonoids and polyphenols compounds presented in the spouted bed dried extract ranges from 4.3 to 12.3 and from 1.2 to 4.7 times the values in the concentrated extract, respectively. Therefore, even considering the losses of bioactive compounds and of AA caused by the drying, the SB dried extracts have advantages compared to the concentrated extract, regarding the concentration bioactive compounds, lower storage volume, and perhaps higher stability.

4. Conclusions

The results here reported indicate that the production of dried extract of *Rosmarinus officinalis* by the spouted bed technology is feasible. The processing conditions should be strictly controlled in order to ensure the product quality with an acceptable dryer performance. The use of experimental design could facilitate the identification of the significant variables and allow the optimization of the processing variables. The drying variables W_s/W_{max} and T_{gi} exert statistically significant effects on the content of bioactive compounds in the dried rosemary extract. The antioxidative properties of rosemary extracts were evaluated by both DPPH* and lipid peroxidation methods. The rosemary extracts caused instantaneous decrease in the absorbance of DPPH* (in a manner similar to that presented by quercetin) and were able to inhibit the lipid peroxidation induced by Fe^{+2}/citrate, indicating their high antioxidant profile. The drying operation caused a slight to moderate decrease in the antioxidant activity of the spouted dried rosemary extracts; but the spouted dried rosemary extracts have higher concentration of bioactive compounds by extract mass (in wet basis) compared to concentrated extract; ranging from 4.3 to 12.3 and from 1.2 to 4.7 times, respectively for T_F and T_p. The spouted bed dried extracts showed powerful antioxidant activity and have potential to be used as a natural antioxidant or phytoactive ingredient in food, cosmetic and pharmaceutical applications. Further work still needs to be carried out in order increase the production yield and to improve product properties to respond to the specifications from the pharmaceutical regulatory bodies.

Author Contributions: C.R.F.S., I.B., V.O.B., R.D.A., A.D., M.L., N.C. and A.S. have contributed to the methodology, formal analysis, investigation, resources, and data curation. C.R.F.S., I.B., V.O.B., E.B.S. and W.P.O. have contributed to the writing of the original manuscript. C.R.F.S., E.B.S. and W.P.O. have contributed to the conceptualization, review and editing of the manuscript, project administration, supervision and funding acquisition. All authors have made a substantial contribution to the work and have approved its publication. All authors have read and agreed to the published version of the manuscript.

Funding: We would like to express our gratitude to the Foundation of Research Support of the São Paulo State (FAPESP) for the funded projects 2011/10333-1, 2012/03427-2 and 2018/26069-0 and for the National Council for Scientific and Technological Development (CNPq) for the financial support. E.B.S. acknowledges the Portuguese Science and Technology Foundation (FCT) for the funded projects M-ERA-NET/0004/2015 (PAIRED) and UIDB/04469/2020 (strategic fund).

Conflicts of Interest: The authors declare no conflict of interest.

References

1. Zielińska, A.; Ferreira, N.R.; Feliczak-Guzik, A.; Nowak, I.; Souto, E.B. Loading, release profile and accelerated stability assessment of monoterpenes-loaded Solid Lipid Nanoparticles (SLN). *Pharm. Dev. Technol.* **2020**, 1–13. [CrossRef]
2. Santini, A.; Novellino, E. Nutraceuticals-shedding light on the grey area between pharmaceuticals and food. *Expert Rev. Clin. Pharm.* **2018**, *11*, 545–547. [CrossRef]

3. Durazzo, A.; Lucarini, M.; Santini, A. Nutraceuticals in Human Health. *Foods* **2020**, *9*, 370. [CrossRef] [PubMed]

4. Santini, A.; Tenore, G.C.; Novellino, E. Nutraceuticals: A paradigm of proactive medicine. *Eur. J. Pharm. Sci.* **2017**, *96*, 53–61. [CrossRef] [PubMed]

5. Daliu, P.; Santini, A.; Novellino, E. From pharmaceuticals to nutraceuticals: Bridging disease prevention and management. *Expert Rev. Clin. Pharm.* **2019**, *12*, 1–7. [CrossRef]

6. Campos, J.R.; Severino, P.; Ferreira, C.S.; Zielinska, A.; Santini, A.; Souto, S.B.; Souto, E.B. Linseed Essential Oil—Source of Lipids as Active Ingredients for Pharmaceuticals and Nutraceuticals. *Curr. Med. Chem.* **2019**, *26*, 4537–4558. [CrossRef] [PubMed]

7. Durazzo, A.; Lucarini, M.; Novellino, E.; Souto, E.B.; Daliu, P.; Santini, A. *Abelmoschus esculentus* (L.): Bioactive Components' Beneficial Properties-Focused on Antidiabetic Role-For Sustainable Health Applications. *Molecules* **2018**, *24*, 38. [CrossRef]

8. Durazzo, A.; Lucarini, M.; Souto, E.B.; Cicala, C.; Caiazzo, E.; Izzo, A.A.; Novellino, E.; Santini, A. Polyphenols: A concise overview on the chemistry, occurrence, and human health. *Phytother. Res.* **2019**, *33*, 2221–2243. [CrossRef]

9. Salehi, B.; Venditti, A.; Sharifi-Rad, M.; Kregiel, D.; Sharifi-Rad, J.; Durazzo, A.; Lucarini, M.; Santini, A.; Souto, E.B.; Novellino, E.; et al. The Therapeutic Potential of Apigenin. *Int. J. Mol. Sci.* **2019**, *20*, 1305. [CrossRef]

10. Santini, A.; Novellino, E.; Armini, V.; Ritieni, A. State of the art of Ready-to-Use Therapeutic Food: A tool for nutraceuticals addition to foodstuff. *Food Chem.* **2013**, *140*, 843–849. [CrossRef]

11. Amarowicz, R.; Carle, R.; Dongowski, G.; Durazzo, A.; Galensa, R.; Kammerer, D.; Maiani, G.; Piskula, M.K. Influence of postharvest processing and storage on the content of phenolic acids and flavonoids in foods. *Mol. Nutr. Food Res.* **2009**, *53*, S151–S183. [CrossRef] [PubMed]

12. Pimentel-Moral, S.; Teixeira, M.C.; Fernandes, A.R.; Arraez-Roman, D.; Martinez-Ferez, A.; Segura-Carretero, A.; Souto, E.B. Lipid nanocarriers for the loading of polyphenols—A comprehensive review. *Adv. Colloid Interface Sci.* **2018**, *260*, 85–94. [CrossRef]

13. Santos, I.S.; Ponte, B.M.; Boonme, P.; Silva, A.M.; Souto, E.B. Nanoencapsulation of polyphenols for protective effect against colon-rectal cancer. *Biotechno. Adv.* **2013**, *31*, 514–523. [CrossRef] [PubMed]

14. Souto, E.B.; Severino, P.; Basso, R.; Santana, M.H. Encapsulation of antioxidants in gastrointestinal-resistant nanoparticulate carriers. *Methods Mol. Bio.* **2013**, *1028*, 37–46. [CrossRef]

15. Durazzo, A.; D'Addezio, L.; Camilli, E.; Piccinelli, R.; Turrini, A.; Marletta, L.; Marconi, S.; Lucarini, M.; Lisciani, S.; Gabrielli, P.; et al. From Plant Compounds to Botanicals and Back: A Current Snapshot. *Molecules* **2018**, *23*, 1844. [CrossRef] [PubMed]

16. Durazzo, A.; Camilli, E.; D'Addezio, L.; Piccinelli, R.; Mantur-Vierendeel, A.; Marletta, L.; Finglas, P.; Turrini, A.; Sette, S. Development of Dietary Supplement Label Database in Italy: Focus of FoodEx2 Coding. *Nutrients* **2019**, *12*, 89. [CrossRef]

17. Souza, C.R.F.; Oliveira, W.P. Spouted bed drying of Bauhinia forficata link extract: The effects of feed atomizer position and operating conditions on equipment performance and product properties. *Braz. J. Chem. Eng.* **2005**, *22*, 239–247. [CrossRef]

18. Runha, F.P.; Cordeiro, D.S.; Pereira, C.A.M.; Vilegas, J.; Oliveira, W.P. Production of Dry Extracts of Medicinal Brazilian Plants by Spouted Bed Process: Development of the Process and Evaluation of Thermal Degradation During the Drying Operation. *Food Bioprod. Process.* **2001**, *79*, 160–168. [CrossRef]

19. Lemos Senna, E.; Petrovick, P.R.; Gonzalez Ortega, G.; Bassani, V.L. Preparation and Characterization of Spray-dried Powders from *Achyrocline satureioides* (Lam.) DC Extracts. *Phytother. Res.* **1997**, *11*, 123–127. [CrossRef]

20. Tran, T.T.A.; Nguyen, H.V.H. Effects of Spray-Drying Temperatures and Carriers on Physical and Antioxidant Properties of Lemongrass Leaf Extract Powder. *Beverages* **2018**, *4*, 84. [CrossRef]

21. Oliveira, W.P.; Bott, R.F.; Souza, C.R.F. Manufacture of Standardized Dried Extracts from Medicinal Brazilian Plants. *Dry. Technol.* **2006**, *24*, 523–533. [CrossRef]

22. Cordeiro, D.S.; Oliveira, W.P. Technical aspects of the production of dried extract of Maytenus ilicifolia leaves by jet spouted bed drying. *Int. J. Pharm.* **2005**, *299*, 115–126. [CrossRef] [PubMed]

23. Passos, M.L.; Massarani, G.; Freire, J.T.; Mujumdar, A.S. Drying Of Pastes In Spouted Beds Of Inert Particles: Design Criteria And Modeling. *Dry. Technol.* **1997**, *15*, 605–624. [CrossRef]

24. Mathur, K.B.; Epstein, N. Dynamics of Spouted Beds. In *Advances in Chemical Engineering*; Drew, T.B., Cokelet, G.R., Hoopes, J.W., Vermeulen, T., Eds.; Academic Press: Cambridge, MA, USA, 1974; Volume 9, pp. 111–191.

25. Durazzo, A.; Lucarini, M. Extractable and Non-Extractable Antioxidants. *Molecules* **2019**, *24*, 1933. [CrossRef] [PubMed]

26. Durazzo, A.; Lucarini, M. A Current shot and re-thinking of antioxidant research strategy. *Braz. J. Anal. Chem.* **2018**, *5*, 9–11. [CrossRef]

27. Durazzo, A. Extractable and Non-extractable polyphenols: An overview. In *Non-Extractable Polyphenols and Carotenoids: Importance in Human Nutrition and Health*; Saura-Calixto, F., Pérez-Jiménez, J., Eds.; Royal Society of Chemistry: London, UK, 2018; pp. 1–37.

28. Cermak, R.; Durazzo, A.; Maiani, G.; Böhm, V.; Kammerer, D.R.; Carle, R.; Wiczkowski, W.; Piskula, M.K.; Galensa, R. The influence of postharvest processing and storage of foodstuffs on the bioavailability of flavonoids and phenolic acids. *Mol. Nutr. Food Res.* **2009**, *53* (Suppl. 2), S184–S193. [CrossRef]

29. Borges, R.S.; Ortiz, B.L.S.; Pereira, A.C.M.; Keita, H.; Carvalho, J.C.T. *Rosmarinus officinalis* essential oil: A review of its phytochemistry, anti-inflammatory activity, and mechanisms of action involved. *J. Ethnopharmacol.* **2019**, *229*, 29–45. [CrossRef]

30. de Oliveira, J.R.; Camargo, S.E.A.; de Oliveira, L.D. *Rosmarinus officinalis* L. (rosemary) as therapeutic and prophylactic agent. *J. Biomed. Sci.* **2019**, *26*, 5. [CrossRef]

31. Souza, C.R.F.; Schiavetto, I.A.; Thomazini, F.C.F.; Oliveira, W.P. Processing of *Rosmarinus officinalis* linne extract on spray and spouted bed dryers. *Braz. J. Chem. Eng.* **2008**, *25*, 59–69. [CrossRef]

32. Lin, J.-Y.; Tang, C.-Y. Determination of total phenolic and flavonoid contents in selected fruits and vegetables, as well as their stimulatory effects on mouse splenocyte proliferation. *Food Chem.* **2007**, *101*, 140–147. [CrossRef]

33. Souza, C.R.F.; Bott, R.F.; Oliveira, W.P. Optimization of the extraction of flavonoids compounds from herbal material using Experimental Design and Multi-response Analysis. *Lat. Am. J. Pharm.* **2007**, *65*, 682–690.

34. Bajaj, K.L.; Devsharma, A.K. A colorimetric method for the determination of tannins in tea. *Microchim. Acta* **1977**, *68*, 249–253. [CrossRef]

35. Tzeng, G.S.; Chen, H.J.; Wang, Y.Y.; Wan, C.C. The effects of roughening on teflon surfaces. *Surf. Coat. Technol.* **1997**, *89*, 108–113. [CrossRef]

36. Cortés-Rojas, D.F.; Oliveira, W.P. Physicochemical Properties of Phytopharmaceutical Preparations as Affected by Drying Methods and Carriers. *Dry. Technol.* **2012**, *30*, 921–934. [CrossRef]

37. Blois, M.S. Antioxidant Determinations by the Use of a Stable Free Radical. *Nature* **1958**, *181*, 1199–1200. [CrossRef]

38. Souto, E.B.; Zielinska, A.; Souto, S.B.; Durazzo, A.; Lucarini, M.; Santini, A.; Silva, A.M.; Atanasov, A.G.; Marques, C.; Andrade, L.N.; et al. (+)-Limonene 1,2-epoxide-loaded SLN: Evaluation of drug release, antioxidant activity and cytotoxicity in HaCaT cell line. *Int. J. Mol. Sci.* **2020**, *21*, 1449. [CrossRef]

39. Souto, E.B.; Souto, S.B.; Zielinska, A.; Durazzo, A.; Lucarini, M.; Santini, A.; Horbańczuk, O.K.; Atanasov, A.G.; Marques, C.; Andrade, L.N.; et al. Perillaldehyde 1,2-epoxide loaded SLN-tailored mAb: Production, physicochemical characterization and in vitro cytotoxicity profile in MCF-7 cell lines. *Pharmaceutics* **2020**, *12*, 161. [CrossRef]

40. Fernandes, M.R.; Azzolini, A.E.; Martinez, M.L.; Souza, C.R.; Lucisano-Valim, Y.M.; Oliveira, W.P. Assessment of antioxidant activity of spray dried extracts of *Psidium guajava* leaves by DPPH and chemiluminescence inhibition in human neutrophils. *BioMed Res. Int.* **2014**, *2014*, 382891. [CrossRef]

41. Rodrigues, T.; Santos, A.C.; Pigoso, A.A.; Mingatto, F.E.; Uyemura, S.A.; Curti, C. Thioridazine interacts with the membrane of mitochondria acquiring antioxidant activity toward apoptosis-potentially implicated mechanisms. *Br. J. Pharm.* **2002**, *136*, 136–142. [CrossRef]

42. Gornall, A.G.; Bardawill, C.J.; David, M.M. Determination of serum proteins by means of the biuret reaction. *J. Biol. Chem.* **1949**, *177*, 751–766.

43. Zainol, M.K.M.; Hamid, A.A.; Bakar, F.A.; Dek, M.S.P. Effect of different drying methods on the degradation of selected flavonoids in Centella asiatica. *Int. Food Res. J.* **2009**, *16*, 531–537.

44. Chaaban, H.; Ioannou, I.; Chebil, L.; Slimane, M.; Gérardin, C.; Paris, C.; Charbonnel, C.; Chekir, L.; Ghoul, M. Effect of heat processing on thermal stability and antioxidant activity of six flavonoids. *J. Food Process. Preserv.* **2017**, *41*, e13203. [CrossRef]

45. Kakkar, S.; Bais, S. A review on protocatechuic Acid and its pharmacological potential. *Isrn Pharmacol.* **2014**, *2014*, 952943. [CrossRef] [PubMed]

46. Van Acker, S.A.B.E.; Van Den Berg, D.-J.; Tromp, M.N.J.L.; Griffioen, D.H.; Van Bennekom, W.P.; Van Der Vijgh, W.J.F.; Bast, A. Structural aspects of antioxidant activity of flavonoids. *Free Radic. Biol. Med.* **1996**, *20*, 331–342. [CrossRef]

47. Sandulachi, E. Water activity concept and its role in food preservation. *Meridian Ing.* **2012**, *4*, 40–48.

48. Labuza, T.P.; Altunakar, L. Water Activity Prediction and Moisture Sorption Isotherms. In *Water Activity in Foods*; Barbosa-Cánovas, G.V., Fontana, A.J., Schmidt, S.J., Labuza, T.P., Eds.; John Wiley & Sons, Inc.: Hoboken, NJ, USA, 2008; Volume 5, pp. 109–154. [CrossRef]

49. Anandharamakrishnan, C.; Padma Ishwarya, S. *Spray Drying Techniques for Food Ingredient Encapsulation*; Wiley-Blackwell, John Wiley & Sons, Inc.: Hoboken, NJ, USA, 2015; pp. 22–23.

50. Cano, A.; Ettcheto, M.; Chang, J.H.; Barroso, E.; Espina, M.; Kuhne, B.A.; Barenys, M.; Auladell, C.; Folch, J.; Souto, E.B.; et al. Dual-drug loaded nanoparticles of Epigallocatechin-3-gallate (EGCG)/Ascorbic acid enhance therapeutic efficacy of EGCG in a APPswe/PS1dE9 Alzheimer's disease mice model. *J. Control. Release* **2019**, *301*, 62–75. [CrossRef]

51. Sánchez-López, E.; Lopez-Machado, A.L.; Bonilla, V.L.; Pizarro, P.G.; Silva, A.M.; Souto, E.B. Lipid nanoparticles as carriers for the treatment of neurodegeneration associated with Alzheimer's disease and glaucoma: Present and future challenges. *Curr. Pharm. Des.* **2020**, *26*, 1235–1250. [CrossRef]

52. Sanchez-Lopez, E.; Egea, M.A.; Davis, B.M.; Guo, L.; Espina, M.; Silva, A.M.; Calpena, A.C.; Souto, E.M.B.; Ravindran, N.; Ettcheto, M.; et al. Memantine-Loaded PEGylated Biodegradable Nanoparticles for the Treatment of Glaucoma. *Small* **2018**, *14*. [CrossRef]

53. Sanchez-Lopez, E.; Ettcheto, M.; Egea, M.A.; Espina, M.; Cano, A.; Calpena, A.C.; Camins, A.; Carmona, N.; Silva, A.M.; Souto, E.B.; et al. Memantine loaded PLGA PEGylated nanoparticles for Alzheimer's disease: In vitro and in vivo characterization. *J. Nanobiotechnol.* **2018**, *16*, 32. [CrossRef]

54. Silva, A.M.; Martins-Gomes, C.; Fangueiro, J.F.; Andreani, T.; Souto, E.B. Comparison of antiproliferative effect of epigallocatechin gallate when loaded into cationic solid lipid nanoparticles against different cell lines. *Pharm. Dev. Technol.* **2019**, *24*, 1243–1249. [CrossRef]

55. Doktorovova, S.; Santos, D.L.; Costa, I.; Andreani, T.; Souto, E.B.; Silva, A.M. Cationic solid lipid nanoparticles interfere with the activity of antioxidant enzymes in hepatocellular carcinoma cells. *Int. J. Pharm.* **2014**, *471*, 18–27. [CrossRef] [PubMed]

56. Niki, E. Assessment of antioxidant capacity in vitro and in vivo. *Free Radic. Biol. Med.* **2010**, *49*, 503–515. [CrossRef] [PubMed]

57. Casagrande, R.; Georgetti, S.R.; Verri, W.A., Jr.; Borin, M.F.; Lopez, R.F.; Fonseca, M.J. In vitro evaluation of quercetin cutaneous absorption from topical formulations and its functional stability by antioxidant activity. *Int. J. Pharm.* **2007**, *328*, 183–190. [CrossRef] [PubMed]

58. Vicentini, F.T.M.C.; Casagrande, R.; Georgetti, S.R.; Bentley, M.V.L.B.; Fonseca, M.J.V. Influence of Vehicle on Antioxidant Activity of Quercetin: A Liquid Crystalline Formulation. *Lat. Am. J. Pharm.* **2007**, *26*, 805.

59. Nieto, G.; Ros, G.; Castillo, J. Antioxidant and Antimicrobial Properties of Rosemary (*Rosmarinus officinalis*, L.): A Review. *Medicines* **2018**, *5*, 98. [CrossRef]

60. Souza, C.R.F.; Georgetti, S.R.; Salvador, M.J.; Fonseca, M.J.V.; Oliveira, W.P. Antioxidant activity and physical-chemical properties of spray and spouted bed dried extracts of Bauhinia forficata. *Braz. J. Pharm. Sci.* **2009**, *45*, 209–218. [CrossRef]

61. Bankole, V.O.; Osungunna, M.O.; Souza, C.R.F.; Salvador, S.L.; Oliveira, W.P. Spray-Dried Proliposomes: An Innovative Method for Encapsulation of *Rosmarinus officinalis* L. Polyphenols. *AAPS PharmSciTech* **2020**, *21*, 143. [CrossRef]

 *agriculture* 

Article

Phenolic Compounds Characterization and Antioxidant Properties of Monocultivar Olive Oils from Northeast Algeria

Soulef Boussahel [1,2], Vita Di Stefano [3,*], Claudia Muscarà [4,5], Mariateresa Cristani [4] and Maria Grazia Melilli [6,*]

[1] Department of Biological Sciences, Faculty of Nature and Life Sciences and Sciences of Earth and Universe, University of Bordj Bou Arreridj, Bordj Bou Arreridj 34030, Algeria; boussahel.soulef@gmail.com

[2] Laboratory of Phytotherapy Applied to Chronic Diseases, Department of Biology and Animal Physiology, Faculty of Nature and Life Sciences, University Setif 1, Setif 19000, Algeria

[3] Department of Biological, Chemical and Pharmaceutical Sciences and Technologies (STEBICEF) University of Palermo, 90123 Palermo, Italy

[4] Department of Chemical, Biological, Pharmaceutical and Environmental Sciences, University of Messina, 98165 Messina, Italy; claudia.muscara@gmail.com (C.M.); mcristani@unime.it (M.C.)

[5] Foundation "Prof. Antonio Imbesi", University of Messina, Piazza Pugliatti 1, 98122 Messina, Italy

[6] Institute for Agricultural and Forest Systems in the Mediterranean, National Council of Research, 95128 Catania, Italy

* Correspondence: vita.distefano@unipa.it (V.D.S.); mariagrazia.melilli@cnr.it (M.G.M.); Tel.: +39-091-23891948 (V.D.S.); +39-0956139916 (M.G.M.)

Received: 23 September 2020; Accepted: 21 October 2020; Published: 23 October 2020

Abstract: In Algeria, the olive tree is one of the main fruit species and plays a very important socioeconomic role. The objective of this study was firstly, to identify and quantify the phenolics of some Algerian olive oils, and secondly, to assess the antioxidant activity of the samples. The olive oils used in this study were derived from Algerian cultivars, including Tefahi, Gelb Elfarroudj, Chemlal, and imported cultivar Manzanilla and Zebboudj. For this purpose, gas chromatography—mass spectrometry (GC-MS) was used to identify olive oil fatty acids profile, while the individual phenolic compounds were assessed by ultra-high-performance liquid chromatography–electrospray ionization–high-resolution mass spectrometry (UHPLC-HESI-MS). To verify the antioxidant capacity, five in vitro free radical assays were used. Questionable values of particular physico-chemical parameters, such as the high value of free acidity and the low concentration of monounsaturated fatty acids in oil from the Zebboudj cultivar, indicate that improvements in olive cultivation and oil production practices are needed. Gelb Elfarroudj, Tefahi, and Manzanilla oils contain quantities of monounsaturated fatty acids in accordance with EU regulations. The oil obtained from the Zebboudj cultivar is not usable for food purposes due to the high value in free acidity and the low concentration of monounsaturated fatty acids. Tefahi and Manzanilla cultivars have given oils with the best antioxidant activity as compared to other studied cultivars; this is attributable to their composition in bioactive phenolic compounds, such as secoiridoids, which play an important role in human health as scavengers of free radicals. The results are interesting for producers and consumers to promote the culture of olive oils derived in particular from the Tefahi cultivar. However, in order to improve the health qualities of this oil, the agronomic techniques essentially linked to the time of harvesting of the olives destined for oil production must be improved.

Keywords: Manzanilla; Tefahi; Gelb Elfarroudj cultivar; secoiridoids; radical scavenging; UHPLC-HESI-MS; phenolics

---

1. Introduction

Olea europaea L., or more commonly olive tree, is largely cultivated for the production of its nutritional and healthy fruits. Extra virgin olive oil (EVOO) is an integral ingredient of the Mediterranean diet and a wide number of analytical techniques were used to identify the chemical composition [1]. These techniques indicated that the fine characteristics, the good health effect, and the biological activity of EVOO are mainly attributed to the presence of the unsaturated fatty acids as major components. They are recognized in olive oils mostly by the presence of the acids: oleic (C18:1), palmitic (C16:0), palmitoleic (C16:1), stearic (C18:0), linoleic (C18:2), and linolenic (C18:3). The high quality of olive oil is also attributed to the presence in its composition of minor components such as phytosterols, carotenoids, tocopherols and hydrophilic phenols. The major phenolic compounds present in olive oil and conferring it the antioxidant activity belong to the class of secoiridoids mainly represented by oleuropein and ligstroside derivatives, which are strong radical scavengers and are also responsible for bitterness and pungency of EVOO [2–4].

Phenolic compounds are used as quality markers for virgin olive oil, and they are of great interest due to their anticancer, antiviral, and anti-inflammatory properties [5,6]. Their content is an important factor when evaluating the EVOO quality because they have been correlated with the oil oxidative stability and, in particular, its resistance to lipid peroxidation [7,8]. Extra virgin olive oil quality production could be influenced by several factors, for instance: olive cultivar, geographical region, environmental factors (seasonal conditions), irrigation, olive ripeness, harvesting, storage, and extraction procedure [9–11]. Light exposure, elevated temperature, and oxygen are all natural adversaries of EVOO and contribute to its deterioration [12].

The Mediterranean countries head the list of olive oil producers; the International Olive Oil Council classed Algeria as the ninth largest producer country of olive oil in the world, with around 87.5 tons in the 2015/16 season [13–15]. The olive oil production in Algeria is continuously increasing. In fact, the country is following a development policy called "National program for agriculture development" which opens the way for financing and supporting the agricultural sector [16]. Within the framework of this program, the government provides financial aid to farmers who plant olive trees and to the producers of olive oil [17]. Even with this, the commercialization of a high quality of olive oil is still presenting a challenge for the producers in Algeria.

The aim of this study was the chemical characterization of monovarietal oils produced in Algeria and the evaluation of their quality. In particular, the study focused on the comparison of the qualitative characteristics of: (a) Tefahi and Gelb Elfarroudj olive oils (cultivars not known in literature); (b) Chemlal and Zebboudj olive oils (known cultivars); (c) Manzanilla oil (imported cultivar). The authors believe that a quantitative characterization of the bioactive constituents of Algerian olive oils could contribute to their enhancement in a constantly growing international market.

For this, firstly, quality control tests were performed on all olive oils under study. Then, phenolic profile, fatty acid methyl esters pattern, and in vitro antioxidant capacity of the olive oils were determined.

2. Materials and Methods

2.1. Plant Material and Olive Oil Samples

Olive oil from four *Olea europaea* cultivars growing in the northeast of Algeria were used in this study (Setif and Batna provinces) (Figure 1). Three cultivars (Chemlal, Tefahi, and Gelb Elfarroudj) are native to the region, while one is locally grown but it is a foreign cultivar (Manzanilla). Furthermore, Zebboudj olive oil (*Olea oleaster*) was bought from the market and included in this study. The cultivars were growing under the same environmental characteristics in a semiarid region. Drop irrigation was used twice a year: in January (every fortnight) and in August (every 10 days). The method used in this study for olive harvest and extraction was the one used by most of Algerian producers. Healthy olive fruits of all cultivars were hand-picked when the skin of the fruit was black. The oils were extracted from 200 kg of the collected fruits using a commercial modern mill located in the same region. The mill originated from Jeha Company (Alexandria, Egypt) and was equipped with a two-phase extraction system. The extraction of olive oil started with leaf stripping and olive cleaning, then olives passed into a hammer-crusher to obtain a homogenized olive paste. The malaxation time and temperature of the olive paste were 30 min at 30 °C. After centrifugation, the oil was separated from the paste and water, and then stored in amber glass bottles at room temperature (15–18 °C) in the dark. Soon after the oil extraction, the samples were used in the chemical analytical methods described below.

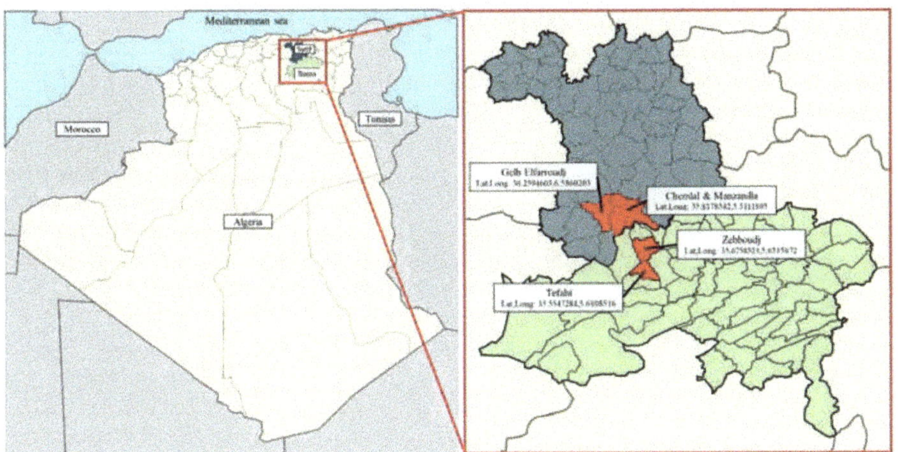

Figure 1. Biogeographical Algerian area where olives are grown.

2.2. Reagents and Standards

Formic acid, water, methanol, acetonitrile (LC-MS grade) were purchased from Biosolve B.V. (Valkenswaard, The Netherlands). All solvents, Folin–Ciocalteu reagent, sodium carbonate (Na_2CO_3), 2,2-diphenyl,1-picrylhydrazyl (DPPH), 2,4,6-tri(2-pyridyl)-1,3,5-triazine (TPTZ), ferric chloride ($FeCl_3 \times 6H_2O$), gallic acid, acetic acid, ferric sulfate ($FeSO_4 \times 7H_2O$), sodium acetate, 2,2′-azino-bis(3-ethylbenzothiazoline-6-sulphonic acid) (ABTS), potassium persulfate, phosphate-buffered saline (PBS), tween 40, β-carotene, Trolox, and chemical standards (cinnamic acid, *p*-coumaric acid, *p*-hydroxybenzoic acid, ferulic acid, syringic acid, vanillic acid, caffeic acid apigenin, apigenin 7-glucoside, diosmetin, hydroxytyrosol, tyrosol, luteolin, oleuropein, vanillin, pinoresinol), were purchased from Sigma-Aldrich (St. Louis, MO, USA).

2.3. Quality Parameters

Free acidity, given as g oleic acid 100 g^{-1} of oil, peroxide value (PV) expressed as meq O_2 kg^{-1} of oil, ΔK and K_{232}, K_{270} extinction coefficients calculated from absorptions at 232 nm and 270 nm were measured as described by the Regulation EC no. 2016/2095 [18].

2.4. FAMEs Composition

Fatty acid methyl esters (FAMEs) were analyzed using the GC-MS method after extraction and hydrolysis of triacyl glycerols using potassium hydroxide in methanol.

A mass of 0.1 g of oil samples were diluted in 1 mL of n-heptane and 0.1 mL of 2N KOH in MeOH solution was added and mixed in a vortex for 2 min. An aliquot of 500 μL of organic phase containing fatty acid methyl esters was diluted with 500 μL n-heptane. Immediately, GC-MS analyses were carried out using a DSQ II single quadrupole system (Thermo Fisher Scientific, Bremen, Germany). The temperature of ion source and injector were 260 °C and 270 °C, respectively. The capillary column used was a ZB-WAX (30 m × 0.25 mm i.d., film thickness 0.25 μm) (Phenomenex, Italy). The oven program temperature started at 165 °C (held for 10 min), it was increased to 200 °C at 1.5 °C/min, then increased to 250 °C at 10 °C/min and kept for 20 min at 250 °C under isothermal conditions. Ionization energy was 70 eV and the mass range scanned was 35–550 m/z. Helium flow rate of 1 mL min^{-1} was used. Monovarietal olive oil samples of 1 μL were injected with a split ratio of 1:100 in triplicate. Fatty acid methyl esters identification was carried out using a mass spectrum database and 37-component fatty acid methyl esters mix (Sigma Aldrich Milan, Milan, Italy).

2.5. Individual Phenolic Compounds Characterization by UHPLC-HESI-MS

Liquid–liquid extraction method was used to isolate the phenolic fraction of olive oils. For this purpose, 2 g of each olive oil was mixed with 5 mL of a solution of methanol–water (80:20 *v/v*); then, the samples were shaken in a vortex for 1 min, placed in an ultrasonic bath for 15 min, and centrifuged at 5000 rpm for 25 min. The resulting aqueous solutions were filtered over PTFE syringe filter 0.45 μm and quickly analyzed by ultra-high-performance liquid chromatography, using a heated electrospray probe and high-resolution mass spectrometer (UHPLC-HESI-MS) for quali-quantitative determination of phenolic compounds, with three different dilution factors (1:2, 1:10, 1:100 *v/v*) to encompass the concentration variability. Triplicate samples of monovarietal olive oil were used for phenolic extractions. Experimental conditions used for qualitative determination of the phenolic constituents have been described in literature [19–21].

Briefly, UHPLC used was a Dionex Ultimate 3000 System equipped with an auto sampler controlled by Chromeleon 7.2 Software (Thermo Fisher Scientific, Bremen, Germany and Dionex Softron GmbH, Germering, Germany). A column Luna C18 50 × 1 mm, 2.5 μm was used. A flow rate of 50 μL min^{-1} was set for separation of the selected compounds. The separation was achieved using eluent A (water with 0.1% acetic acid (*v/v*) pH 3.2) and eluent B (acetonitrile). The gradient elution program was: 0–2 min 5% B; 2–4.5 min linear increase to 10% B; 4.5–16 min linear increase to 25% B; 16–29 min linear increase to 95% B; 29–31 min decrease to 5% B; 31–33 min hold 5% B coming back to the initial conditions and being equilibrated. The column temperature was set at 35 °C and the injection volume at 1 μL. Heated electrospray ion source (HESI) was used for the ionization. HESI parameters were optimized as follows: sheath gas flow rate 35 arbitrary units; auxiliary gas unit flow rate 4 arbitrary units; capillary temperature 250 °C; auxiliary gas heater temperature 259 °C; spray voltage 3.5 kV; and S lens RF level 30. MS in negative mode was selected for analysis of low-molecular phenolic compounds [19,21]. Detection of phenolic compounds was performed using a quadrupole Orbitrap mass spectrometer (Q Exactive; Thermo Scientific, Bremen, Germany). Full scan (100–800 m/z) acquisition method and a targeted single ion monitoring (SIM) analysis were performed using the mass inclusion list of the target analytes.

Phenolic compounds were quantified using solutions containing all commercial standards at six different concentration levels (5, 2.5, 1, 0.5, 0.25, 0.1 μg mL^{-1}). Each point of the external calibration graph corresponded to the average of five independent injections.

In the case of secoiridoids derivatives, where commercial standards were not available, their equivalent values were estimated. Monovarietal olive oil contents of 3,4-DHPEA-EA (oleuropein aglycon and stereoisomers), 3,4-DHPEA-EDA (oleacein), p-HPEA-EA (ligstroside aglycon and stereoisomers), deacetoxy-10-hydroxy oleuropein aglycon (DAc-10-OH Ole Agly), and elenolic acid (EA) were expressed as a p-HPEA-EDA (oleocanthal) equivalent (mg kg^{-1} of oil) [18]. Isolation of p-HPEA-EDA (oleocanthal) from olive oil was carried out according to a reported procedure developed in literature [19–21]. Briefly, 500 mL of olive oil was mixed with 250 mL hexane and 250 mL methanol. The mixture was sonicated for 15 min and after partition, the methanolic phase was centrifuged at 3000 rpm for 10 min. The hexane phase was extracted with 200 mL of methanol again. Combined methanolic phases were evaporated and the oily residue was extracted with 25 mL methanol–water (1:1) and 50 mL of hexane.

The aqueous MeOH layer was used for isolation of oleocanthal. Preparative chromatography was performed on an HPLC Agilent 1100 binary pump equipped with a UV-Vis detector. The chromatographic column was a reverse-phase Supelcosil LC 318 column, 25 cm, 4.6 mm, 3 μm particle size. HPLC elution was performed at 5 mL/min with a binary gradient system with water (solvent A) and acetonitrile (solvent B). The gradient was: 0–5 min, 30% B; 5–30 min, 100% B; 35–40 min, 30% B. The eluate was monitored at 278 nm and fractions of about 3 mL were collected from the detector. After fractions collection, the solvents were evaporated under reduced pressure using a rotary evaporator, and finally the residue was stored at −20 °C. The obtained fractions were analyzed by UHPLC-ESI-MS/MS and those with similar composition were combined for further preparative chromatography in the same conditions. Identity and purity of the oleocanthal were verified through spectroscopic techniques such as ^{1}HNMR, ^{13}C NMR, and UHPLC-HESI-MS.

Oleocanthal was used as external standard to prepare further calibration solutions (10, 5, 2.5, 1, 0.5, 0.25, 0.1 μg mL^{-1}). The deprotonated molecule [M-H]$^-$ detection was based on calculated exact mass and on retention time of target compounds.

2.6. Olive Oil Extracts for Determination of Total Phenolic Content and for Antioxidant Assays

The oil samples (2 g) were added to 5 mL of a methanol/water (80:20, *v/v*) mixture in a centrifuge tube. After vigorous mixing, they were centrifuged for 25 min at 5000 rpm. The hydroalcoholic phase was collected, washed with 2 mL of n-hexane to remove the residual oil, and then concentrated under reduced pressure using a rotary evaporator at 35 °C. The issued phenolic fraction extracts (PFEs) were fully dried using a vacuum desiccator. The oil PFEs were dissolved in the methanol/water mixture to use them in the determination of total phenolic content and in antioxidant assays.

2.7. Total Phenols Content

The total phenol compounds of the oil's phenolic fraction extracts (PFE), was determined by the Folin–Ciocalteu colorimetric assay [22]. Briefly, fifty microliter of the solutions containing different concentrations (5 mg mL^{-1}, 3 mg mL^{-1}, 1 mg mL^{-1}, 0.5 mg mL^{-1}, and 0.25 mg mL^{-1}) oil's PFE to be tested were added to 450 μL of deionized water, 500 μL of Folin-Ciocalteau reagent, and 500μL of 10% aqueous sodium carbonate solution. Samples were then maintained at room temperature for 1 h and absorbance was measured at 786 nm (UV-Vis Spectrophotometer, Shimadzu Japan) against blank containing 50 μL of the methanol (sample solvent). Gallic acid at different concentrations (300 μg mL^{-1}, 150 μg mL^{-1}, 75 μg mL^{-1}, and 25 μg mL^{-1}) was used as standard. The experiment was performed in triplicate and the results were expressed as μg gallic acid equivalents/mg of phenolic fraction extracts (μg GAE mg^{-1} PFE).

2.8. Antioxidant Activity

2.8.1. 2,2-Diphenyl-1-Picrylhydrazyl Test

The free-radical-scavenging capacity of the oil's phenolic fraction extracts (PFEs) was determined by the 2,2-diphenyl-1-picrylhydrazyl (DPPH) assay as reported in literature [23]. A volume of 1.5 mL of DPPH$^\bullet$ solution (100 mM in methanol) was mixed with 37.5 μL of various concentrations (87 mg mL^{-1}, 60 mg mL^{-1}, 50 mg mL^{-1}, 45 mg mL^{-1}, 30 mg mL^{-1}, 25 mg mL^{-1}, 10 mg mL^{-1}, 5 mg mL^{-1}) of oil's PFE, or with methanol as control. After 20 min of incubation at room temperature, the absorbance was recorded at 517 nm with a UV-Vis spectrophotometer (Shimadzu, Japan). Results were expressed as mmol Trolox equivalents mg^{-1} of phenolic fraction extracts (mmol TE mg^{-1} PFE), using the calibration curve prepared with Trolox as standard using different concentrations (1.5 μg mL^{-1}, 1 μg mL^{-1}, 0.5 μg mL^{-1}, 0.25 μg mL^{-1}, 0.125 μg mL^{-1}). Each determination was performed in triplicate.

2.8.2. Trolox Equivalents Antioxidant Capacity

The Trolox equivalents antioxidant capacity (TEAC), also known as the ability of the oil's PFEs to scavenge the 2,2'-azinobis-(3-ethylbenzothiazine-6-sulfonic acid radical (ABTS$^+$), was evaluated as previously described [24]. The ABTS$^+$ radical cation was produced by the oxidation of ABTS$^+$ (1.7 mM) with potassium persulfate (4.3 mM) in water. The mixture was allowed to stand in the dark at room temperature and the ABTS$^+$ solution was diluted with phosphate-buffered saline (PBS) at pH 7.4 to give absorbance of 0.7 ± 0.02 at 734 nm. An aliquot of 50 µmL of a solution containing different concentrations (50–0.5 mg mL^{-1}) of the oil's PFE or methanol (blank) was added to 2 mL of the ABTS$^+$ solution, and the absorbance was recorded at 734 nm in a UV/Vis spectrophotometer (Shimadzu, Japan) after allowing the reaction to stand for 6 min in the dark at room temperature. Results were expressed as mmol Trolox equivalents/milligram of phenolic fraction extracts (mmol TEmg^{-1} PFE) using the calibration curve prepared with Trolox standard at different concentrations (0.215 µg mL^{-1}, 0.0625 µg mL^{-1}, 0.0312 µg mL^{-1}, 0.015 µg mL^{-1},). Each determination was repeated at least three times.

2.8.3. Ferric-Reducing/Antioxidant Power

The ferric-reducing/antioxidant power (FRAP) of the phenolic fraction extracts was evaluated as previously described [25]. The fresh working solution ferric-reducing/antioxidant power (FRAP) reagent was prepared by mixing 2.5 mL of 10 mM 2,4,6-tripyridyl-s-triazine solution (prepared in 40 mM HCl) with 25mL of 0.3 M acetate buffer (pH 3.6) and 2.5 mL of 20 mM FeCl$_3$·6H$_2$O solution, and then preheated at 37 °C before use. A volume of 50 µL of a methanolic solution containing different concentrations (25–1 mg mL^{-1}) of the oil's PFE to be tested, or of the solvent, were added to 1 mL of FRAP reagent, and the absorbance was measured at 593 nm in a spectrophotometer (Shimadzu, Japan) after incubation at 20 °C for 4 min. A standard curve was prepared using various concentrations (1 µg mL^{-1}, 0.5 µg mL^{-1}, 0.25 µg mL^{-1}, 0.1 µg mL^{-1}) of FeSO$_4$ × 7H$_2$O. Each determination was performed in triplicate; the results were expressed as mmol Fe^{2+}/milligram of phenolic fraction extracts (mmol Fe^{2+}/mg of PFE).

2.8.4. Beta-Carotene Bleaching Assay

The potential of the oil's PFE in the β-carotene bleaching assay (BCB) was determined as described in literature [26]. A mass of 1 mg of β-carotene was dissolved in 10 mL of chloroform. Then, 5 mL of solution was added to 40 µL of linoleic acid and 400 µL of Tween 40. Chloroform was removed using a rotary evaporator and 100 mL of distilled water was added.

A volume of 5mL was added to 200 µL of a solution containing different concentrations (25–0.5 mg mL^{-1}) of oil's PFE, whereas 5 mL of DMSO was used as blank. The absorbance was measured at 470 nm and results were expressed as 50% mean inhibition concentration (IC 50) with confidence limits (CL) at 95% calculated by Litchfield and Wilcoxon test. Each determination was carried out in triplicate and repeated at least three times.

2.9. Statistical Analysis

The results are expressed as mean values ± standard deviation (SD) from three separate experiments. The results of total phenolic content and antioxidant activity were analyzed using Student's *t* test. Differences were considered to be statistically significant at $p < 0.05$.

3. Results

3.1. Quality Parameters

The qualitative parameters of the oils studied, obtained using analytical methods described by the International Olive Council [27], highlight some defects of Algerian monovarietal oils.

The results are summarized in Table 1. The oil obtained from the Zebboudj cultivar had a free acidity value (1.25 g oleic acid 100 g^{-1} of oil) higher than the established limit, whereas the oils obtained from the Manzanilla and Gelb Elfarroudj cultivars had free acidity values 0.80 and 0.75 respectively; the first value was equal to the maximum allowed limit and the second one was very close to it. The peroxide values were also quite high even if below the maximum allowed limit; in particular, they were between 12.75 meq O_2 kg^{-1} and 15.50 meq O_2 kg^{-1}. The conjugated trienes K_{270} and dienes K_{232} showed values in the range of 0.12–0.21 and 1.47–2.30, respectively, while ΔK results were in the range of −0.007–0.11.

Table 1. Results of qualitative parameters of monovarietal Algerian olive oils under study.

Cultivars	Free Acidity (%)	PV (meq O_2/kg)	K_{232}	K_{270}	ΔK
Tefahi	0.48 ± 0.03 c	12.68 ± 1.23 b	1.98 ± 0.17 b	0.17 ± 0.02 a	−0.002 ± 0.01 c
Chemlal	0.50 ± 0.05 c	12.75 ± 1.13 b	2.29 ± 0.28 a	0.19 ± 0.01 a	0.001 ± 0.01 b
Gelb Elfarroudj	0.75 ± 0.06 b	15.50 ± 1.53 a	1.97 ± 0.26 b	0.21 ± 0.03 a	0.003 ± 0.03 b
Manzanilla	0.80 ± 0.04 b	15.50 ± 0.24 a	1.47 ± 0.24 c	0.12 ± 0.01 b	−0.007 ± 0.01 d
Zebboudj	1.25 ± 0.11 a	12.51 ± 0.32 b	0.39 ± 0.03 d	0.11 ± 0.05 b	0.11 ± 0.05 a

Free acidity = g oleic acid 100 g^{-1} of oil, peroxide value (PV) = meq O_2 kg^{-1} of oil. Results were expressed as mean ± SD. Different letters in columns indicate statistical differences at $p < 0.05$ (Student's t test) among cultivars.

3.2. Fatty Acid Composition

Using GC-MS, 12 fatty acids were detected as shown in Table 2. The composition of the samples was a combination of four saturated fatty acids (palmitic acid C16:0, margaric acid C17:0, stearic acid C18:0, and arachidic acid C20:0); six mono-unsaturated fatty acids (palmitoleic acid C16:1 (7-cis) (ω-9), palmitoleic acid C16:1 (9-cis) (ω-7), heptadecenoic acid C17:1, oleic acid C18:1 (ω-7), oleic acid C18:1 (ω-9), and 11-eicosenoic acid C20:1 (ω-9)), and two poly-unsaturated fatty acids (linoleic acid C18:2 (ω-6) and α-linolenic acid C18:3 (ω-3)). The monounsaturated fatty acids were higher in the monovarietal oils of Gelb Elfarroudj, Tefahi, and Manzanilla, due to the high content of oleic acid C18:1 (ω-9). Polyunsaturated fatty acids content was highest in Zebboudj oil mainly due to the high content of linoleic acid. The percentages of fatty acids in our work are very close to those of other Algerian cultivars [28]. Our results in the concentrations of palmitic acid, linoleic acid, and α-linolenic are also in agreement with those reported in other studies about Chemlal Algerian olive oil [29].

Table 2. Fatty acid methyl esters (in percentage) identified by GC-MS in the monovarietal olive oils under study.

Fatty Acids	Tefahi	Chemlal	Gelb Elfarroudj	Manzanilla	Zebboudj
			Cultivars		
C16:0	12.27 ± 0.04 b	15.75 ± 0.02 a	12.70 ± 0.05 b	9.97 ± 0.31 c	11.59 ± 0.09 b
C16:1(ω-7)	0.14 ± 0.01 b	0.35 ± 1.5 a	0.16 ± 0.01 b	0.20 ± 0.01 b	nf
C16:1 (ω-9)	1.12 ± 0.01 b	1.05 ± 0.22 a	1.21 ± 0.05 b	0.57 ± 0.01 c	0.10 ± 0.01 d
C17:0	0.03 ± 0.01 b	0.91 ± 1.56 a	0.08 ± 0.02 b	0.05 ± 0.04 b	0.08 ± 0.01 b
C17:1	0.05 ± 0.01 b	0.15 ± 0.09 a	0.10 ± 0.01 b	0.08 ± 0.01 b	0.05 ± 0.01 b
C18:0	2.46 ± 0.04 c	4.95 ± 0.88 a	2.68 ± 0.05 c	2.39 ± 0.03 c	4.27 ± 0.01 b
C18:1(ω-7)	3.17 ± 0.11 a	2.37 ± 0.55 b	3.48 ± 0.39 a	2.31 ± 0.11 b	1.55 ± 0.03 c
C18:1(ω-9)	66.93 ± 0.26 b	59.97 ± 1.11 c	66.03 ± 0.49 b	74.9 ± 0.35 a	26.65 ± 0.08 c
C18:2(ω-6)	12.43 ± 0.17 b	9.34 ± 0.33 c	12.10 ± 0.09 b	8.18 ± 0.06 c	48.51 ± 0.02 a
C18:3(ω-3)	0.84 ± 0.09 b	0.83 ± 0.01 b	0.97 ± 0.10 b	0.81 ± 0.01 b	6.35 ± 0.01 a
C20:0	0.26 ± 0.04 a	0.32 ± 0.01 a	0.33 ± 0.03 a	0.27 ± 0.02 a	0.33 ± 0.01 a
C20:1 (ω-9)	0.27 ± 0.07 b	0.25 ± 0.01 b	0.24 ± 0.05 b	0.31 ± 0.02 a	0.16 ± 0.01 c
ΣSFA [†] %	15.02 b	21.93 a	15.69 b	12.58 c	16.27 b
ΣMFA [‡] %	71.7 a	64.14 b	71.2 a	78.4 a	28.5 c
ΣPFA [§] %	13.27 b	10.17 c	13.07 b	8.99 c	54.86 a

The results are expressed as mean % ± SD of total fatty acid methyl esters (n = 3). [†]: saturated fatty acid; [‡]: monounsaturated fatty acid; [§]: polyunsaturated fatty acid; nf = not found. Different letters for each fatty acid indicate statistical differences at $p < 0.05$ (Student's t test) among cultivars.

3.3. Individual Phenolic Compounds Identification

In our study, UHPLC-HESI-MS was used for identification of the phenolic compounds. Table 3 shows the calibration data of reference compounds used as external standards in the UHPLC-HESI-MS characterization.

Table 3. Calibration data of reference compounds used as external standards for identification of phenolic compounds.

Phenolic Compounds	Molecular Formula	Experimental m/z $[M-H]^-$	Retention Time	Linear Regression	(r^2)
Tyrosol	$C_8H_{10}O_2$	137.05933	12.05	$Y = -124{,}926 + 8556 \times X$	0.9937
Hydroxytyrosol	$C_8H_{10}O_3$	153.05438	9.15	$Y = -5.29 \times 10^6 + 7.31 \times 10^7 \times X$	0.9974
p-Coumaric Acid	$C_9H_8O_3$	163.03847	18.30	$Y = -5.83 \times 10^6 + 8.72 \times 10^7 \times X$	0.9959
Caffeic Acid	$C_9H_8O_4$	179.03392	14.90	$Y = -2.65 \times 10^6 + 8.46 \times 10^7 \times X$	0.9750
Ferulic Acid	$C_{10}H_{10}O_4$	193.05026	19.10	$Y = -5.99 \times 10^6 + 5.15 \times 10^7 \times X$	0.9983
Vanillic Acid	$C_8H_8O_4$	167.03388	14.30	$Y = -412{,}147 + 2.85 \times 10^6 \times X$	0.9812
p-Hydroxybenzoic Acid	$C_7H_6O_3$	137.02304	13.50	$Y = 648{,}298 + 6.49 \times 10^7 \times X$	0.9982
Syringic Acid	$C_9H_{10}O_5$	197.04519	14.56	$Y = 746{,}927 + 5.54 \times 10^6 \times X$	0.9852
Cinnamic Acid	$C_9H_8O_2$	147.04049	3.50	$Y = -1.69 \times 10^6 + 2.42 \times 10^6 \times X$	0.9888
Gallic Acid	$C_7H_6O_5$	169.01319	7.45	$Y = -1.94 \times 10^6 + 4.30 \times 10^7 \times X$	0.9968
Protocatechuic Acid	$C_7H_6O_4$	153.01763	10.35	$Y = -2.85 \times 10^6 + 2.09 \times 10^7 \times X$	0.9929
Oleocanthal	$C_{17}H_{20}O_5$	303.12387	23.50	$Y = -3.19 \times 10^6 + 3.09 \times 10^7 \times X$	0.9984
Vanillin	$C_8H_8O_3$	151.03867	17.22	$Y = -318{,}221 + 5.02 \times 10^6 \times X$	0.9994
Pinoresinol	$C_{20}H_{22}O_6$	357.13445	23.05	$Y = -61{,}525.2 + 634911 \times X$	0.9988
Oleuropein	$C_{25}H_{32}O_{13}$	539.17732	20.60	$Y = -3.11 \times 10^6 + 3.10 \times 10^7 \times X$	0.9985
Luteolin	$C_{15}H_{10}O_6$	285.04037	23.30	$Y = 3.61 \times 10^6 + 1.24 \times 10^8 \times X$	0.9992
Diosmetin	$C_{16}H_{12}O_6$	299.05621	24.05	$Y = 1.435 \times 108 \times X$	0.9853
Apigenin	$C_{15}H_{10}O_5$	269.04565	24.00	$Y = 2.37 \times 10^7 + 2.18 \times 10^8 \times X$	0.9989
Apigenin-7-glucoside	$C_7H_6O_5$	431.09854	22.90	$Y = 9.18 \times 10^6 + 7.56 \times 10^7 \times X$	0.9954

The UHPLC-HESI-MS analysis allowed the identification of 13 phenolic compounds and the class of secoiridoids is the most represented as shown in Table 4.

Table 4. Phenolic compounds ($mgkg^{-1}$) identified in the monovarietal olive oils by UHPLC-ESI-MS.

Phenolic Compound		Cultivars				
		Tefahi	Chemlal	Gelb Elfarroudj	Manzanilla	Zebboudj
Phenolic alcohol	Hydroxytyrosol	0.41 a	nf	nf	0.24a	Nf
Phenolic acids	*p*-Coumaric acid	0.54 b	nf	0.61 a	nf	0.41 b
	Caffeic acid	0.07 c	0.98 c	1.22 a	0.8 b	0.72 a
	Ferulic acid	0.41 a	0.55 c	0.68 b	nf [†]	nf
	Vanillic acid	nf	nf	0.70 c	nf [†]	1.43 b
	p-Hydroxy benzoic acid	nf	nf	0.42 a	0.32 c	0.45 a
Secoiridoids	Oleacein (3,4-DHPEA-EDA)	8.01 a	2.50 b	2.30 a	25.16 b	nf
	Oleocanthal (p-HPEA-EDA)	2.07 b	3.54 c	1.13 b	4.20 a	nf
	Oleuropein aglycon (3,4-DHPEA-EA)	49.65 a	7.28 a	13.71 a	62.67 a	0.13 a
	Deacetoxy-10-hydroxy oleuropein aglycon (DAc-10-OH Ole Agly);	0.61 a	0.18 a	nf	0.84 b	nf
	Elenolic acid	16.88 b	3.70 b	1.72 a	3.18 b	nf
	Ligstroside aglycon (p-HPEA-EA)	nf	8.09 c	3.46 c	12.86 a	0.22 a
Flavonoids	Luteolin	0.83 b	0.54 a	0.51 b	0.41 a	0.14 b

The results are given by the mean value ($n = 3$) of independent determinations, including extraction and injection. Different letters for each phenolic compound indicate statistical differences at $p < 0.05$. nf = not found.

Monovarietal oil samples showed a very similar phenolic composition. Differences from a quantitative point of view were recorded, in particular for the portion of the secoiridoids where

3,4-DHPEA-EA (oleuropein aglycon) was the main compound found. With regard to flavonoid derivatives, luteolin was detected in all samples, contrary to diosmetin and apigenin-7-glucoside.

The main phenolic compound was 3,4-DHPEA-EA (oleuropein aglycon) with contents varying from 62.67 mg kg^{-1} to 0.13 mg kg^{-1}, together with p-HPEA-EA (ligstroside aglycon) that ranged from 12.86 mg kg^{-1} to 0.22 mg kg^{-1}; 3,4-DHPEA-EDA (oleacein) was found in a range of 2.30 mg kg^{-1} to 25.16 mg kg^{-1}; p-HPEA-EDA (oleocanthal) showed the highest content (3.54 mg kg^{-1}) in Chemlal and the lowest content (2.07 mg kg^{-1}) in Tefahi.

In the group of phenolic acids, the amounts of the compounds were less abundant in olive oils than the amount of the secoiridoids group. p-coumaric acid, caffeic acid, ferulic acid, vanillic acid, and p-hydroxybenzoic acid were the phenolic acids found in some of the oils with small amounts.

3.4. Antioxidant Activity and Total Content of Phenolic Compounds

The results of the in vitro antioxidant assays performed on the olive oil phenolic fraction extracts are summarized in Table 5. They revealed that the oil PFE issued from the Tefahi cultivar was the most active, while Zebboudj was the least active and did not give any result in 3/5 of the assays (DPPH, FRAP, and TEAC). There was no significant difference observed between the oil PFE of Tefahi and Manzanilla cultivars in the results obtained with FRAP, TEAC, and BCB assays and much closer results were observed in the Folin–Ciocalteu test between the oils of Manzanilla, Chemlal, and Gelb Elfarroudj.

Table 5. Antioxidant activity and total phenolic content of the olive oil phenolic fraction extracts as measured by means of five in vitro assays.

Varieties	Folin–Ciocalteu µg GAEmg^{-1} ($\pm$SD)	DPPH mmol TEmg^{-1} ($\pm$SD)	FRAP mmol Fe^{2+} Emg^{-1} ($\pm$SD)	TEAC mM TEmg^{-1} ($\pm$SD)	BCB IC$_{50\%}$ mgml^{-1} (CL95)
Tefahi	237.19 ± 23.70	0.70 ± 0.08	0.84 ± 0.12	0.31 ± 0.07	1.84 (1.52–2.15)
Chemlal	59.24 ± 8.99 **	0.05 ± 0.001 **	0.12 ± 0.02 **	0.008 ± 0.001 **	14.81 (12.74–17.21)*
Gelb Elfarroudj	48.94 ± 5.65 **	0.03 ± 0.004 **	0.044 ± 0.002*	0.020 ± 0.003 **	8.46 (7.41–9.66)*
Manzanilla	46.32 ± 9.10 **	0.22 ± 0.06 **	0.34 ± 0.20	0.37 ± 0.02	6.54 (5.36–7.99)
Zebboudj	37.34 ± 4.71 **	-	-	-	7.44 (5.62–9.84)

Results are expressed as mean ± SD of three experiments for DPPH, FRAP, and TEAC tests, and as 50% mean inhibition concentration (IC$_{50}$) with confidence limits (CL) at 95% for the BCB assay. GAE: Gallic acid equivalents; TE: Trolox equivalents. * $p < 0.05$; ** $p < 0.01$ versus Tefahi oil phenolic fraction extract.

4. Discussion

The qualitative parameters of the studied oils highlight very high values which bring into question their quality. In particular, Zebboudj, Manzanilla, and Gelb Elfarroudj oils had high values of free acidity (0.75–1.25 g of oleic acid 100 g^{-1} of oil), which could be connected with the prolonged ripening before the olive harvest.

The amount of phenolic compounds in EVOO is an important factor when evaluating its quality, given that the natural phenols improve its resistance to oxidation, and to a certain extent, are responsible for its sharp bitter taste.

In this study, for the first time, the composition of secoiridoids in olive oils of some Algerian cultivars was analyzed with UHPLC-HESI-MS. The most abundant secoiridoids of the olive oil analyzed were the aglycones of oleuropein and ligstroside, and these results agree with previous data found for Algerian Chemlal olive oils [30,31]. In general, all the samples showed lower concentrations of oleocanthal, if compared to the other cultivars of the countries of the Mediterranean basin [19,20].

We suggest that the composition in phenolic compounds of the oils under study could be higher if the harvesting time is done earlier; previous studies demonstrated that olives have the highest phenolic compound content at the phase between green and darker skin [13,32]. The extraction of olive oil from black olive fruits was justified by the Algerian producers in that it allows them to get better olive oil yield [33].

The phenolic compounds are known to contribute to antioxidant/radical scavenging activity; their concentrations have a relationship with the percentage of radical inhibition [34]. Consequently, the lack of antioxidant activity in some experiments when testing Zebboudj oil PFE could be explained by the low concentration of phenolic compounds. However, Zebboudj oil PFE demonstrated good BCB results, probably due to the presence of vanillic acid, since previous studies reported its antioxidant and protective effects on the peroxidation of lipids [35].

The antioxidant activity of Tefahi oil PFE may be related to its composition in secoiridoids mainly due to oleuropein aglycon (3,4-DHPEA-EA) and elenolic acid. The antioxidant activity of Manzanilla oil PFE is probably due to the presence also of high amounts of secoiridoids, mostly oleuropein aglycon (3,4-DHPEA-EA) and ligstroside aglycon (p-HPEA-EA). We noticed that in both Tefahi and Manzanilla, there were high amounts of oleuropein aglycon (3,4-DHPEA-EA), a hydrophilic secoiridoid that demonstrated in other works a metal-chelation and a free-radical-scavenging action [36]. Some other phenolics may contribute to the antioxidant activity with their remarkable presence such as the phenolic acid oleacein (3,4-DHPEA-EDA) in both Tefahi and Manzanilla oil PFE.

In another study about the Manzanilla cultivar growing in Australia, the authors reported amounts of caffeic acid (0.46 mg kg^{-1}) less than the amounts found in this study (0.8 mg kg^{-1}), whereas they detected other compounds that we could not find such as vanillic and ferulic acids. The same study reported that a significant gradual decrease was noted in major polyphenolic compounds in the later harvest stage [37].

Other authors tested the antioxidant activity of Algerian olive oils (cultivar Chemlal and *Olea europaea* L. subsp. Oleaster) extracted using laboratory technics instead of extraction in the mill, and the results of antioxidant activity were higher with respect to our results [30,38]. Unfortunately, there are no studies on the composition and antioxidant activity of olive oils from Algerian cultivars growing in the same region of our study (Setif and Batna); for this reason, it is difficult to compare the results. Certainly the region of cultivation of the various cultivars plays a decisive role in the chemical characteristics of the oils produced [39].

The technics used in the commercial mills for the extraction of olive oil may play also a crucial role in the olive oil quality such as the malaxation temperature. In literature it was reported that a significant increment of total phenols concentration was found with a maximum at 27 °C, whereas for higher temperatures (30–36 °C), a progressive decrement was observed [40]. The malaxation temperature used in the commercial mill in this study was 30 °C; we suggest that reducing this temperature may give better olive oil quality.

In our study, the olive oils were obtained with the method used from local producers in Algeria. This method did not give us a very high olive oil quality. In general, to produce EVOOs with high quality which reinforces its enhancement in the international market, a better knowledge of the local cultivars and the best agronomic techniques are necessary.

5. Conclusions

Antioxidant activity as well as fatty acids composition, phenolic acids, and secoiridoids content of some Algerian monovarietal olive oils were investigated. The results indicate that the quality parameters are acceptable for all oils, except for Zebboudj oil. The most important results from the chemical point of view were the phenolic portion. Manzanilla reported the greatest quantity of secoiridoids in particular as regards oleuropein aglycon, ligstroside aglycon, and oleacein (62.67 mg kg^{-1}, 12.86 mg kg^{-1}, and 25.16 mg kg^{-1}, respectively). Tefahi oil showed quality parameters in line with international regulations, an appreciable content of monounsaturated fatty acids (66.93%) but modest levels of oleuropein aglycon (49.65 mg kg^{-1}) and other secoiridoids if compared to those of other oils of the Mediterranean areas. The antioxidant activity of Tefahi oil was found to be the best among the cultivars used. This study constitutes a starting point for the diffusion and marketing of Tefahi monovarietal oils due to their phenolic content and antioxidant capacity.

This is just a preliminary report since only a single sample from each cultivar was investigated. However, a more in-depth study must be done on agronomic techniques and ripening stage before the olive harvest takes place, which has been shown to have great influence on the quality of the oils produced.

Author Contributions: Conceptualization, V.D.S. and M.G.M.; methodology, V.D.S. and S.B.; software, M.C.; validation, V.D.S., S.B., C.M., and M.C.; formal analysis, V.D.S., S.B., C.M., and M.C.; investigation, V.D.S., S.B., C.M., and M.C.; resources, V.D.S. and M.C.; data curation, M.C., V.D.S., C.M., and S.B.; writing—original draft preparation, S.B., V.D.S., and M.C.; writing—review and editing, M.G.M. and V.D.S.; visualization, S.B.; supervision, V.D.S. and M.C.; project administration, M.G.M.; funding acquisition, M.G.M. All authors have read and agreed to the published version of the manuscript.

Funding: This work has been partially supported by National Research Council, CNR—DISBA project NutrAge (project nr.7022).

Acknowledgments: The authors are grateful to Loucif Ayache, an investor in agriculture, owner of the modern mill and of some of the cultivars. He provided olive oil samples and allowed the inquiry at the mill. The authors thank Dott.ssa Lucia Sollima for her administrative support.

Conflicts of Interest: The authors declare that there is no conflict of interest.

References

1. Gomez Caravaca, A.M.; Maggio, R.M.; Cerretani, L. Chemometric applications to assess quality and critical parameters of virgin and extra-virgin olive oil. *Anal. Chim. Acta* **2016**, *913*, 1–21. [CrossRef]
2. Ouni, Y.; Taamalli, A.; Maria Gómez-Caravaca, A.; Segura-Carretero, A.; Fernández-Gutiérrez, A.; Zarrouk, M. Characterisation and quantification of phenolic compounds of extra-virgin olive oils according to their geographical origin by a rapid and resolutive LC-ESI-TOF MS method. *Food Chem.* **2011**, *127*, 1263–1267. [CrossRef]
3. Deng, J.; Xu, Z.; Xiang, C.; Liu, J.; Zhou, L.; Li, T.; Yang, Z.; Ding, C. Comparative evaluation of maceration and ultrasonic-assisted extraction of phenolic compounds from fresh olives. *Ultrason. Sonochem.* **2017**, *37*, 328–334. [CrossRef]
4. Genovese, A.; Yang, N.; Linforth, R.; Sacchi, R.; Fisk, I. The role of phenolic compounds on olive oil aroma release. *Food Res. Int.* **2018**, *112*, 319–327. [CrossRef]
5. Ghanbari, R.; Anwar, F.; Alkharfy, K.M.; Gilani, A.H.; Saari, N. Valuable Nutrients and Functional Bioactives in Different Parts of Olive (*Olea europaea* L.). *Int. J. Mol. Sci.* **2012**, *13*, 3291–3340. [CrossRef]
6. Reboredo-Rodríguez, P.; Varela-López, A.; Forbes-Hernández, T.Y.; Gasparrini, M.; Afrin, S.; Cianciosi, D.; Zhang, J.; Pia Manna, P.; Bompadre, S.; Quiles, J.L.; et al. Phenolic Compounds Isolated from Olive Oil as Nutraceutical Tools for the Prevention and Management of Cancer and Cardiovascular Diseases. *J. Funct. Foods* **2018**, *19*, 2305. [CrossRef]
7. García-Rodríguez, R.; Belaj, A.; Romero-Segura, C.; Sanz, C.; Pérez, A.G. Exploration of genetic resources to improve the functional quality of virgin olive oil. *J. Funct. Foods* **2017**, *38*, 1–8. [CrossRef]
8. Oueslati, I.; Krichene, D.; Manaï, H.; Taamalli, W.; Zarrouk, M.; Flamini, G. Monitoring the volatile and hydrophilic bioactive compounds status of fresh and oxidized Chemlali. *Food Res. Int.* **2018**, *112*, 425–433. [CrossRef]
9. Jiménez, B.; Sánchez-Ortiz, A.; Lorenzo, M.L.; Rivas, A. Influence of fruit ripening on agronomic parameters, quality indices, sensory attributes and phenolic compounds of Picudo olive oils. *Food Res. Int.* **2013**, *54*, 1860–1867. [CrossRef]
10. Ammar, S.; Kelebek, H.; Zribi, A.; Abichou, M.; Selli, S.; Bouaziz, M. LC-DAD/ESI-MS/MS characterization of phenolic constituents in Tunisian extra-virgin olive oils: Effect of olive leaves addition on chemical composition. *Food Res. Int.* **2017**, *100*, 477–485. [CrossRef]
11. Natasa, P.K.; Reza, A.; Nikolaos, S.T. Application of an advanced and wide scope non-target screening workflow with LC-ESI-QTOF-MS and chemometrics for the classification of the Greek olive oil varieties. *Food Chem.* **2018**, *256*, 53–61.
12. Salvo, A.; Rotondo, A.; La Torre, G.L.; Cicero, N.; Dugo, G. Determination of 1,2/1,3-diglycerides in Sicilian extra-virgin olive oils by ^{1}H-NMR over a one-year storage period. *Nat. Prod. Res.* **2017**, *31*, 822–828. [CrossRef]

13. Bakhouche, A.; Lozano Sánchez, J.; Bengana, M.; Fernández Gutiérrez, A.; Segura Carretero, A. Time course of Algerian Azeradj extra virgin olive oil quality during olive ripening. *Eur. J. Lipid Sci. Technol.* **2015**, *117*, 389–397. [CrossRef]

14. Yakhlef, W.; Arhab, R.; Romero, C.; Brenes, M.; de Castro, A.; Medina, E. Phenolic composition and antimicrobial activity of Algerian olive products and by-products. *LWT Food Sci. Technol.* **2018**, *93*, 323–328. [CrossRef]

15. Abdul-Hussain, K.H.; Abdul-Hussain, M.S. Influence of the gibberelic acid on the germination of the seeds of olive-tree *Olea europaea* L. *J. Cent. Eur. Agric.* **2004**, *5*, 1–4.

16. International conventions and agreements-laws and decrees, judgments, decisions, notices, communications and announcements. *Off. J. People's Democr. Repub. Alger.* **2008**, *5*, 46.

17. Algerian Ministry of Agriculture. Fixation des Conditions D'éligibilité au Soutien sur le Compte D'affectation Spéciale no.302-139 Intitulé "Fonds National de Développement Agricole" ligne 1 "Développement de L'investissement Agricole", Ainsi que les Modalités de Paiement des Subventions; Decision no. 414; 2014. Available online: https://eur-lex.europa.eu/legal-content/EN/ALL/?uri=CELEX:32016R2095 (accessed on 22 October 2020).

18. European Commission. Regulation no. 2016/2095 amending regulation no. 2568/91 on the characteristics of olive oil and olive-residue oil and on the relevant methods of analysis. Available online: https://eur-lex.europa.eu/legal-content/EN/ALL/?uri=CELEX:32016R2095 (accessed on 1 December 2016).

19. Di Stefano, V.; Melilli, M.G. Effect of storage on quality parameters and phenolic content of Italian extra-virgin olive oils. *Nat. Prod. Res.* **2020**, *34*, 78–86. [CrossRef]

20. Grilo, F.; Novara, M.E.; D'Oca, M.C.; Rubino, S.; Lo Bianco, R.; Di Stefano, V. Quality evaluation of extra-virgin olive oils from Sicilian genotypes grown in a high-density system. *Int. J. Food Sci. Nutr.* **2020**, *71*, 397–409. [CrossRef]

21. Di Stefano, V.; Avellone, G.; Bongiorno, D.; Indelicato, S.; Massenti, R.; Lo Bianco, R. Quantitative evaluation of the phenolic profile in fruits of six avocado (Persea americana) cultivars by ultra-high-performance liquid chromatography-heated electrospray-mass spectrometry. *Int. J. Food Prop.* **2017**, *6*, 1–27. [CrossRef]

22. Napoli, E.; Siracusa, L.; Ruberto, G.; Carrubba, A.; Lazzara, S.; Speciale, A.; Cimino, F.; Saija, A.; Cristani, M. Phytochemical profiles, phototoxic and antioxidant properties of eleven Hypericum species. A comparative study. *Phytochemistry.* **2018**, *152*, 162–173. [CrossRef]

23. Ouerghemmi, S.; Sebei, H.; Siracusa, L.; Ruberto, G.; Saija, A.; Cimino, F.; Cristani, M. Comparative study of phenolic composition and antioxidant activity of leaf extracts from three wild Rosa species grown in different Tunisia regions: Rosa canina L., Rosa moschata Herrm. And *Rosa sempervirens* L. *Ind. Crops Prod.* **2016**, *94*, 167–177. [CrossRef]

24. Dehimi, K.; Speciale, A.; Saija, A.; Dahamna, S.; Raciti, R.; Cimino, F.; Cristani, M. Antioxidant and anti-inflammatory properties of Algerian Thymelaeamicrophylla Coss. And Dur. extracts. *Pharmacogn. Mag.* **2016**, *12*, 203–210.

25. Boussahel, S.; Speciale, A.; Dahamna, S.; Amar, Y.; Bonaccorsi, I.; Cacciola, F.; Cimino, F.; Donato, P.; Ferlazzo, G.; Harzallah, D.; et al. Flavonoid profile, antioxidant and cytotoxic activity of different extracts from Algerian Rhamnus alaternus L. bark. *Pharmacogn. Mag.* **2015**, *11*, 102–109.

26. Martorana, M.; Arcoraci, T.; Rizza, L.; Cristani, M.; Bonina, F.P.; Saija, A.; Trombetta, D.; Tomaino, A. In vitro antioxidant and in vivo photoprotective effect of pistachio (Pistacia vera L. variety Bronte) seed and skin extracts. *Fitoterapia* **2013**, *85*, 41–48. [CrossRef]

27. International Olive Council. Trade Standard Applying to Olive Oils and Olive Pomace Oils COI/T.15/NC no. 3/Rev 13. Available online: https://www.internationaloliveoil.org/wp-content/uploads/2019/11/COI-T.15-NC. -No-3-Rev.-13-2019-Eng.pdf (accessed on 23 October 2020).

28. Boudour Benrachou, N.; Plard, J.; Pinatel, C.; Artaud, J.; Dupuy, N. Fatty Acid Compositions of Olive Oils from Six Cultivars from East and South-Western Algeria. *Adv. Food Technol. Nutri. Sci.* **2017**, *3*, 1–5. [CrossRef]

29. Laincer, F.; Iaccarino, N.; Amato, J.; Pagano, B.; Pagano, A.; Tenore, G.; Tamendjari, A.; Rovellini, P.; Venturini, S.; Bellan, G.; et al. Characterization of monovarietal extra virgin olive oils from the province of Béjaïa (Algeria). *Food Res. Int.* **2016**, *89*, 1123–1133. [CrossRef]

30. Laincer, F.; Laribi, R.; Tamendjari, A.; Arrar, L.; Rovellini, P.; Venturini, S. Olive oils from Algeria: Phenolic compounds, antioxidant and antibacterial activities. *Grasas Aceites* **2014**, *65*, 1–10.

31. Louadj, L.; Giuffre, A.M. Analytical characteristics of olive oil produced with three different processes in Algeria. *Riv. Ital. Sostanze Grasse.* **2010**, *87*, 186–195.

32. Bengana, M.; Bakhouche, A.; Lozano Sánchez, J.; Amir, Y.; Youyou, A.; Segura Carretero, A.; Fernández Gutiérrez, A. Influence of olive ripeness on chemical properties and phenolic composition of "Chemlal" extra-virgin olive oil. *Food Res. Int.* **2013**, *54*, 1868–1875. [CrossRef]

33. Yangui, A.; Costa-Font, M.; Gil, J.M. The effect of personality traits on consumers' preferences for extra virgin olive oil. *Food Qual Prefer.* **2016**, *51*, 27–38. [CrossRef]

34. McDonald, S.; Prenzler, P.D.; Antolovich, M.; Robards, K. Phenolic content and antioxidant activity of olive extracts. *Food Chem.* **2001**, *73*, 73–84. [CrossRef]

35. Chou, T.H.; Ding, H.Y.; Hung, W.J.; Liang, C.H. Antioxidative characteristics and inhibition of α-melanocyte-stimulating hormone-stimulated melanogenesis of vanillin and vanillic acid from Origanumvulgare. *Exp. Dermatol.* **2010**, *19*, 742–750. [CrossRef] [PubMed]

36. Visioli, F.; Poli, A.; Galli, C. Antioxidant and Other Biological Activities of Phenols from Olives and Olive Oil. *Med. Res. Rev.* **2002**, *22*, 65–75. [CrossRef] [PubMed]

37. Alowaiesh, B.; Singha, Z.; Fangc, Z.; Kailisd, S.G. Harvest time impacts the fatty acid compositions, phenolic compounds and sensory attributes of Frantoio and Manzanilla olive oil. *Sci. Hortic.* **2018**, *234*, 74–80. [CrossRef]

38. Bouarroudja, K.; Tamendjaria, A.; Larbat, R. Quality, composition and antioxidant activity of Algerian wild olive (OleaeuropaeaL. subsp.Oleaster) oil. *Ind. Crop. Prod.* **2016**, *83*, 484–491. [CrossRef]

39. Mailer, R.J.; Ayton, J.; Graham, K. The Influence of Growing Region, Cultivar and Harvest Timing on the Diversity of Australian Olive Oil. *J. Am. Oil Chem. Soc.* **2010**, *87*, 877–884. [CrossRef]

40. Parenti, A.; Spugnoli, P.; Masella, P.; Calamai, L. The effect of malaxation temperature on the virgin olive oil phenolic profile under laboratory-scale conditions. *Eur. J. Lipid Sci. Technol.* **2008**, *110*, 735–741. [CrossRef]

Publisher's Note: MDPI stays neutral with regard to jurisdictional claims in published maps and institutional affiliations.

Article

Phenolics and Antioxidant Activity of Green and Red Sweet Peppers from Organic and Conventional Agriculture: A Comparative Study

Rosa Guilherme [1,2], **Alfredo Aires** [3], **Nuno Rodrigues** [4], **António M. Peres** [4] and **José Alberto Pereira** [4,*]

1   Instituto Politécnico de Coimbra, ESAC, Coimbra,
    Portugal & CERNAS-Centro de Estudos de Recursos Naturais, Ambiente e Sociedade, Bencanta,
    3045-601 Coimbra, Portugal; rguilherme@esac.pt
2   GeoBioTec, Faculdade de Ciências e Tecnologia, Universidade Nova de Lisboa, 2825-149 Caparica, Portugal
3   Centre for the Research and Technology for Agro-Environment and Biological Sciences, CITAB, UTAD,
    Quinta de Prados, 5000-801 Vila Real, Portugal; alfredoa@utad.pt
4   Centro de Investigação de Montanha (CIMO), ESA, Instituto Politécnico de Bragança,
    Campus de Santa Apolónia, 5300-253 Bragança, Portugal; nunorodrigues@ipb.pt (N.R.);
    peres@ipb.pt (A.M.P.)
*   Correspondence: jpereira@ipb.pt; Tel.: +351-2733030818

Received: 1 November 2020; Accepted: 16 December 2020; Published: 21 December 2020

Abstract: Today, consumers are very concerned regarding food quality, nutritional composition and positive health effects of consumed foods. In this context, the preference and consumption of organic products has been increasing worldwide. In the present work, sweet peppers in two maturation stages (i.e., green and red peppers) from organic and conventional production systems were evaluated in regards to phenolic composition and antioxidant activity. Nine phenolic compounds were identified and quantified by a high-performance liquid chromatography-diode-array detector (HPLC-DAD), namely resveratrol, meta-coumaric acid, ortho-coumaric acid, clorogenic acid, caffeic acid, myricetin, rutin, luteolin-7-O-glucoside and quercitin-3-O-rhamnoside. In contrast to the production system, the maturation stage showed a pronounced significant effect on the phenolic composition of the studied sweet peppers; in general, green peppers possessed higher contents than red ones. Meta-coumaric acid, ortho-coumaric acid and quercitin-3-O-rhamnoside were more abundant in green conventional peppers and chlorogenic acid, caffeic acid and rutin were found in higher levels in red organic peppers. Regarding the antioxidant activity, green conventional peppers showed the highest DPPH, ABTS$^{\bullet+}$ and total reducing capacities, while red conventional peppers had higher TEAC values. Finally, principal component analysis showed that the phenolic composition together with the antioxidant capacities could be used to differentiate the production system and the maturation stage of sweet peppers. This finding confirmed that both factors influenced the peppers' phenolic composition and antioxidant capacity, allowing their possible use as maturation–production biomarkers.

Keywords: phenolic compounds; resveratrol; linear discriminant analysis; production–maturation mode discrimination

---

1. Introduction

Sweet pepper (*Capsicum annuum* L.), is one of the most popular and consumed vegetables around the world. The high diversity of fruit forms and colors, in many cases related to its maturation degree, and also its pungency, specific taste and/or distinct aroma make sweet peppers very popular and an excellent ingredient to be included in many types of diets and dishes with high attractiveness for several types of consumers [1].

In the last decades, there has been an increasing concern by consumers for healthier, safer and high-quality foods produced under environmentally friendly practices and economically fair modes. In this sense, the worldwide demand for organic products has increased, and is expected to undergo a sharper increase in the coming years [2–4]. Consumers believe that organic products are of better quality, tastier, with high amounts of vitamins and other healthy compounds and are consequently more nutritious, and these perceptions are the main driver of the observed increase in preference for organic products [5]. This perception is usually related to the fact that the use of chemical fertilizers or synthetic plant protection products are not permitted in organic farming [6–8], being in-line with the reported higher levels of bioactive compounds reported for organic compared to conventional sweet peppers [8,9]. Sweet peppers are, in general, recognized as a potential food source of vitamins, phenolic compounds, carotenoids and flavonoids, which possess known positive health effects [8,10]. Besides the agronomic production system [11–16], the richness in bioactive compounds (e.g., carotenoids and phenolic compounds) and the related antioxidant capacities greatly depend on the sweet peppers' cultivar [17–20] and on the fruits' maturation stage [10,21]. Several phenolic compounds have been detected and quantified in sweet peppers. The phenolics have been detected in both free and bound forms [20]. Depending on the cultivar, production system, maturation stage, environmental-climatic conditions and geographical origin, several phenolics have been found, including flavonoids and hydroxycinnamic acids. In fact, apigenin, caffeic, chlorogenic, ferulic, p-courmaric, p-hydroxybenzoic, rosmarinic, sinapic and vanillic, acids, naringenin, quercetin-3-*O*-glucoside and luteolin have been found in different levels, not always considering the effect of some of the above mentioned factors [16,18–27].

Therefore, it still is of utmost interest to assess the specific phenolic composition of different sweet pepper cultivars, as well as to establish the possible effects of the production system and maturation stage. Indeed, it has been previously reported that the phenolic profiles alone or coupled with other chemical data (e.g., volatile fraction composition and antioxidant activity data, among others) allowed for differentiating peppers' cultivars [26,28], such as red pepper cultivars grown under different shade and controlled-temperature conditions [20], fresh and cooked sweet peppers of two cultivars [27], sweet and hot peppers [29] as well as fresh and dried red peppers grown under conventional or organic systems [16]. Recently, the research team has shown that chemical-sensory data and potentiometric signal profiles, recorded using a lab-made electronic tongue could be satisfactorily used as biomarkers of sweet peppers' agronomic production system (organic versus conventional) and the maturation stage (green versus red fruits) [30]. In the present work, we intended to evaluate the effect of the production system (organic and conventional) and maturation stage (green and red colors) on the phenolic composition and on the antioxidant activity of sweet peppers from the Entinas variety. These effects were further evaluated based on unsupervised (principal component analysis, PCA) and/or supervised (linear discriminant analysis, LDA) pattern recognition techniques. To limit/overcome the known influence of non-controlled external factors (e.g., sweet pepper varieties, agro-climatic conditions, soil compositions, harvest time-periods, among others), the study was limited to a single pepper variety grown by two producers within the same geographical area. This option allowed a deep evaluation of the two factors under study (production system and maturation stage), although it posed some additional difficulties in establishing an optimal and general study model.

2. Materials and Methods

2.1. Sweet Peppers Production System and Sampling

For this study, two different sweet peppers (*Capsicum annuum* L.) producers were selected. Both producers, one organic and one conventional, were located near Coimbra, in the center region of Portugal with around one kilometer of distance between fields. The organic producer followed the organic production European Commission guidelines [6], and the conventional producer followed the conventional rules of agricultural production without the limitations of use of pesticides and fertilizers. The peppers' producers were selected taking into account some aspects, namely, soil with similar physical and chemical characteristics. Soil of both fields was analyzed before the experiment. The organic field presented a pH value of 6.4, organic matter 1.5%, 105 mg P_2O_5 kg^{-1} (available P) and 156 mg K_2O kg^{-1} (available K), whereas the values for the conventional field were 6.0 for pH, 1.8% for organic matter, 181 mg P_2O_5 kg^{-1} for available P and 134 mg K_2O kg^{-1} for available K. The production occurred in open field conditions. The planting of sweet pepper seedlings, from the Entinas variety, was performed in the last days of May 2018. The number of plants per hectare was 22 222 (75 cm between rows, 60 between plants). In both fields, plants were drip-irrigated with water from Mondego River, with a flow rate of 2.66 L s^{-1} per emitter during 20 min every two or three days according to the climatic conditions. In the organic field, nutritional requirements were supplied by the incorporation of 29.6 t ha^{-1} (5.8 g of N kg^{-1}, 2.8 g P_2O_5 kg^{-1} and 5.3 g K_2O kg^{-1}) of horse manure on soil before planting. In the conventional field, chemical fertilizers were used at a rate of 700 kg ha^{-1} before sweet pepper planting. The fertilizer was composed of 7% N, 14% P_2O_5, 14% K_2O, 3% CaO, 2% MgO, 9% SO_3 and 0.02% B; this was reinforced in late July with 300 kg ha^{-1} of a fertilizer composed of 27% N and 4% CaO. A phytosanitary treatment was applied against aphids (100 g L^{-1} of deltamethrin) in the conventional field. The harvest of sweet peppers occurred in the second week of September 2018, and from each production system (organic and conventional) two different maturation stages were selected. The first, green, corresponds to completely developed fruits that have reached the requirements to be collected; and the red corresponds to an advancement of maturation when fruits are completely red. For each production system and each color, five independent 2 kg samples were collected, for a total of 20 independent samples. Afterward, the fruits were transported to the laboratory, washed, had all non-edible parts removed, frozen at -20 °C and lyophilized until analysis.

2.2. Sweet Peppers Phenolic Compounds Evaluation by High-Performance Liquid Chromatography-Diode-Array Detector (HPLC-DAD)

The evaluation of the phenolic composition of sweet pepper samples was performed using a solid–liquid extraction followed by a high-performance liquid chromatography (HPLC)-diode-array detector (DAD) injection to identify each phenolic present in each sample. The extraction of phenolic samples was performed by mixing 40 mg dried powder (dw) of each sample with 950 µL of 70% methanol and 50 µL of internal standard (naringin). Each mixture was agitated thoroughly in a vortex and placed in a thermoblock at 70 °C for 45 min, and then centrifuged (Centrifuge 5804 R, Eppendorf, Hamburg, Germany) at 4000 rpm for 15 min. The extracts were then filtered (Fisherbrand Ø 90 mm) and supernatants transferred to amber vials and stored at -20 °C (Chromacol 2-SVWK(A)ST-CPK, ThermoScientific, Langerwehe, Germany) until further analysis. The HPLC-DAD system used was a Gilson HPLC (Villers-le-bel, France) equipped with a Finnigan/Surveyor DAD (Thermo Electron, San Jose, CA, USA, C18 column (250 $\times$ 4.6 mm, 5 µm) (ACE, Aberdeen, Scotland) with an eluent composed of water with 0.1% of trifluoroacetic acid (TFA) (solvent A) and acetonitrile with 0.1% TFA (solvent B) and a flow rate of 1 mL min^{-1}. The gradient used started from 0% solvent B at 0 min, 0% solvent B at 5 min, 20% solvent B at 15 min, 50% solvent B at 30 min, 100% solvent B at 45 min, 100% solvent B at 50 min, 0% solvent B at 55 min and 0% solvent B at 60 min. The chromatograms were recorded at 254, 280, 320 and 370 nm. The identification of phenolics was based on their peak retention times, UV spectra and UV max absorbance bands in comparison with commercial standards (resveratrol, meta-coumaric acid, ortho-coumaric acid, chlorogenic acid, caffeic acid, myricetin, rutin, luteolin-7-*O*-glucoside, quercitin-3-*O*-ramnoside; Extrasynthese, Genay, Rhône, France) and literature. Phenolics were quantified using the internal standard method and the results expressed in µg g^{-1} dw as the mean ± standard error of three replicates.

2.3. Sweet Peppers Antioxidant Activity Assays

2.3.1. Preparation of Extracts

The different samples were split according to the production system (organic and conventional) and maturation stage (green and red). All peppers were washed, cut, cleaned of seeds, frozen in plastic bags and then freeze-dried. Afterward, the extraction was performed using 600 mg of freeze-dried sample in 150 mL of water–methanol solution (70:30, *v/v*), at 70 °C for 45 min. Afterward, methanol was removed using a rotary evaporator (Stuart Re300), and the remaining water was removed by freeze-drying until a dry extract was obtained. Typical electron transfer antioxidant-based assays were performed, including DPPH, ABTS$^{\bullet+}$ and TEAC.

2.3.2. Determination of the Blocking Effect of 2,2-Diphenyl-1-Picrilhhydrazyl Free Radicals (DPPH)

The evaluation method used to detect the ability to block DPPH free radicals from pepper extracts was described by Hatano et al. [31]. Thus, 0.3 mL of extract with predetermined concentrations of each sample was mixed with 2.7 mL of a methanolic solution containing DPPH radicals (6×10^{-5} mol L^{-1}) (Sigma–Aldrich, St Louis, MO, USA). The mixture was vigorously stirred and left to stand in the dark at room temperature for 60 min, until stable absorbance values were obtained. The spectrophotometric reading was done at 517 nm in a UV-Visible UV-1280 spectrophotometer model Shimadzu and the results were presented as a percentage of DPPH discoloration, using the following equation:

$$\% \text{ Blocking effect} = [(ADPPH - AA)/ADPPH] \times 100$$

The antioxidant activity values corresponding to the absorbance of the solution with the sample extract and the ADPPH values the absorbance of the DPPH solution (white).

2.3.3. Radical Scavenging Activity (ABTS$^{\bullet+}$)

The formation of the ABTS radical [2,2′-azinobis-(3-ethylbenzothiazoline-6-sulfonic acid)] is the basis of one of the spectrophotometric methods that has been applied to measure the antioxidant activity of products. The peppers' radical scavenging activity was carried out following the method of Sánchez et al. [32] with some modifications described below. To prepare the solution, 25 mL of ABTS$^{\bullet+}$ solution was used. The radical was generated using 0.440 mL of the potassium persulfate solution; after stirring, it was placed in the dark for 12 to 16 h. The solution was diluted with absolute ethanol until an absorbance of 0.70 ± 0.02 was obtained, read at $\lambda = 734$ nm. Once the radical was formed, 2 mL of ABTS$^{\bullet+}$ solution was mixed with 0.1 mL of the sample with the previously determined concentration. After 6 min of reaction, each sample was read at 734 nm on a Shimadzu UV-Visible UV-1280 spectrophotometer. The results were presented in percentage of ABTS$^{\bullet+}$ discoloration using the same equation presented in the free radical blocking effect (DPPH) method.

2.3.4. Reducing Power

To perform the assessment of the reducing power of the extracts, the method described by Berker et al. was used [33]. Thus, 1 mL of the extract solution was used, 2.5 mL of 0.2 M sodium phosphate buffer solution with a pH of 6.6 and 2.5 mL of potassium ferrocyanide ($K_3Fe(CN)_6$) to 1%. The formed mixture was stirred and incubated at 50 °C for 20 min in a water bath. After cooling the samples, 2.5 mL of 10% (*w/v*) trichloroacetic acid was added and stirred vigorously. After this, 2.5 mL of the mixture supernatant was removed and 2.5 mL of distilled water and 0.5 mL of 0.1% iron (III) chloride were added. After the mixture with all the necessary reagents was ready, we waited for 2 min and the reading to assess the reducing capacity was made at 700 nm absorbance. The results obtained were expressed in mg Trolox per g of sample.

2.4. Statistical Analysis

One-way ANOVA was applied to discuss the statistical significance of the agronomic production system–maturation stage effect on the individual and total phenolic composition as well as on the antioxidant radical scavenging activity. In the case that a statistically significant effect was detected (i.e., p-value < 0.05) the post-hoc multi-comparison Tukey's test was further used to identify which levels were or were not significantly different. Boxplots were used to visualize the statistical results. As usual, the 1st, 2nd (median) and 3rd quartiles were plotted and the box bars represented the values that are within the 1st and 3rd quartiles. Additionally, whiskers were plotted (vertical lines) from the middle of the top and bottom edges of each box. The whiskers were 1.5 times the inner quartile spread in length, being measured from the median. The whiskers provided an arbitrary cutoff point to identify data points that were possible outside values. Minimum and maximum values that fell outside the whisker range were plotted (dot symbols) and symbolized possible extreme values or outliers.

In addition, the differentiation/discrimination of the agronomic production system and/or fruit's maturation stage was also evaluated (unsupervised and supervised classifications) using, respectively, the principal component analysis (PCA) or the linear discriminant analysis (LDA). The unsupervised differentiation performance was evaluated using 3D-plots of the first three principal components (PCs). The supervised classification technique was implemented together with the simulated annealing (SA) algorithm (i.e., a variable selection meta-heuristic algorithm) to choose the non-redundant variables with the most discrimination potential and to minimize noise effects [34,35]. The predictive performances of the LDA-SA models were checked using two cross-validation (CV) variants, namely the leave-one-out cross-validation (LOO-CV) and the repeated K-fold-CV (with 10 repeats and K set equal to 4, allowing that 25% of the data were used for validation purposes at each iteration). For both variants, the percentage of correct classifications (i.e., the model's sensitivity) was calculated. The statistical analysis was performed using the Subselect [35] and MASS [36] packages of the open source statistical program R (version 2.15.1), at a 5% significance level.

3. Results and Discussion

3.1. Phenolic Composition of Entinas Sweet Peppers

Nine phenolic compounds (caffeic, chlorogenic, m-coumaric and o-coumaric acids, luteolin-7-*O*-glucoside, myricetin, resveratrol, rutin and quercetin-3-*O*-rhamnoside) were detected by HPLC-DAD in all the studied Entinas sweet pepper samples grown under both conventional and organic systems and at the two maturation stages (green and red peppers). The contents (in $\mu g\ g^{-1}$ dw) of the nine phenolic compounds quantified in Entinas peppers are shown in Figure 1 as well as in Table S1, according to the maturation stage (green and red peppers) and the agronomic production system (conventional and organic systems). The less abundant phenolic was m-coumaric acid (red organic peppers with a mean content of 1.15 $\mu g\ g^{-1}$ dw) and the most abundant one was luteolin-7-*O*-glucoside (green conventional peppers with a mean content of 458.54 $\mu g\ g^{-1}$ dw). Although the cultivar, production system, fruits' maturation stage, environmental-climatic conditions and post-harvest treatments significantly influence the peppers' phenolic composition and contents, in general, the identified phenolics and respective levels found for the Entinas peppers were in-line with the wide range of values previously reported for other sweet pepper cultivars [16,19–21,23,25–27,37]. As can be visualized, for each phenolic compound, the contents varied within each production system–maturation stage group. Still, significant statistical differences (p-values < 0.05, for one-way ANOVA) were found among the contents of the individual phenolic compounds of green and red peppers grown under conventional or organic systems. Green peppers had higher contents of chlorogenic, m-coumaric and o-coumaric acids, as well as of resveratrol, myricetin, luteolin-7-*O*-glucoside and quercetin-3-*O*-rhamnoside compared to red peppers; these two latter phenolics were the most abundant ones. This finding pointed out that, with the exception of caffeic acid and rutin, maturation promoted a significant decrease of the abundance of the majority of the individual phenolic compounds. Similar trends were reported for total phenolic contents decreasing with the ripening type in different pepper cultivars [38,39]. The observed decreasing trend could be tentatively attributed to the synthesis of amino acids (e.g., phenylalanine, tyrosine, among others), in the early stage of fruit ripening, which enter the metabolic pathway of the shikimic acid, acting as precursors of phenolic acid formation, leading to possible higher contents in green fruits. However, other works also showed that the phenolic content trend (decrease or increase) with the maturation stage could be cultivar-dependent [40]. Indeed, an increasing trend of the total phenolic contents (TPC) with the ripening time was described for green and red bell peppers [41].

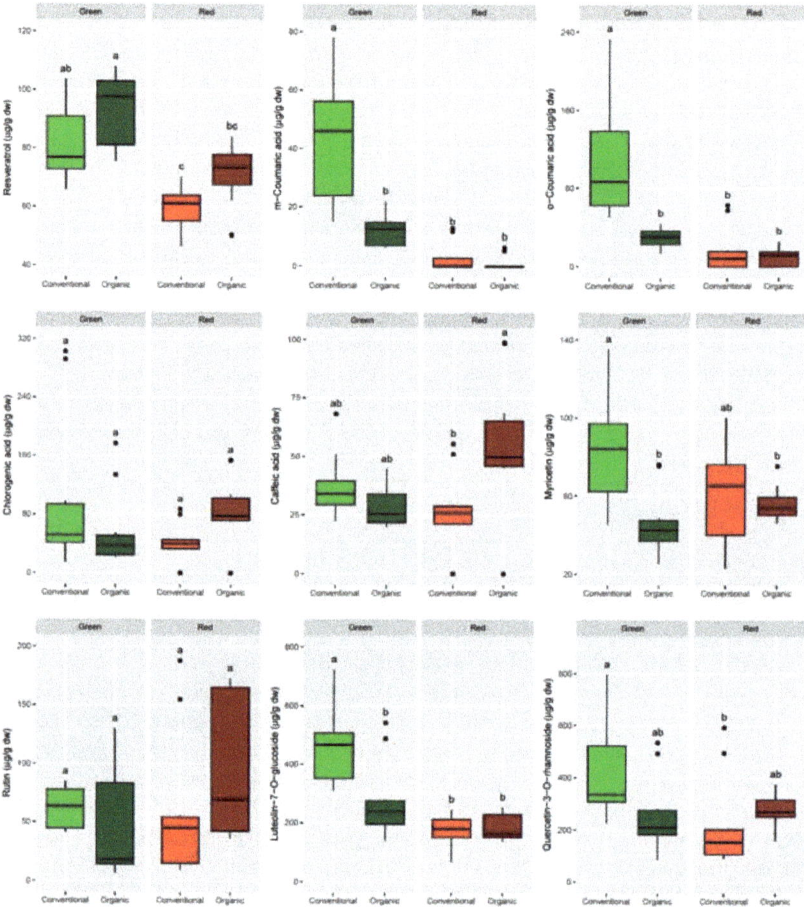

Figure 1. Individual phenolic compounds quantified (in µg g^{-1} dw), by High-Performance Liquid Chromatography-Diode-Array Detector (HPLC-DAD), in sweet pepper samples according to the maturation stage (green or red peppers) and agronomic production system (conventional or organic systems). Box bars represent the values that are within the 1st and 3rd quartiles and the horizontal line represents the 2nd quartile (median). Vertical lines from the middle of the top and bottom edges of each box represent the whiskers. Dot points represent minimum or maximum values that fall outside the whisker range (extreme values or outliers). Different lower case letters mean significant statistical differences (*p*-value < 0.05) among maturation–production levels.

On the other hand, although several studies reported that peppers grown under the organic system were richer in phenolics, namely in total phenol content compared to those grown under the conventional system [8,9,14,16], the results of the present study (Figure 1) show that, sometimes, conventional peppers possessed greater contents of some individual phenolic compound.

In fact, Entinas peppers produced in the conventional system had higher contents of m-coumaric and o-coumaric acids, myricetin, luteolin-7-*O*-glucoside and quercetin-3-*O*-rhamnoside, but lower contents of caffeic acid, resveratrol and rutin, compared to those grown under the organic system. However, Marín et al. [22] observed slight differences among the contents of individual and total phenolic compounds of green and red peppers cultivated under organic, integrated or soil-less systems.

The opposite findings reported in the literature, together with those of the present study clearly show the difficulty in attempting to establish a priori the effects of the agronomic production system and/or maturation stage on the phenolic composition of sweet peppers, pointing out the relevance of factors such as cultivar, environmental-climatic conditions and post-harvest treatments. If an optimal and general evaluation model was envisaged, a broader study would be needed, which should take into account different sweet pepper varieties grown in different geographical regions under different agronomic practices and subjected to different climatic conditions. However, several evaluation difficulties would arise from such a wide-ranging approach, leading to a complex data analysis where main conclusions could be hard to identify and further extrapolate to other practical cases.

3.2. Antioxidant Activity and Total Phenolic Content of Entinas Sweet Peppers

The total phenolic contents (TPC), calculated as the sum of the individual contents of the detected phenolics, as well as the antioxidant activities (DPPH, in %; ABTS$^{\bullet+}$, in %; and TEAC, in mg Trolox g^{-1}) of the Entinas sweet peppers are shown in Figure 2, according to the production system (conventional and organic) and maturation stage (green and red). The mean TPC values varied from 656 to 1400 µg g^{-1} dw for red and green conventional peppers, respectively, which is in agreement with the wide range of literature values determined using either spectrophotometric (Folin–Ciocalteu based-method: 300–7700 µg g^{-1}) [21,27,41–43] or chromatographic assays (LC coupled or not with MS: 120–4000 µg g^{-1}) [16,20,27]. These values greatly depend on several factors, including, pepper cultivar, production system, environmental-climatic conditions, maturation stage and post-harvest treatments. Similarly, red and green conventional peppers showed, respectively, the lowest and highest mean DPPH (varying from 55% to 71%) and ABTS$^{\bullet+}$ (ranging between 45% and 61%) radical scavenging activities. An opposite finding was observed for TEAC, for which red conventional peppers had the greater values (11 mg Trolox g^{-1}) and green conventional peppers had the lower ones (8 mg Trolox g^{-1}). It should be noted that the DPPH values found are in accordance with the values reported by several research teams (varying from 45% to 87%) for different pepper cultivars at different maturation stages and grown under different agronomic systems [43–45].

Overall, green peppers had significantly greater DPPH and ABTS$^{\bullet+}$ activities compared to red peppers, probably due to the higher TPC, but showed lower TEAC. Thus, the TPC and antioxidant activity were favored by early maturation stage of Entinas peppers. By contrast, Lutz et al. [42] and Cisternas-Jamet et al. [41] observed an increase in the DPPH activity with ripening. Regarding the oxygen radical antioxidant capacity (ORAC), a similar increasing trend was found by Lutz et al. [42], although no clear effect of maturity was observed by Cisternas-Jamet et al. [41]. In addition, Martí et al. [46] did not find significant changes in the total antioxidant activity between green and red peppers of different cultivars. It should be noted that DPPH and ABTS$^{\bullet+}$ radical scavenging activity as well as the ferric reducing antioxidant power assay (FRAP) are greatly cultivar-dependent [21,27,43,47,48], which can partially explain the different trends found in the present study as well as in the literature. Concerning the agronomic production system effect on the radical scavenging activity, conventional peppers showed the highest TPC and TEAC but lower DPPH and ABTS$^{\bullet+}$ activities, compared to organic peppers.

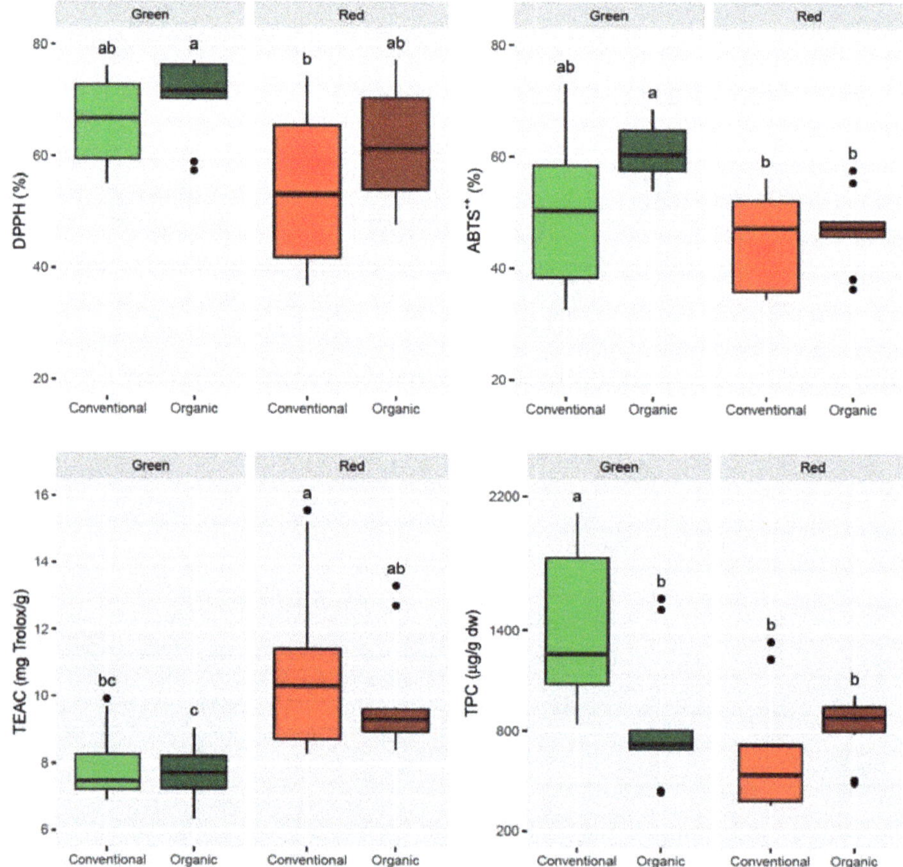

Figure 2. Antioxidant activities (DPPH, in %; ABTS$^{\bullet+}$, in %; and, TEAC, in mg Trolox g^{-1}) and total phenolic contents (TPC) (sum of individual phenolics, in µg g^{-1} dw) of sweet peppers according to the maturation stage (green or red peppers) and agronomic production system (conventional or organic systems). Box bars represent the values that are within the 1st and 3rd quartiles and the horizontal line represents the 2nd quartile (median). Vertical lines from the middle of the top and bottom edges of each box represent the whiskers. Dot points represent minimum or maximum values that fall outside the whisker range (extreme values or outliers). Different lower case letters mean significant statistical differences (*p*-value < 0.05) among maturation–production levels.

3.3. Unsupervised and Supervised Entinas Peppers Differentiation Based on the Phenolic Profile and Antioxidant Radical Scavenging Data

The possible use of the phenolic composition (individual and total contents) and/or the related radical scavenging activity (DPPH, ABTS$^{\bullet+}$ and TEAC) data to act as possible biomarkers of Entinas peppers production system/maturation stage was further evaluated using unsupervised (PCA) and supervised (LDA) multivariate pattern recognition techniques. Previously these statistical techniques have been applied, for example, to differentiate pepper cultivars/maturation stage based on the phenolic composition [26], although different cultivars/maturation stages were grouped into the same clusters; or to distinguish fresh and dried peppers grown under conventional or organic systems [16], although in this case, the use of both phenolic and aroma compounds was required. Recently, the production system–maturation stage of Entinas peppers could be identified using chemical-sensory data or potentiometric signal profiles, recorded using a lab-made taste sensor device [30]. As reported by Guilherme et al. [30], LDA classification models could be established using non-redundant independent parameters selected by the simulated annealing (SA) algorithm, allowing the correct classification of 80–90% of the studied samples (leave-one-out cross-validation, LOO-CV). In the present study, the possibility of assessing the production system–maturation stage of Entinas peppers, based on the phenolic composition and antioxidant capacity was further evaluated using PCA and LDA-SA approaches.

Figure 3 shows the 3D-PCA plots based on the individual (caffeic, chlorogenic, m-coumaric and O-coumaric acids, luteolin-7-O-glucoside, myricetin, resveratrol, rutin and quercetin-3-O-rhamnoside) and total phenolic contents (TPC) together with the radical scavenging activities (DPPH, ABTS$^{\bullet+}$ and TEAC). As can be inferred, the three first principal components (1st, 2nd and 3rd PCs), which explained 80% of the total data variance, allowed a satisfactory differentiation of the studied Entinas peppers either considering simultaneously the production system–maturation stage (Figure 3a), or taking into account each factor separately (Figure 3b,c, for production system or maturation stage, respectively). Moreover, from a qualitative point of view, it is also clear that the abovementioned data would enable a better recognition of the maturation stage compared to the production system, pointing out that the phenolics composition and related antioxidant activities are more influenced by peppers' maturation. To identify the parameters that had the higher discriminant power, LDA-SA models were established for each aim, i.e., to simultaneously discriminate the production system–maturation stage or each factor per se.

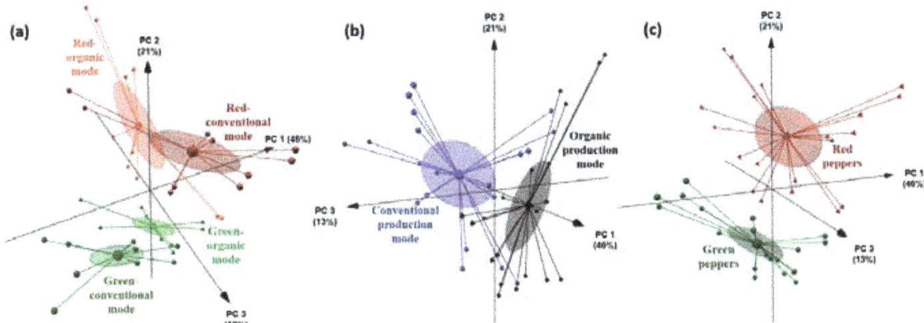

Figure 3. Unsupervised differentiation (3D principal component analysis (PCA) plots) of Entinas sweet peppers based on the nine individual phenolics detected by HPLC-DAD (caffeic, chlorogenic, m-coumaric and o-coumaric acids, luteolin-7-O-glucoside, myricetin, resveratrol, rutin and quercetin-3-O-rhamnoside, µg g^{-1} dw), total phenolic content (TPC, µg g^{-1} dw) and radical scavenging activity data (DPPH, in %; ABTS$^{\bullet+}$, in %; and, TEAC, in mg Trolox g^{-1}): (**a**) according to the agronomic production system (organic and conventional) and the maturation stage (green and red peppers); (**b**) according to the agronomic production system (organic and conventional) independently of the maturation stage; (**c**) according to the maturation stage (green and red peppers) independently of the agronomic production system.

For the simultaneous discrimination of the four production system–maturation stage levels a LDA-SA model with three discriminant functions was established based on seven non-redundant parameters (resveratrol, m-coumaric acid, chlorogenic acid, myricetin, quercetin-3-O-rhamnoside, TPC and TEAC) selected by the SA algorithm. The classification model allowed the correct classification of 100%, 75% and 76 ± 16% of the peppers, for the original data grouped (training), LOO-CV and repeated K-fold-CV (predictive internal validation variants). Although 5 of 20 peppers were misclassified, it should be remarked that, the misclassification occurred between the production system within the same maturation stage (i.e., green and red peppers were always correctly classified). Therefore, to further confirm this finding, LDA-SA models were also developed for predicting the peppers' production system or maturation stage. For the production system, classification models with one discriminant function were obtained based on four non-redundant parameters (o-coumaric acid, chlorogenic acid, myricetin and DPPH). This model had sensitivities (i.e., correct classifications) of 100%, 90% and 89 ± 11% for the original data grouped, LOO-CV and repeated K-fold-CV. In this case, only 2 of the 20 peppers were misclassified, with one sample of each production system misclassified. Finally, for the maturation stage, a model with one discriminant function was also established based on only three phenolic compounds (resveratrol, m-coumaric acid and myricetin), which allowed 100% of correct classifications for training and both predictive cross-validation variants. The overall results clearly show that maturation stage could be easily predicted based on the peppers' phenolic composition; this was also reliable for the assessment of the production system. Thus, phenolic and antioxidant data could be used as a preliminary peppers classification tool. However, for taking into account both factors simultaneously (i.e., production system and maturation stage) a data fusion approach, using other physicochemical data would be required.

4. Conclusions

The study carried out confirmed that the production system, as well as the maturation stage has a significant effect on the phenolic profile and on the antioxidant capacity of Entinas sweet peppers. Regarding the Entinas cultivar, it was observed that a lower maturation degree (i.e., green peppers) promoted the increase of the total phenolic content, the DPPH-radical scavenging activity and the bleaching of the ABTS radical cation, but a lower TEAC. Regarding the individual phenolic levels, the maturation effect greatly depended on the specific phenolic compound. Concerning the agronomic production system, the conventional system seemed to enhance the overall phenolic-antioxidant richness of the studied Entinas peppers, although this trend was not observed for all individual phenolics. Finally, the chemometric evaluation performed allowed us to verify that Entinas peppers' phenolic-antioxidant levels could be satisfactorily used as discrimination biomarkers for both production system and maturation stage; this finding is more visible for the maturation stage recognition. Nevertheless, it should be emphasized that the abovementioned conclusions were established for a specific sweet pepper variety grown within a narrow geographical region and so, any extrapolation should be carefully made if different varieties grown under different agro-climatic conditions are considered in the future. Indeed, to establish an optimal and general model, other external factors must be included in a future study.

Supplementary Materials: The following are available online at http://www.mdpi.com/2077-0472/10/12/652/s1, Table S1: Individual phenolic compounds and total phenolc compounds quantified (in µg g^{-1} dw), by HPLC-DAD, and antioxidante activitires (DPPH, in %; ABTS•+, in %; and, TEAC, mg Trolox/g), mean ± standard deviation of 10 individual samples, in sweet pepper samples according to the maturation stage (green or red peppers) and agronomic production system (conventional or organic systems).

Author Contributions: Conceptualization, R.G. and J.A.P.; methodology, R.G., A.A., N.R. and A.M.P.; investigation, all authors; resources, A.A. and J.A.P.; writing—original draft preparation, R.G., A.A., N.R., A.M.P. and J.A.P.; writing—review and editing, A.M.P. and J.A.P. All authors have read and agreed to the published version of the manuscript.

Funding: The authors are grateful to the Foundation for Science and Technology (FCT, Portugal) for financial support by national funds FCT/MCTES to CIMO (UIDB/00690/2020). N.R. gives thanks to National funding by FCT–Foundation for Science and Technology, through the institutional scientific employment program–contract.

Conflicts of Interest: The authors declare no conflict of interest.

References

1. Eggink, P.M.; Maliepaard, C.; Tikunov, Y.; Haanstra, J.P.W.; Bovy, A.G.; Visser, R.G.F. A taste of sweet pepper: Volatile and non-volatile chemical composition of fresh sweet pepper (*Capsicum annuum*) in relation to sensory evaluation of taste. *Food Chem.* **2012**, *132*, 301–310. [CrossRef] [PubMed]
2. Massey, M.; O'Cass, A.; Otahal, P. A meta-analytic study of the factors driving the purchase of organic food. *Appetite* **2018**, *125*, 418–427. [CrossRef] [PubMed]

3.   Willer, H.; Lernoud, J. *The World of Organic Agriculture. Statistics and Emerging Trends 2019*; Research Institute of Organic Agriculture (FiBL): Frick, Switzerland; IFOAM–Organics International: Bonn, Germany, 2019; pp. 1–351.

4.   Denver, S.; Jensen, J.D.; Olsen, S.B.; Christensen, T. Consumer Preferences for 'Localness' and Organic Food Production. *J. Food Prod. Mark.* **2019**, *25*, 668–689. [CrossRef]

5.   Jensen, M.M.; Jørgensen, H.; Lauridsen, C. Comparison between conventional and organic agriculture in terms of nutritional quality of food—A critical review. *CAB Rev.* **2013**, *8*. [CrossRef]

6.   Council Regulation (EC) No 834/2007 of 28 June 2007 on organic production and labelling of organic products and repealing Regulation (EEC) No 2092/91. *Off. J. Eur. Union.* **2007**, *L189*, 1–23.

7.   Baudry, J.; Péneau, S.; Allès, B.; Touvier, M.; Hercberg, S.; Galan, P.; Amiot, M.-J.; Lairon, D.; Méjean, C.; Kesse-Guyot, E. Food choice motives when purchasing in organic and conventional consumer clusters: Focus on sustainable concerns (The NutriNet-Santé Cohort Study). *Nutrients* **2017**, *9*, 88. [CrossRef]

8.   Hallmann, E.; Marszałek, K.; Lipowski, J.; Jasińska, U.; Kazimierczak, R.; Średnicka-Tober, D.; Rembiałkowska, E. Polyphenols and carotenoids in pickled bell pepper from organic and conventional production. *Food Chem.* **2019**, *278*, 254–260. [CrossRef]

9.   Barański, M.; Średnicka-Tober, D.; Volakakis, N.; Seal, C.; Sanderson, R.; Stewart, G.B.; Benbrook, C.; Biavati, B.; Markellou, E.; Giotis, C.; et al. Higher antioxidant and lower cadmium concentrations and lower incidence of pesticide residues in organically grown crops: A systematic literature review and meta-analyses. *Br. J. Nutr.* **2014**, *112*, 794–811. [CrossRef]

10.   Cisternas-Jamet, J.; Salvatierra-Martínez, R.; Vega-Gálvez, A.; Uribe, E.; Goñi, M.G.; Stoll, A. Root inoculation of green bell pepper (*Capsicum annum*) with *Bacillus amyloliquefaciens* BBC047: Effect on biochemical composition and antioxidant capacity. *J. Sci. Food Agric.* **2019**, *99*, 5131–5139. [CrossRef]

11.   Ritota, M.; Marini, F.; Sequi, P.; Valentini, M. Metabolomic characterization of Italian sweet pepper (*Capsicum annum* L.) by means of HRMAS-NMR Spectroscopy and multivariate analysis. *J. Agric. Food Chem.* **2010**, *58*, 9675–9684. [CrossRef]

12.   Sim, K.H.; Sil, H.Y. Antioxidant activities of red pepper (*Capsicum annuum*) pericarp and seed extracts. *Int. J. Food Sci. Technol.* **2008**, *43*, 1813–1823. [CrossRef]

13.   Sun, T.; Xu, Z.; Wu, C.-T.; Janes, M.; Prinyawiwatkul, W.; No, H.K. Antioxidant activities of different colored sweet bell peppers (*Capsicum annuum* L.). *J. Food Sci.* **2007**, *72*, S98–S102. [CrossRef] [PubMed]

14.   Del Amor, F.M.; Serrano-Martínez, A.; Fortea, M.A.; Núñez-Delicado, E. Differential effect of organic cultivation on the levels of phenolics, peroxidase and capsidiol in sweet peppers. *J. Sci. Food Agric.* **2008**, *88*, 770–777. [CrossRef]

15.   Flores, P.; López, A.; Fenoll, J.; Hellín, P.; Kelly, S. Classification of organic and conventional sweet peppers and lettuce using a combination of isotopic and bio-markers with multivariate analysis. *J. Food Compos. Anal.* **2013**, *31*, 217–225. [CrossRef]

16.   Guclu, G.; Keser, D.; Kelebek, H.; Keskin, M.; Emre Sekerli, Y.; Soysal, Y.; Selli, S. Impact of production and drying methods on the volatile and phenolic characteristics of fresh and powdered sweet red peppers. *Food Chem.* **2021**, *338*, 128129. [CrossRef]

17.   Chassy, A.W.; Bui, L.; Renaud, E.N.C.; van Horn, M.; Mitchell, A.E. Three year comparison of the content of antioxidant microconstituents and several quality characteristics in organic and conventionally managed tomatoes and bell peppers. *J. Sci. Food Agric.* **2006**, *54*, 8244–8252. [CrossRef]

18.   Morales-Soto, A.; Gómez-Caravaca, A.M.; García-Salas, P.; Segura-Carretero, A.; Fernández-Gutiérrez, A. High-performance liquid chromatography coupled to diode array and electrospray time-of-flight mass spectrometry detectors for a comprehensive characterization of phenolic and other polar compounds in three pepper (*Capsicum annuum* L.) samples. *Food Res. Int.* **2013**, *51*, 977–984. [CrossRef]

19.   Materska, M. Bioactive phenolics of fresh and freeze-dried sweet and semi-spicy pepper fruits (*Capsicum annuum* L.). *J. Funct. Foods* **2014**, *7*, 269–277. [CrossRef]

20.   Lekala, C.S.; Madani, K.S.H.; Phan, A.D.T.; Maboko, M.M.; Fotouo, H.; Soundy, P.; Sultanbawa, Y.; Sivakumar, D. Cultivar-specific responses in red sweet peppers grown under shade nets and controlled-temperature plastic tunnel environment on antioxidant constituents at harvest. *Food Chem.* **2019**, *275*, 85–94. [CrossRef]

21. Alu'datt, M.H.; Rababah, T.; Alhamad, M.N.; Gammoh, S.; Ereifej, K.; Al-Karaki, G.; Tranchant, C.C.; Al-Duais, M.; Ghozlan, K.A. Contents, profiles and bioactive properties of free and bound phenolics extracted from selected fruits of the Oleaceae and Solanaceae families. *LWT Food Sci. Technol.* **2019**, *109*, 367–377. [CrossRef]

22. Marín, A.; Gil, M.I.; Flores, P.; Hellín, P.; Selma, M.V. Microbial quality and bioactive constituents of sweet peppers from sustainable production systems. *J. Agric. Food Chem.* **2008**, *56*, 11334–11341. [CrossRef] [PubMed]

23. Hallmann, E.; Rembiałkowska, E. Characterisation of antioxidant compounds in sweet bell pepper (*Capsicum annuum* L.) under organic and conventional growing systems. *J. Sci. Food Agric.* **2012**, *92*, 2409–2415. [CrossRef] [PubMed]

24. Zhuang, C.; Chen, L.; Sun, L.; Cao, J. Bioactive characteristics and antioxidant activities of nine peppers. *J. Funct. Foods* **2012**, *4*, 331–338. [CrossRef]

25. Rodrigues, C.A.; Nicácio, A.E.; Jardim, I.C.S.F.; Visentainer, J.V.; Maldaner, L. Determination of Phenolic Compounds in Red Sweet Pepper (*Capsicum annuum* L.) Using using a Modified Modified QuEChERS Method and UHPLC-MS/MS Analysis and Its Relation to Antioxidant Activity. *J. Braz. Chem. Soc.* **2019**, *30*, 1229–1240. [CrossRef]

26. Fratianni, F.; D'acierno, A.; Cozzolino, A.; Spigno, P.; Riccardi, R.; Raimo, F.; Pane, C.; Zaccardelli, M.; Lombardo, V.T.; Tucci, M.; et al. Biochemical characterization of traditional varieties of sweet pepper (*Capsicum annuum* L.) of the Campania region, southern Italy. *Antioxidants* **2020**, *9*, 556. [CrossRef]

27. Kelebek, H.; Sevindik, O.; Uzlasir, T.; Selli, S. LC-DAD/ESI MS/MS characterization of fresh and cooked Capia and Aleppo red peppers (*Capsicum annuum* L.) phenolic profiles. *Eur. Food Res. Technol.* **2020**, *246*, 1971–1980. [CrossRef]

28. Sora, G.T.S.; Haminiuk, C.W.I.; da Silva, M.V.; Zielinski, A.A.F.; Gonçalves, G.A.; Bracht, A.; Peralta, R.M. A comparative study of the capsaicinoid and phenolic contents and in vitro antioxidant activities of the peppers of the genus *Capsicum*: An application of chemometrics. *J. Food Sci. Technol.* **2015**, *52*, 8086–8094. [CrossRef]

29. Valle, A.D.; Dimmito, M.P.; Zengin, G.; Pieretti, S.; Mollica, A.; Locatelli, M.; Cichelli, A.; Novellino, E.; Ak, G.; Yerlikaya, S.; et al. Exploring the nutraceutical potential of dried pepper *capsicum annuum* L. on market from Altino in Abruzzo region. *Antioxidants* **2020**, *9*, 400. [CrossRef]

30. Guilherme, R.; Rodrigues, N.; Marx, Í.M.G.; Dias, L.G.; Veloso, A.C.A.; Ramos, A.C.; Peres, A.M.; Pereira, J.A. Sweet peppers discrimination according to agronomic production mode and maturation stage using a chemical-sensory approach and an electronic tongue. *Microchem. J.* **2020**, *157*, 105034. [CrossRef]

31. Hatano, T.; Kagawa, H.; Yasuhara, T.; Okuda, T. Two new flavonoids and other constituents in licorice root: Their relative astringency and radical scavenging effects. *Chem. Pharm. Bull.* **1988**, *36*, 2090–2097. [CrossRef]

32. Sánchez, C.S.; González, A.M.T.; Parrilla, M.C.G.; Granados, J.J.Q.; Serrana, H.L.G.D.L.; Martínez, M.C.L. Different radical scavenging tests in virgin olive oil and their relation to the total phenol content. *Anal. Chim. Acta* **2007**, *593*, 103–107. [CrossRef] [PubMed]

33. Berker, K.I.; Güçlü, K.; Tor, İ.; Apak, R. Comparative evaluation of Fe (III) reducing power-based antioxidant capacity assays in the presence of phenanthroline, batho-phenanthroline, tripyridyltriazine (FRAP), and ferricyanide reagents. *Talanta* **2007**, *72*, 1157–1165. [CrossRef] [PubMed]

34. Bertsimas, D.; Tsitsiklis, J. Simulated annealing. *Stat. Sci.* **1993**, *8*, 10–15. [CrossRef]

35. Cadima, J.; Cerdeira, J.O.; Minhoto, M. Computational aspects of algorithms for variable selection in the context of principal components. *Comput. Stat. Data Anal.* **2004**, *47*, 225–236. [CrossRef]

36. Venables, W.N.; Ripley, B.D. *Modern Applied Statistics with S (Statistics and Computing)*, 4th ed.; Springer: New York, NY, USA, 2002.

37. Jeong, W.Y.; Jin, J.S.; Cho, Y.A.; Lee, J.H.; Park, S.; Jeong, S.W.; Kim, Y.-H.; Lim, C.-S.; Abd El-Aty, A.M.; Kim, G.-S.; et al. Determination of polyphenols in three *Capsicum annuum* L. (bell pepper) varieties using high-performance liquid chromatography-tandem mass spectrometry: Their contribution to overall antioxidant and anticancer activity. *J. Sep. Sci.* **2011**, *34*, 2967–2974. [CrossRef] [PubMed]

38. Ghasemnezhad, M.; Sherafati, M.; Payvast, G.A. Variation in phenolic compounds, ascorbic acid and antioxidant activity of five coloured bell pepper (*Capsicum annum*) fruits at two different harvest times. *J. Funct. Foods* **2011**, *3*, 44–49. [CrossRef]

39. Ekinci, M.; Turan, M.; Yildirim, E.; Güneş, A.; Kotan, R.; Dursun, A. Effect of plant growth promoting rhizobacteria on growth, nutrient, organic acid, amino acid and hormone content of cauliflower (*Brassica oleracea* L. var. *botrytis*) transplants. *Acta Sci. Pol. Hortorum Cultus* **2014**, *13*, 71–85.
40. Howard, L.R.; Talcott, S.T.; Brenes, C.H.; Villalon, B. Changes in phytochemical and antioxidant activity of selected pepper cultivars (*Capsicum* species) as influenced by maturity. *J. Agric. Food Chem.* **2000**, *48*, 1713–1720. [CrossRef]
41. Cisternas-Jamet, J.; Salvatierra-Martínez, R.; Vega-Gálvez, A.; Stoll, A.; Uribe, E.; Goñi, M.G. Biochemical composition as a function of fruit maturity stage of bell pepper (*Capsicum annum*) inoculated with Bacillus amyloliquefaciens. *Sci. Hortic.* **2020**, *263*, 109107. [CrossRef]
42. Lutz, M.; Hernández, J.; Henríquez, C. Phenolic content and antioxidant capacity in fresh and dry fruits and vegetables grown in Chile. *CYTA J. Food* **2015**, *13*, 541–547.
43. Hamed, M.; Kalita, D.; Bartolo, M.E.; Jayanty, S.S. Capsaicinoids, polyphenols and antioxidant activities of *Capsicum annuum*: Comparative study of the effect of ripening stage and cooking methods. *Antioxidants* **2019**, *8*, 364. [CrossRef] [PubMed]
44. Zhang, D.; Hamauzu, Y. Phenolic compounds, ascorbic acid and antioxidant properties of green, red and yellow bell peppers. *J. Food Agric. Environ.* **2003**, *1*, 22–27.
45. Wong, P.Y.; Tan, S.T. Comparison of total phenolic content and antioxidant activities in selected coloured plants. *Br. Food J.* **2020**, *122*, 3193–3201. [CrossRef]
46. Martí, M.C.; Camejo, D.; Vallejo, F.; Romojaro, F.; Bacarizo, S.; Palma, J.M.; Sevilla, F.; Jiménez, A. Influence of Fruit Ripening Stage and Harvest Period on the Antioxidant Content of Sweet Pepper Cultivars. *Plant Foods Hum. Nutr.* **2011**, *66*, 416–423. [CrossRef] [PubMed]
47. Aslani, L.; Mobli, M.; Keramat, J. Comparison of anti-oxidant activities and fruit quality attributes in four sweet bell pepper (*Capsicum annuum* L.) cultivars grown in two soilless media. *J. Hortic. Sci. Biotechnol.* **2016**, *91*, 497–505. [CrossRef]
48. Alam, M.A.; Saleh, M.; Mohsin, G.M.; Nadirah, T.A.; Aslani, F.; Rahman, M.M.; Roy, S.K.; Juraimi, A.S.; Alam, M.Z. Evaluation of phenolics, capsaicinoids, antioxidant properties, and major macro-micro minerals of some hot and sweet peppers and ginger land-races of Malaysia. *J. Food Process. Preserv.* **2020**, *44*, e14483. [CrossRef]

Publisher's Note: MDPI stays neutral with regard to jurisdictional claims in published maps and institutional affiliations.

MDPI

Article

Grain Quality of Maize Cultivars as a Function of Planting Dates, Irrigation and Nitrogen Stress: A Case Study from Semiarid Conditions of Iran

Maryam Rahimi Jahangirlou [1,2], Gholam Abbas Akbari [1,*], Iraj Alahdadi [1], Saeid Soufizadeh [3] and David Parsons [2]

[1] Department of Agronomy and Plant Breeding Sciences, College of Aburaihan, University of Tehran, Pakdasht 1417466191, Iran; maramrahimi204@ut.ac.ir (M.R.J.); alahdadi@ut.ac.ir (I.A.)
[2] Department of Agricultural Research for Northern Sweden, Swedish University of Agricultural Sciences, 90183 Umeå, Sweden; david.parsons@slu.se
[3] Department of Agroecology, Environmental Sciences Research Institute, Shahid Beheshti University, Tehran P.O. Box 19835-196, Iran; s_soufizadeh@sbu.ac.ir
* Correspondence: ghakbari@ut.ac.ir; Tel.: +98-2136-040-902

Abstract: Maize grain is an important source of human and animal feed, and its quality can be affected by management practices and climatic conditions. This study aimed to evaluate the concentration and composition of starch, protein and oil in grain of maize cultivars in response to different planting dates (20 June and 21 July), irrigation (12-day and 6-day intervals) and nitrogen rates (0 and 184 kg N ha^{-1}). The first two principal components (PCs) accounted for 84.5% of the total variation. High N fertilization increased protein (by 6.0 and 10.9 g kg^{-1}) and total nonessential amino acids (by 3.4 and 2.4 g kg^{-1}) during 2018 and 2019, respectively. With the high irrigation rate, the high N rate increased oil, total unsaturated fatty acids, and starch and amylopectin, whereas with the low irrigation rate, there was no effect of the N rate. With earlier planting, total saturated fatty acids were higher. The findings highlight the complicated relationship between the different factors and how they affect quality characteristics of maize grain. There was a large impact of year, which to a great extent cannot be controlled, even in this environment where water supply was controlled and rainfall did not affect the results.

Keywords: amino acids; fatty acids; oil; protein; starch; *Zea mays* L.

Citation: Rahimi Jahangirlou, M.; Akbari, G.A.; Alahdadi, I.; Soufizadeh, S.; Parsons, D. Grain Quality of Maize Cultivars as a Function of Planting Dates, Irrigation and Nitrogen Stress: A Case Study from Semiarid Conditions of Iran. *Agriculture* **2021**, *11*, 11. http://dx.doi.org/10.3390/agriculture11010011

Received: 6 November 2020
Accepted: 21 December 2020
Published: 27 December 2020

Publisher's Note: MDPI stays neutral with regard to jurisdictional claims in published maps and institutional affiliations.

1. Introduction

Dent maize (*Zea mays* L. var. indentata) is cultivated because of its various end uses including livestock nutrition, human consumption and ethanol production [1].

The mature grain of dent maize is composed of storage components including 60% to 72% starch [2], 8 to 11% protein [3] and 4 to 6% oil [1]. Amylose and amylopectin are glucose storage polymers that can alter physicochemical properties of starch [4]. The optimal amylose–amylopectin ratio differs depending on the purpose of use. While high-amylose maize grain is a good source to supply resistant starch for food industries [5], high-amylopectin grain can be appropriate for livestock feeding because it is easier to digest [6]. For poultry diets, amylose content is significantly correlated with lower digestibility or higher resistant starch content because of a very compact physical structure [7].

Protein of cereals is critically important with regards to nutritional quality. Protein is relevant in terms of its overall concentration, and the essential and nonessential amino acids and their ratios [8]. Diets with essential amino acids as the only N source are used less efficiently than diets with an optimal ratio of essential to nonessential amino acids [9], which affects the amount of N losses from poultry excrement [10]. Saturated and unsaturated fatty acids and their distribution in triacylglycerol molecules are also important compositional factors of crude fat or oil to estimate metabolizable energy values in animal diets and oil nutritional value for human consumption [11].

The abovementioned compositional factors, as substantial dimensions of cereal grain quality are strongly affected by the genetic potential, the growing environment and agricultural practices [12,13]. Among agronomic practices, key strategies include cultivar selection [14]; water regimes [15]; nutritional status, especially N utilization [8]; and edaphic and climatic conditions, in particular, temperature changes due to year, location and planting date [16]. The effect of these factors on grain quality is more complex than yield. As an example, apart from interspecies differences, reports have shown that higher N rates reduced rice cooking quality by decreasing amylose [17], and in wheat grain, higher N doses caused an increase in different fractions of proteins and amino acids [18]. Lower temperatures resulted in smaller starch granule formation and decreased amylose content of maize grain starch [19]. Concentrations of amino acids increased with canopy warming in winter wheat [20]. Soil water deficit decreased the contents of amylose and starch in wheat grain [21] and of oil and linolenic acid in maize grain [22]. Water stress caused an increase in amino acid concentrations of maize grain [23]. An increase in starch [11] and a decrease in protein concentration [24] were reported for sorghum grain grown under high water-deficit conditions. Interactions between factors can also affect grain quality. For example, variations in nitrogen-based compounds like protein and amino acids can be a consequence of the dilution effect—decreasing the concentration of elements in plant tissue due to changes in environmental conditions—in response to irrigation and N interactions [25]. As a result, cultivar, water regime, N rate and planting date impact the qualitative characteristics of cereal grains; however, more knowledge is needed on how these factors work in an interactive way.

Thus, in this study, we assessed the interactive effects of planting date, irrigation timing and N fertilization on various measures of maize grain quality. The aim of the research was to provide a basis of knowledge for defining future research and for developing guidelines for improving maize grain quality.

2. Materials and Methods

2.1. Experiment Site

Field experiments were performed in 2018 and 2019 in Pakdasht city (35.4669° N, 51.6861° E), Tehran province, Iran. The geographical location of Pakdasht city in Iran is shown in Figure 1. This region has a semiarid climate with relatively cold winters and hot summers. Annual precipitation in Pakdasht is approximately 160 mm and concentrated in late autumn and winter. Irrigated summer maize in this region is typically cultivated after winter cereals like wheat and barley. Summaries of the climatic parameters during the summer maize growing season for two years of the experiment are presented in Table 1. The overall means of daily temperatures were 25.3 and 24.9 °C for 2018 and 2019, respectively, and the total precipitation was less than 1.5 mm in each growing season. The highest temperatures occurred in July for both years. Soil sampling was conducted before the first cultivation year from two depths (0–30 and 30–60 cm). The samples were air dried; passed through a 2-mm sieve; and tested for organic carbon using the Walkley–Black method [26], total nitrogen (N) using the Kjeldahl method [27], for available phosphorus (P) by the Olsen procedure [28] and for available Potassium (K) by using a flame photometer [29]. The soil is characterized as clay-loam, and the initial physicochemical characteristics in 0−30 cm soil depth were as follows: organic carbon, 1.09%; total N, 0.12%; P, 90.17 mg kg^{-1}; and K, 453 mg kg^{-1}. Moisture release curve was used to determine soil water [30,31]. Water content at field capacity and permanent wilting point were 0.346 and 0.115 g·cm^{-3}, respectively.

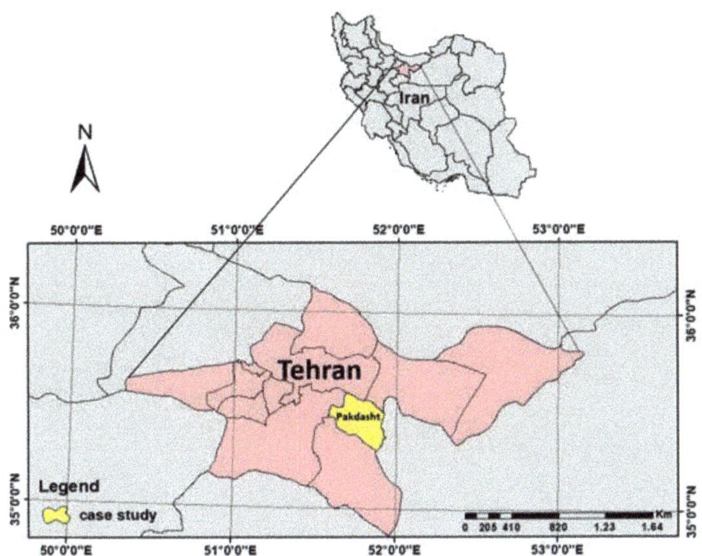

Figure 1. The geographical location of Pakdasht city (35.4669° N, 51.6861° E), Tehran province, Iran.

Table 1. Average minimum and maximum daily air temperatures and monthly total precipitation during the summer maize growing season from June through November of 2018 and 2019 in Pakdasht city, Iran.

Month	Temperature (°C)				Precipitation (mm)	
	Minimum		Maximum			
	2018	2019	2018	2019	2018	2019
June	20.3	20.8	38.7	39.1	0.1	0.0
July	24.4	24.2	42.4	41.9	0.0	0.0
August	22.5	21.0	40.1	39.3	0.0	0.0
September	17.8	16.9	35.1	35.3	0.0	0.0
October	12.3	12.2	26.6	26.9	0.3	0.8
November	6.7	4.4	16.5	15.6	0.9	0.6
Total	17.3	16.6	33.2	33.0	0.24	0.23

2.2. Experimental Design

In mid-June 2018, experimental strips and plots were designated at a farm where no crops had been cultivated for the previous three years. The field was divided into two strips (24-m length) to apply different irrigation timings (12-day and 6-day intervals). Two planting dates (20 June and 21 July) were randomized within the irrigation treatments. Two ditches crossed the strips, partitioning each of them into three plots. Thus, the experiment comprised twelve main plots. The four combinations of cultivars (KSC704 and KSC260) and N rates (0 and 184 kg ha^{-1}) were randomized to subplots within each plot. Thus, a total of 12 main plots containing a total of 48 subplots were obtained from three replications in each cultivation year. Individual subplots were 6 m in length and consisted of 6 rows sown at a density of six and eight plants per square meter for KSC704 and KSC260, respectively. Treatments were applied to the same plots in each year.

KSC704 is a locally popular high-yield late maturity hybrid which has been cultivated in Iran since 1980 and accounts for 80% of the maize growing areas of Iran [32]. KSC260, an early maturity maize cultivar, was introduced in 2008 and has shown good performance in various experiments comparing new early-maturing cultivars [33]. Harvesting date, maturity period and the total growing degree days (GDDs) of cultivars in the present study are shown in Table 2. GDD was calculated using the GDD calculator program [34]. For this study, 10 °C was set as the base temperature, 34 °C was set as the optimum temperature and 40 °C was set as the maximum temperature threshold, adapted from Cutforth and Shaykewich [35]. The two planting dates were chosen to create different temperature conditions during grain filling. The variability of daily average temperature during the reproductive stage (VT-R6) for both planting dates is shown in Figure 2. There were clear differences between the recorded temperatures for the planting dates in each year and differences between the two years. The second year was a warmer year during the reproductive period.

Table 2. Planting dates, harvesting dates, maturity period and growing degree days (GDDs) of cultivars in 2018 and 2019.

Year	Maize Hybrids	Planting Date	Harvesting Date	Maturity Period (Days)	GDDs (°C)
2018	KSC704	20 June	7 November	140	2424
		21 July	26 November	138	1839
	KSC260	20 June	18 October	120	2311
		21 July	11 November	118	1796
2019	KSC704	20 June	2 November	135	2370
		21 July	28 November	130	1855
	KSC260	20 June	12 October	120	2220
		21 July	14 November	116	1704

The temperature thresholds used to calculate GDDs were 10 °C (base temperature), 34 °C (optimum) and 40 °C (maximum) (Cutforth and Shaykewich, 1990).

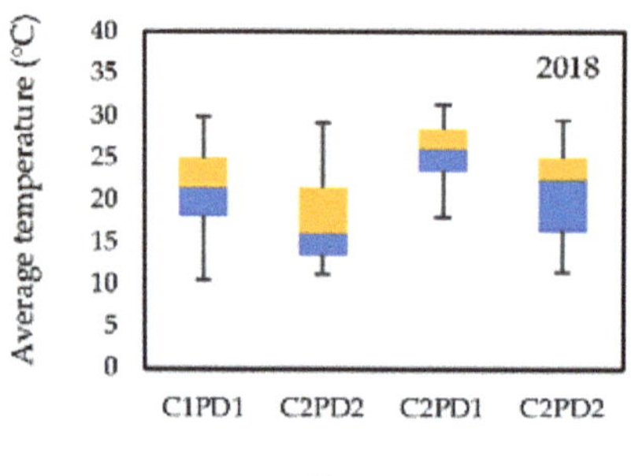

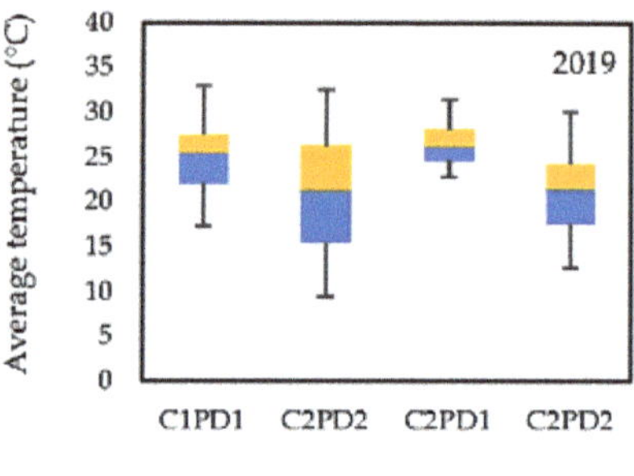

Figure 2. Variability of average daily temperatures (°C) during the reproductive stage of maize for conventional and late planting dates in 2018 and 2019: the upper and lower hinges of the box indicate the 75th percentile and 25th percentile of the data set, respectively. The line in the box indicates the median value of the data, and the upper and lower whiskers represent the maximum and minimum of the data, respectively. PD1, planting date 21 June; PD2, planting date 22 July; C1, KSC704; C2, KSC260.

An irrigation regime with 6-day intervals and N rate of 184 kg ha^{-1} were specified as typical non-stressed growing conditions. Furrow irrigation was used for irrigation. The closed-end furrows were constructed with a ditcher. The ridge and furrow widths were 40 cm and 20 cm, respectively, and the depth of the furrow was 15 cm. Based on the flow rate per furrow (approximately 1.3 L/s), water was delivered until the furrows were completely full and the duration of each irrigation event was recorded. The known flow rate of the irrigation pump was then used to estimate irrigation volumes.

In total, the KSC704 cultivar received about 9200 m^3 ha^{-1} of water with 6-day irrigation timing and about 5800 m^3 ha^{-1} with 12-day irrigation timing. KSC260 received less water due to a shorter growing period and less irrigation events: 7800 m^3 ha^{-1} with 6-day irrigation timing and 4800 m^3 ha^{-1} with 12-day irrigation timing. The Food and Agriculture Organization (FAO) estimates the water requirement of late-maturing maize cultivars in arid and semiarid regions to be around 7000 to 8000 m^3 ha^{-1} [36]. Urea was used as the N source and top-dressed in equal proportions at two stages (pre-planting and V4–V6). Fresh irrigation water was used from the main ditch for each replication, and excess water was not reused.

2.3. Procedures of Sampling and Laboratory

2.3.1. Sample Preparation

At the R6 (physiological maturity) stage, based on the Hanway standard [37], 15 plants in an area of approximately 11 m^2 (4-m row length $\times$ 0.7-m row spacing $\times$ 4 rows) were randomly cut from each subplot. Ears were threshed and dried at 60 °C to a constant weight. Seeds were separated from ears and were weighed and ground. The milled samples were maintained at 5 $\pm$ 1 °C until the start of laboratory procedures.

2.3.2. Analysis of Starch Content and Composition

Total starch and its composition (amylose and amylopectin) were determined using a Spectrophotometer (V-M5 model, BEL Engineering, Monza, Italy) set at 510 nm according to the amylose/amylopectin Megazyme procedure [38].

2.3.3. Analysis of Protein Content and Composition

Crude protein, essential amino acids (composed of methionine, threonine, valine, isoleucine, leucine, phenylalanine, histidine, arginine, lysine and tryptophan) and nonessential amino acids (composed of cysteine, asparagine, serine, glutamine, proline, glycine and alanine) were estimated using Near-Infrared Reflectance Spectroscopy (NIRS) (NIRS-XDS model, Foss, Hilleroed, Denmark) in the 1100–2500 nm wavelength range at five-nm intervals. A dent maize library of global origin samples that previously had been assayed by wet chemistry or non-NIR methods was used for model calibration.

2.3.4. Analysis of Oil Concentration and Fatty Acid Composition

Oil extraction was conducted using a Soxhlet extractor for approximately 4 h with hexane as the solvent, with a solid to solvent ratio of 1:7 m v^{-1}. The fatty acid methyl esters (FAME) were extracted according to AOAC 996.06 protocol [39]. The synthesized FAME was injected into gas chromatography (GC) (CP-Sil 88 model, Varian, Walnut Creek, CA, USA) to detect fatty acid composition by curve and retention time. The GC was equipped with a flame ionization detector (FID) (Column: CP-Sil 88 (100 m $*$ 250 μm $*$ 0.2 μm) removable phase: Nitrogen, 28.8 min, heat injection chamber: 270 °C, heat detector: 260 °C).

2.4. Statistical Analysis

Principal component analysis (PCA) scores were derived using Minitab, v.19 [40] after standardizing the variables by using the correlation matrix. Results from the two cultivation years were analyzed separately, using mixed model procedures with PROC MIXED in SAS, v.9.4 software [41]. The mixed-effects model included fixed effects of irrigation, planting date, cultivar, N, their interactions, and the random effects of irrigation $\times$ planting date and irrigation $\times$ planting date $\times$ block interactions. Because of the experimental design, irrigation effects have to be interpreted with caution—possible meaningful effects of irrigation and its interactions can be either caused by differences between the planned irrigation treatments or by differences in the two sections of the field where the irrigation treatments were applied. Tukey's statistic was used to test differences ($p \leq 0.05$) among means.

3. Results

3.1. Correlations between Variables and Treatments

The PCA comprising the first two principal components accounted for 84.5% of the total variance. The loading plot (Figure 3A) shows the eigenvectors for nine variables: starch, amylose, amylopectin, protein, total essential amino acids (ΣEAA), total nonessential amino acids (ΣNEAA), oil, total saturated fatty acids (ΣSFA) and total unsaturated fatty acids (ΣUSFA).

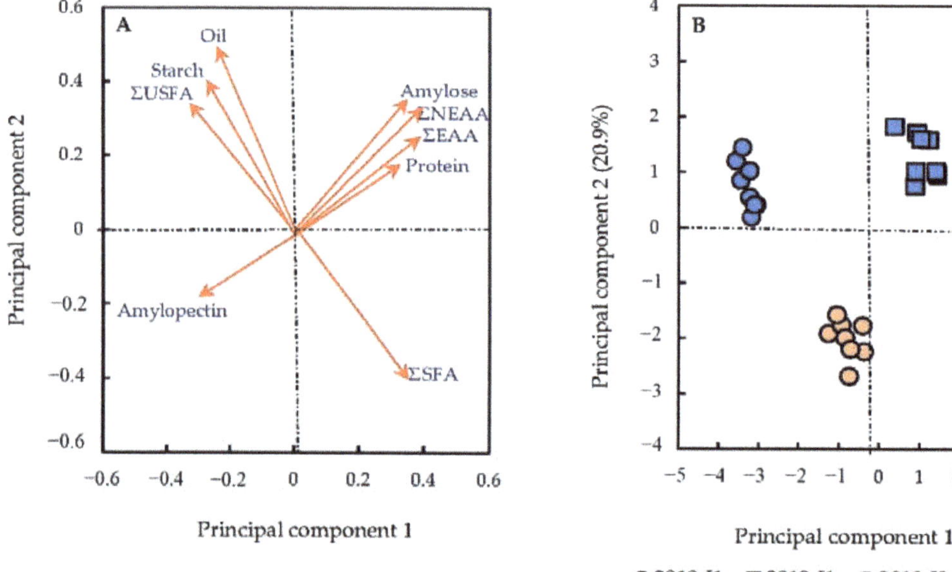

Figure 3. Plots of principal component 1 versus principal component 2 based on measured maize grain quality characteristics: the loading plot (**A**) shows the eigenvector of each characteristic. The score plot (**B**) shows the means, grouped by year and irrigation treatment. Each marker represents the mean of replicates in the field. Abbreviations; I1, Irrigation at 12-day intervals; I2, Irrigation at 6 day-intervals; ΣEAA, total essential amino acids; ΣNEAA, total non-essential amino acids; ΣSFA, total saturated fatty acids; ΣUSFA, total unsaturated fatty acids.

The cosine of the angle between two vectors estimates the correlation between them; therefore, clustered points are highly correlated with each other. There are two clusters of variables that are strongly correlated with each. The first cluster includes protein concentration, protein composition (ΣEAA and ΣNEAA) and starch composition (amylose concentration) of the grain. The eigenvector for amylopectin points in the opposite direction; thus, the first cluster is highly negatively correlated with amylopectin (e.g., the correlation between ΣNEAA and amylopectin is −0.917). Running perpendicular to the first cluster, the second cluster of highly correlated variables includes oil, starch and ΣUSFA. This cluster is highly negatively correlated with ΣSFA (e.g., the correlation between starch and ΣSFA is −0.663).

The score plot means are shown in Figure 3B. To illustrate the most noticeable trends, points are labelled according to combinations of year and irrigation treatment. The first component separates the first year of the experiment from the second year—markers for the first year (2018) are located in the left of the score plot, whereas markers for the second year (2019) are in the right. The second component separates well-irrigated samples from those under potential water-stress. Markers for well-watered samples are distributed toward the top, while markers from water-limited plots are mostly located toward the bottom.

However, the differences of irrigation timing are more clearly separated in the first year of the experiment than second.

The data points presented in score plots coincide with the directions of change in maize grain compositional variables in the loading plots. The two clusters of eigenvectors and their negative correlations define two axes, making an x shape. The first axis (from the southwest to northeast quadrants) separates 2018 from 2019. i.e., 2019 samples were higher in amylose, ΣEAA, ΣNEAA and protein. The second axis (from the northwest to southeast quadrants) separates the irrigation treatments. Irrigated plots were higher in starch, oil and ΣUSFA.

3.2. Treatment Effects on Starch Content and Composition

After calculating *p*-values of all main effects and interactions (Supplementary Materials S1), the treatment main effect means are presented in Table 3 and significant interactions were plotted (unless no pairs of means were significantly different or the effect could not be clearly interpreted).

Table 3. Least square means, significances and standard errors of maize grain starch, amylopectin, amylose, protein, total nonessential amino acids (ΣNEAA), total essential amino acids (ΣEAA), oil, total unsaturated fatty acids (ΣUFA) and total saturated fatty acids (ΣSFA) in response to treatment main effects (irrigation, planting date, cultivar and nitrogen rate) in 2018 and 2019.

Treatment	Starch		Amylopectin		Amylose		Protein		ΣNEAA		ΣEAA		Oil		ΣUSFA		ΣSFA	
	g kg^{-1}																	
	2018	2019	2018	2019	2018	2019	2018	2019	2018	2019	2018	2019	2018	2019	2018	2019	2018	2019
I1	668.10	682.50	549.17	494.23	100.48	187.36	105.00	121.40	42.20	61.90	31.80	43.30	31.30	31.90	26.98	27.20	4.24	4.60
I2	750.80	684.00	637.77	509.22	128.26	173.99	97.50	116.10	39.00	58.50	31.50	41.60	48.70	41.70	42.71	36.33	5.75	5.30
p-value	0.056	0.937	0.043	0.565	0.035	0.133	0.206	0.337	0.042	0.137	0.012	0.334	0.068	0.022	0.065	0.022	0.083	0.005
SEM	7.31	15.51	6.00	18.44	1.50	2.83	2.52	3.11	2.14	7.36	0.05	9.42	1.86	3.37	1.62	3.20	0.19	0.55
PD1	705.90	685.60	594.67	496.33	111.31	187.59	101.50	115.10	40.40	60.70	31.40	43.00	43.00	37.20	37.43	31.95	5.43	5.06
PD2	713.00	680.90	592.26	507.12	117.43	173.77	101.00	122.50	40.80	59.70	31.90	41.80	37.00	36.50	32.26	31.57	4.56	4.84
p-value	0.509	0.814	0.757	0.663	0.154	0.129	0.898	0.254	0.309	0.400	0.007	0.421	0.191	0.291	0.194	0.447	0.143	0.016
SEM	7.31	15.51	6.00	18.44	1.50	2.83	2.52	3.11	2.14	7.36	0.05	9.42	1.86	3.37	1.62	3.20	0.19	0.55
C1	710.40	688.50	593.72	499.36	112.94	187.52	99.20	117.50	40.00	60.30	31.20	43.10	40.00	36.40	34.78	31.38	5.02	4.89
C2	708.50	677.90	593.22	504.09	115.80	173.84	103.40	120.00	41.20	60.20	32.10	41.80	40.00	37.20	34.91	32.15	4.97	5.02
p-value	0.628	0.011	0.895	0.251	0.015	<0.001	0.056	0.212	<0.001	0.866	<0.001	0.012	0.955	0.188	0.881	0.181	0.697	0.194
SEM	3.89	3.90	3.75	40.41	1.09	1.89	2.10	1.88	1.51	2.24	0.35	4.96	1.02	6.38	0.88	5.64	0.13	0.95
N1	711.50	682.90	589.59	493.17	119.25	190.18	98.10	111.80	38.90	59.10	31.20	41.70	40.20	33.90	34.99	29.10	5.08	4.67
N2	707.50	683.50	597.35	510.28	109.50	171.17	104.40	125.70	42.30	61.40	32.00	43.10	39.80	39.80	34.70	34.43	4.91	5.24
p-value	0.316	0.878	0.049	<0.001	<0.001	<0.001	0.006	<0.001	<0.001	<0.001	<0.001	0.009	0.669	<0.001	0.743	<0.001	0.214	<0.001
SEM	3.89	3.90	3.75	40.41	1.09	1.89	2.10	1.88	1.51	2.24	0.35	4.96	1.02	6.38	0.88	5.64	0.13	0.95

I1, irrigation at 12-day intervals; I2, irrigation at 6 day-intervals; PD1, planting date 21 June; PD2, planting date 22 July; C1, cultivar KSC704; C2, cultivar KSC260; N1, zero-nitrogen; N2, 184 kg ha^{-1} nitrogen.

In 2019, average starch concentration in KSC704 was higher than KSC260 by 11 g kg^{-1} (Table 3). In 2018, there was an interactive effect of irrigation and N rate on starch (Figure 4A). With the high irrigation rate, the high N rate increased starch, whereas with low irrigation, the high N rate decreased starch. Grain starch was higher with the high irrigation rate, regardless of the N rate. In 2019, high N increased amylopectin concentration by 17 g kg^{-1} (Table 3). In 2018, there was an interactive effect of irrigation and N rate on amylopectin (Figure 4B). With the high irrigation rate, the high N rate increased amylopectin (649 g kg^{-1}), whereas with low irrigation, there was no N effect. Grain amylopectin was higher with the high irrigation rate, regardless of the N rate.

In 2018, there were interactive effects of irrigation and N rate (Figure 4C); irrigation rate and cultivar (Figure 4D); and N rate, planting date and cultivar (Figure 4E) on amylose concentration. For either irrigation rate, high N decreased amylose (Figure 4C). Amylose was higher with the high irrigation rate (with either N rate) than with the low irrigation rate (Figure 4C). With the high irrigation rate, there was no cultivar effect, whereas with low irrigation, KSC260 (103 g kg^{-1}) had higher amylose than KSC704 (97 g kg^{-1}) (Figure 4D). Grain amylose was higher with the high irrigation rate regardless of the cultivar (Figure 4D). With the zero N rate, KSC704 planted late had higher amylose than that planted earlier. With the high N rate, KSC260 planted late had higher amylose than that planted earlier. In general, amylose was higher with zero N fertilizer than with the high N rate (Figure 4E).

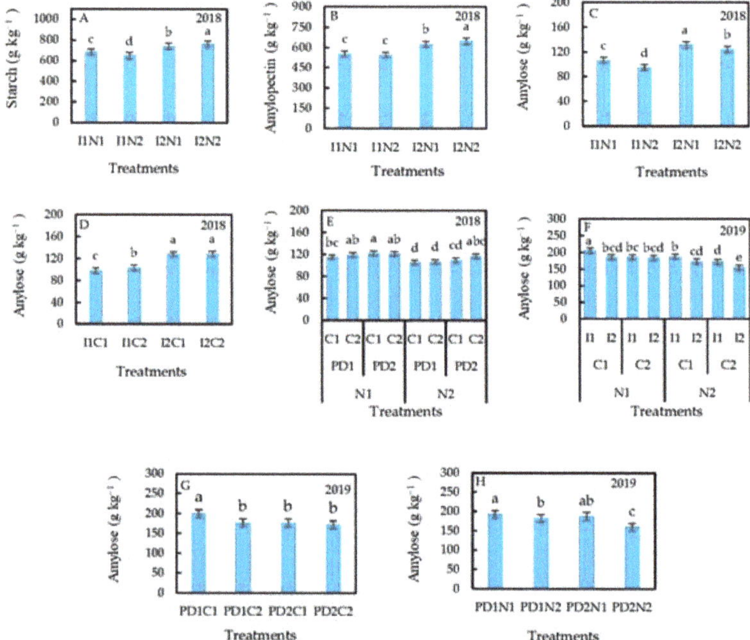

Figure 4. Least square means of maize grain starch (**A**), amylopectin (**B**) and amylose (**C**) in response to interaction effects of irrigation and nitrogen, and least square means of maize grain amylose in response to interaction effects of irrigation and cultivar (**D**); nitrogen, planting date and cultivar (**E**); nitrogen, cultivar and irrigation (**F**); planting date and cultivar (**G**); and planting date and nitrogen (**H**). I1, irrigation at 12-day intervals; I2, irrigation at 6 day-intervals; PD1, planting date 21 June; PD2, planting date 22 July; C1, KSC704; C2, KSC260; N1, 0; N2, 184 kg N ha^{-1}. Least square means labelled with the same letter do not differ significantly at $p < 0.05$ based on Tukey's test. Vertical bars represent the 95% confidence interval.

In 2019, there were interactive effects of N rate, cultivar and irrigation rate (Figure 4F); planting date and cultivar (Figure 4G); and planting date and N rate (Figure 4H) on amylose concentration. Low irrigation rate and zero N rate in KS704 had the greatest amylose concentration (205.15 g kg^{-1}), whereas high irrigation and N rate in KSC260 decreased amylose to its lowest value (154.53 g kg^{-1}) (Figure 4F). With an early planting date, KSC704 had higher amylose than KSC260 (198.98 g kg^{-1}), whereas with a late planting date, there was no cultivar effect (Figure 4G). For both planting dates, N application decreased amylose concentration; however, the effect was greater with the late planting date (Figure 4H).

3.3. Treatment Effects on Protein Content and Composition

Irrigation, planting date and cultivar did not influence protein concentration in either year. However, high N increased proteins by 6.0 g kg^{-1} in 2018 and 10.9 g kg^{-1} in 2019. Moreover, high N increased ΣNEAA concentrations by 3.4 g kg^{-1} in 2018 and 2.4 g kg^{-1} in 2019 (Table 3). In 2018, low irrigation rate increased ΣNEAA by 3.2 g kg^{-1}, and KSC260 had higher ΣNEAA than KSC704 by 1.2 g kg^{-1} (Table 3). In 2019, there was an interactive effect of planting date and cultivar on ΣNEAA (Figure 5A). With the early planting date, KSC704 had a higher ΣNEAA than KSC260, whereas with late planting, KSC260 had a higher ΣNEAA than KSC704.

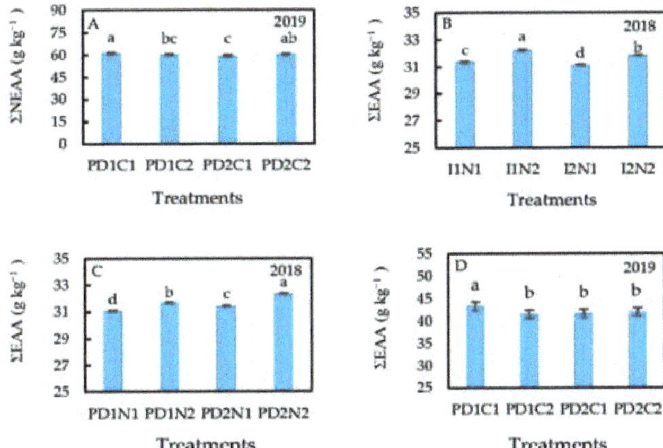

Figure 5. Least square means of maize grain for total nonessential amino acids (ΣNEAA) in response to interaction effects of planting date and cultivar (**A**) and least square means of maize grain total essential amino acids (ΣEAA) in response to interaction effects of irrigation and nitrogen (**B**), planting date and nitrogen (**C**), and planting date and cultivar (**D**). I1, irrigation at 12-day intervals; I2, irrigation at 6 day-intervals; PD1, planting date 21 June; PD2, planting date 22 July; C1, KSC704; C2, KSC260; N1, 0; N2, 184 kg N ha^{-1}. Least square means labelled with the same letter do not differ significantly at $p < 0.05$ based on Tukey's test. Vertical bars represent the 95% confidence interval.

In 2018, KSC260 had higher ΣEAA than KSC704 by 0.8 g kg^{-1} (Table 3). In 2019, high N increased the ΣEAA concentration by 1.4 g kg^{-1} (Table 3). In addition, in 2018, there was an interactive effect of irrigation and N (Figure 5B), and planting date and N (Figure 5C) on ΣEAA concentration. For either irrigation rate, a high N rate increased the ΣEAA concentration (Figure 5B); however, the effect was greater with a low irrigation rate. For either planting date, the high N rate increased the ΣEAA concentration (Figure 5C); however, high N had a greater effect on ΣEAA concentration (32.35 g kg^{-1}) for the later planting date. In 2019, there was an interactive effect of planting date and cultivar on ΣEAA (Figure 5D). With the early planting date, KSC704 had higher ΣEAA than KSC260, whereas with the late planting date, there was no cultivar effect.

3.4. Treatment Effects on Oil Content and Composition

Planting date and cultivar did not affect oil and ΣUSFA concentration in either year. In both years, there was an interactive effect of irrigation and N rate on oil (Figure 6A,B) and ΣUSFA (Figure 6C,D). In 2018, with the high irrigation rate, there was no N effect on oil and ΣUSFA, whereas with low irrigation, the high N rate decreased oil and ΣUSFA (Figure 6A,C). In 2019, with the high irrigation rate, the high N rate increased oil and ΣUSFA, whereas with low irrigation, there was no N effect (Figure 6B,D). In both years, oil and ΣUSFA were higher in the high irrigated plots (with either N rate) than with the low irrigated ones. In 2019, ΣSFA concentration with the early planting date was higher (by 0.22 g kg^{-1}) than with the late planting date (Table 3). In both years, there was an interactive effect of irrigation and N rate on ΣSFA (Figure 6E,F). In 2018, with the high irrigation rate, there was no N effect on ΣSFA, whereas with low irrigation, the high N rate decreased ΣSFA (Figure 6E). In 2019, with the high irrigation rate, the high N rate increased ΣSFA, whereas with low irrigation rate, the high N rate decreased ΣSFA (Figure 6F).

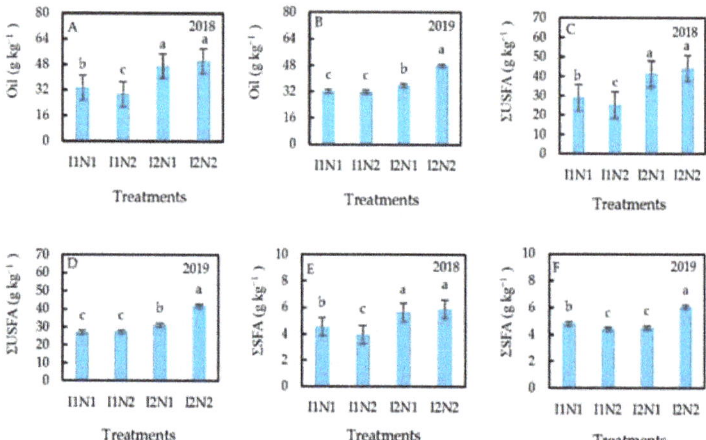

Figure 6. Least square means of maize grain oil (**A,B**), total unsaturated fatty acids (ΣUSFA) (**C,D**) and total saturated fatty acids (ΣSFA) (**E,F**) in response to interaction effects of irrigation and nitrogen. I1, irrigation at 12-day intervals; I2, irrigation at 6 day-intervals; PD1, planting date 21 June; PD2, planting date 22 July; C1, KSC704; C2, KSC260; N1, 0; N2, 184 kg N ha^{-1}. Least square means labelled with the same letter do not differ significantly at $p < 0.05$ based on Tukey's test. Vertical bars represent the 95% confidence interval.

4. Discussion

4.1. Correlations between Variables and Treatments

The results suggest that year and irrigation were the most influential effects in this experiment. The findings are well supported by Hammac et al. [42], who reported that temperature and water changes are more effective than soil nutrient status for changing rapeseed composition. In this study, there are two main potential sources for the differences among the trend of the years: (a) daily temperature differences among years and (b) performing the experiment after three years of fallow. The grain filling period is an enzyme-dependent stage of accumulating storage materials, primarily starch and protein and is sensitive to factors affecting photosynthesis, especially temperature and soil moisture, and nutritional status [43].

Many studies point to the negative impact of elevated temperatures and water deficit on oil, starch and dry matter accumulation in cereal grain [44,45]. Examples from the literature report 76.8% of the changes in oat oil content [46] and 52% of the changes in canola oil content [47] explained by climatic variation, especially water and temperature variability among years. However, Riccardi et al. [48] found that water stress induces the expression of proteins not specifically related to this stress but rather to reactions against cell damage. This may be the reason why the protein and its composition did not show a clear response to further irrigation compared to starch and oil, according to the PCA results. Castro et al. [49] reported that slight heat stress increased protein of wheat grain by shortening the grain filling period and by increasing the rate of N remobilization to grain. This is a potential reason for the long eigenvector of protein concentration in the second year.

The great impact of irrigation amount on the starch, protein and oil content of maize grain was reported by Kresović et al. [15]. Moreover, other studies point to the large effect of water availability [50] and genotype–environment interactions [51] on compositional attributes of maize and wheat grain. On the potential effect of fallow, fallows improve soil fertility, organic matter and physical properties to supply essential nutrients needed for assimilation and changes in grain composition. In contrast, continuous cultivation may result in some elements being deficient in grain, if they are not provided through fertilization [52].

4.2. Treatment Effects on Starch Content and Composition

The results showed that the average starch concentration for KSC704 was higher than for KSC260. One potential reason for this result is genetic differences between the two cultivars [53]. There is a positive correlation between grain weight and starch content in cereals [54,55]. Considering that the 100-grain weight in KSC704, as a late maturity hybrid, is higher than KSC260 [53], a higher starch concentration is expected. The highest concentration of starch was achieved by interactive effects of higher N and irrigation rate. Starch accumulation in grains is a physiological process of transportation and conversion of photosynthetic assimilates into starch and is expected to increase under irrigation and adequate N supply [56]. Moreover, the results showed that high N application with low irrigation decreased starch concentration. One potential reason for this result is the inverse starch and protein relationship in grain with high N application [57].

In the first year, the highest value of amylopectin was achieved by the interactive effect of high irrigation and N rate. In the second year, a high N rate increased amylopectin concentration. These results are in agreement with Jiyun et al. [58] and Kaplan et al. [59], who reported that amylopectin contents of maize hybrids increased under high irrigation and N rate in a similar trend to starch. The effect of N on amylose in the present experiment was that high N application was associated with reduced grain amylose. This is in close agreement with Kaufman et al. [60], who reported that high N application reduces type A granules in sorghum grain compared to type B or C, [61], suggesting that high N application reduces amylose by decreasing type A granules.

In addition, the results of the first year suggest that amylose was higher with the high irrigation rate regardless of N rate. The results are not in agreement with Kaplan et al. [59], who reported that the amylose content of maize grain decreased with an increase in irrigation rate, and it increased with an increase in N rate. Potential reasons for the discrepancy between results are different N and irrigation rates, and climatic conditions. The second-year results suggest that planting KSC704 earlier significantly increased amylose concentration. Few studies have investigated the simultaneous effects of maturity group and planting date on amylose content of grain. However, the increase in amylose in response to earlier planting [19] and lengthening maize kernel maturation [62,63] has been suggested.

4.3. Treatment Effects on Protein Content and Composition

Higher protein content in plots with higher N rates was not surprising. These results are similar to the results obtained by Saint Pierre et al. [64], Yang et al. [65], and Cao et al. [17], who reported there is a positive correlation between N application and protein content of cereal grains. This is because N stimulates the activity of panel enzymes involved in protein biosynthesis [66]. Increasing the crude protein concentration in grain can be achieved via two scenarios: (a) increasing N utilization and (b) sustaining higher partitioning of N to grain (nitrogen harvest index-NHI) [67]. Due to the nonsignificant differences between cultivars in terms of protein, it can be concluded that NHI did not differ between cultivars.

Our findings on the significant effect of N, irrigation and their interaction on ΣNEAA and ΣEAA are in agreement with Zhang et al. [8], who reported that ΣEAA in wheat grain increased in response to high N rate and low frequency irrigation. They also reported an interactive effect of irrigation and N rate on wheat grain amino acids but did not have a consistent response during the three years of their experiment. Since N is one of the basic element of amino acids and protein compounds in grain [68], higher amino acids in N-contained treatments is expected. In the first year, ΣNEAA was lower in plots with more frequent irrigation. One potential reason for a decrease in amino acids in high irrigation conditions is due to yield dilution effects on N-containing compounds in the grain [8].

The first-year results showed that KSC260 had higher ΣNEAA and ΣEAA than KSC704. In the second year, amino acids differed between cultivars and changing planting dates. In general, the inconsistency of cultivars amino acids among years and planting dates could suggest a large impact of exogenous factors in addition to the maturity group and genetic potentials on grain quality [69], which can further be associated with the complexities of genotype–environment–management interactions. The results are in close agreement with Huang et al. [70] in rice, who found that, with earlier planting dates, early maturity cultivars enter the reproductive phase earlier than late maturity ones, potentially causing flowering to coincide with high summer temperatures and consequently reducing amino acids in early maturity cultivars. Accordingly, we infer that the absence of ΣEAA differences between maturity classes in the late planting date may be because of lower temperatures during grain filling.

4.4. Treatment Effects on Oil Content and Composition

The results demonstrated that, for any year of cultivation, oil concentrations were affected by interaction effects of irrigation and N rate. This observation gives weight to the results from Aguirrezábal et al. [71] and Kaplan et al. [22], who reported that the interaction effects of irrigation and N rate are very influential on grain oil. In any year of cultivation, grain oil was higher with the high irrigation rate, irrespective of N rate. This implies that water was more critical than N for increasing oil concentration and quality. This is because water deficiency reduces grain oil by decreasing N uptake [72] and germ growth and by reducing the enzyme activity responsible for lipid biosynthesis [73].

The results suggest that, for any year of cultivation, ΣUSFA was higher with high irrigation, regardless of N rate. Our findings for irrigation effects on fatty acids were somewhat different with Kaplan et al. [22]. These authors reported that high irrigation decreased linoleic acid, as the most abundant unsaturated fatty acids in maize grain. Higher irrigation regimes, lower air temperature and higher precipitation during grain filling are potential reasons for decreasing linoleic acid under higher irrigation rates in the study conducted by Kaplan et al. [22].

The results showed a negative impact of high N rate on ΣSFA under the low irrigation rate. This is in close agreement with the study of Ali and Ullah [74], who reported that a high N rate (225 kg ha^{-1}) decreased palmitic acid and stearic acid in sunflower hybrids. In the second year, ΣSFA concentration was higher with earlier planting. Similar results were reported by Obeng et al. [75] in camelina cultivars. A possible reason for the difference between planting dates in the second year is a higher air temperature during the reproductive stage of maize growth. Some studies proposed that, as the average daily temperature rises during grain filling, the crop tends to produce more saturated fatty acids in sunflower [76] and oilseed crops [77]. This may be related to high-temperature impacts on lipid profiles by destabilizing enzymes effective in unsaturated fatty acid synthesis; as a result, saturated fatty acids increase in the grain [78].

There are some commonalities between trends from PCA and ANOVA results; however, one possible reason for some discrepancy between ANOVA and PCA trends is that the PCA plots only present the data in two dimensions (principle components 1 versus 2), whereas the correlation matrix values take into account all dimensions [79].

5. Conclusions

This study aimed to provide insights into understanding the relationship among quality characteristics of maize grain. Applying principal components analysis, the first two PCs accounted for 84.5% of the total variation. Year and irrigation had the greatest effect on yield and quality. Plots with high irrigation were associated with higher starch, oil and total unsaturated fatty acids (ΣUSFA), and data points from 2019, a warmer year, were associated with higher amylose, protein and amino acids. Analysis of variance results revealed more details on the effects of other factors and their interactions on maize grain components. In any year of cultivation, N application significantly increased protein and ΣNEAA values. A combination of high irrigation and N rate often increased oil and fatty acids values, whereas with the low irrigation rate, increased N had no effect. The cultivar KSC704 had a higher starch concentration, and KSC260 had a higher amino acids concentration. With earlier planting, ΣSFA was higher. The study was limited to two cultivars, and although there were clear differences between them, further studies that include additional cultivars would provide more confidence in the results. However, year-to-year variations in the effects of factors on amylose, amylopectin and amino acids suggest that the response of cultivars to the environment plays an important role in the final composition of starch and protein. The findings highlight the complicated relationship between the experimental factors and the large impacts of growing season conditions on quality attributes of maize grain.

Supplementary Materials: The following are available online at https://www.mdpi.com/2077-047 2/11/1/11/s1, S1: *P*-values of maize grain starch, amylopectin, amylose, protein, total non-essential amino acids (ΣNEAA), total essential amino acids (ΣEAA), oil, total unsaturated fatty acids (ΣUFA) and total saturated fatty acids (ΣSFA) in response to treatment (irrigation, planting date, cultivar and nitrogen rate) effects in 2018 and 2019.

Author Contributions: Conceptualization, S.S., M.R.J. and D.P.; data curation, S.S., G.A.A. and I.A.; formal analysis, D.P. and M.R.J.; investigation, M.R.J.; methodology, S.S., D.P. and M.R.J.; software, D.P. and M.R.J.; writing—original draft, M.R.J.; writing—review and editing, D.P., S.S., G.A.A. and I.A. visualization, D.P. and M.R.J. All authors have read and agreed to the published version of the manuscript.

Funding: This research received no external funding.

Data Availability Statement: Data are available by contacting the authors.

Acknowledgments: The authors would like to thank the University of Tehran (UT); the Animal research institute of Iran (ASRI); the Agricultural Research, Education & Extension Organization of Iran (AREEO); the Strategic Technologies Laboratory Network of Iran; and the Department of Agricultural Research for Northern Sweden for their financial contributions. The authors also would like to thank Akbar Yaghoubfar for his valuable guidance with the chemical analyses.

Conflicts of Interest: The authors declare no conflict of interest.

References

1. Darrah, L.L.; McMullen, M.; Zuber, M. Breeding, Genetics and Seed Corn Production. In *Corn*; Serna-Saldivar, S., Ed.; Elsevier: Amsterdam, The Netherlands, 2019; pp. 19–41.
2. Rausch, K.D.; Hummel, D.; Johnson, L.A.; May, J.B. Wet Milling: The Basis for Corn Biorefineries. In *Corn*; Serna-Saldivar, S., Ed.; Elsevier: Amsterdam, The Netherlands, 2019; pp. 501–535.
3. Ewu, Y.; Messing, J. Proteome balancing of the maize seed for higher nutritional value. *Front. Plant Sci.* **2014**, *5*, 240. [CrossRef]
4. Moran, J.E.T. Starch: Granule, Amylose-Amylopectin, Feed Preparation, and Recovery by the Fowl's Gastrointestinal Tract. *J. Appl. Poult. Res.* **2019**, *28*, 566–586. [CrossRef]
5. Singh, N.; Singh, S.; Shevkani, K. Maize: Composition, Bioactive Constituents, and Unleavened Bread. In *Flour and Breads and their Fortification in Health and Disease Prevention*; Preedy, V., Watson, R., Patel, V., Eds.; Elsevier: Amsterdam, The Netherlands, 2019; pp. 111–121.
6. Copeland, L.; Blazek, J.; Salman, H.; Tang, M.C. Form and functionality of starch. *Food Hydrocoll.* **2009**, *23*, 1527–1534. [CrossRef]
7. Cai, C.; Zhao, L.; Huang, J.; Chen, Y.; Wei, C. Morphology, structure and gelatinization properties of heterogeneous starch granules from high-amylose maize. *Carbohydr. Polym.* **2014**, *102*, 606–614. [CrossRef]
8. Zhang, P.; Ma, G.; Wang, C.; Lu, H.; Li, S.; Xie, Y.; Ma, D.; Zhu, Y.; Guo, T. Effect of irrigation and nitrogen application on grain amino acid composition and protein quality in winter wheat. *PLoS ONE* **2017**, *12*, e0178494. [CrossRef]
9. Allen, N.K.; Baker, D.H. Quantitative Evaluation of Nonspecific Nitrogen Sources for the Growing Chick. *Poult. Sci.* **1974**, *53*, 258–264. [CrossRef]
10. Lenis, N.P.; Van Diepen, H.T.M.; Bikker, P.; Jongbloed, A.W.; Van Der Meulen, J. Effect of the ratio between essential and nonessential amino acids in the diet on utilization of nitrogen and amino acids by growing pigs. *J. Anim. Sci.* **1999**, *77*, 1777–1787. [CrossRef] [PubMed]
11. Peng, L.-P.; Men, S.-Q.; Liu, Z.; Tong, N.-N.; Imran, M.; Shu, Q.Y. Fatty Acid Composition, Phytochemistry, Antioxidant Activity on Seed Coat and Kernel of Paeonia ostii from Main Geographic Production Areas. *Foods* **2019**, *9*, 30. [CrossRef]

12. Cook, J.P.; McMullen, M.D.; Holland, J.B.; Tian, F.; Bradbury, P.; Ross-Ibarra, J.; Buckler, E.S.; Flint-Garcia, S.A. Genetic Architecture of Maize Kernel Composition in the Nested Association Mapping and Inbred Association Panels. *Plant Physiol.* **2011**, *158*, 824–834. [CrossRef] [PubMed]

13. Jensen, L. Animal Manure Fertiliser Value, Crop Utilisation and Soil Quality Impacts. *Anim. Manure Recycl.* **2013**, 295–328. [CrossRef]

14. Silva, R.R.; Zucareli, C.; Fonseca, I.C.D.B.; Riede, C.R.; Gazola, D. Nitrogen management, cultivars and growing environments on wheat grain quality. *Rev. Bras. Eng. Agrícola Ambient.* **2019**, *23*, 826–832. [CrossRef]

15. Kresovic, B.; Gajic, B.; Tapanarova, A.; Dugalić, G. How Irrigation Water Affects the Yield and Nutritional Quality of Maize (*Zea mays* L.) in a Temperate Climate. *Pol. J. Environ. Stud.* **2018**, *27*, 1123–1131. [CrossRef]

16. Struik, P.C. *Effect of Temperature on Development, Dry-Matter Production, Dry-Matter Distribution and Quality of Forage Maize (Zea mays L.)*; Veenman & Zonen: Mariastraat, The Netherlands, 1983.

17. Cao, X.; Sun, H.; Wang, C.; Ren, X.; Liu, H.; Zhang, Z. Effects of late-stage nitrogen fertilizer application on the starch structure and cooking quality of rice. *J. Sci. Food Agric.* **2018**, *98*, 2332–2340. [CrossRef]

18. Asthir, B.; Jain, D.; Kaur, B.; Bains, N. Effect of nitrogen on starch and protein content in grain influence of nitrogen doses on grain starch and protein accumulation in diversified wheat genotypes. *J. Environ. Biol.* **2017**, *38*, 427–433. [CrossRef]

19. Medic, J.; Abendroth, L.J.; Elmore, R.W.; Jane, J.-L.; Slukova, M. Effect of planting date on maize starch structure, properties, and ethanol production. *Starch Stärke* **2015**, *68*, 476–487. [CrossRef]

20. Wang, J.; Hasegawa, T.; Li, L.; Lam, S.K.; Zhang, X.; Liu, X.; Pan, G. Changes in grain protein and amino acids composition of wheat and rice under short-term increased [CO2] and temperature of canopy air in a paddy from East China. *New Phytol.* **2019**, *222*, 726–734. [CrossRef]

21. Dai, Z.-M.; Yin, Y.-P.; Zhang, M.; Li, W.-Y.; Yan, S.-H.; Cai, R.; Wang, Z. Distribution of Starch Granule Size in Grains of Wheat Grown Under Irrigated and Rainfed Conditions. *Acta Agron. Sin.* **2008**, *34*, 795–802. [CrossRef]

22. Kaplan, M.; Kale, H.; Karaman, K.; Unlukara, A. Influence of different irrigation and nitrogen levels on crude oil and fatty acid composition of maize (*Zea mays* L.). *Grasas Aceites* **2017**, *68*, 207. [CrossRef]

23. Anwar, S.; Iqbal, M.; Akram, H.M.; Niaz, M.; Rasheed, R. Influence of Drought Applied at Different Growth Stages on Kernel Yield and Quality in Maize (*Zea mays* L.). *Commun. Soil Sci. Plant Anal.* **2016**, *47*, 2225–2232. [CrossRef]

24. Impa, S.; Perumal, R.; Bean, S.R.; Sunoj, V.J.; Jagadish, S.V.K. Water deficit and heat stress induced alterations in grain physico-chemical characteristics and micronutrient composition in field grown grain sorghum. *J. Cereal Sci.* **2019**, *86*, 124–131. [CrossRef]

25. Bicego, B.; Sapkota, A.; Torrion, J.A. Differential Nitrogen and Water Impacts on Yield and Quality of Wheat Classes. *Agron. J.* **2019**, *111*, 2792–2803. [CrossRef]

26. Walkley, A.; Black, I.A. An examination of the degtjareff method for determining soil organic matter, and a proposed modification of the chromic acid titration method. *Soil Sci.* **1934**, *37*, 29–38. [CrossRef]

27. Bremner, J.M. Determination of nitrogen in soil by the Kjeldahl method. *J. Agric. Sci.* **1960**, *55*, 11–33. [CrossRef]

28. Olsen, S.R. *Estimation of Available Phosphorus in Soils by Extraction with Sodium Bicarbonate*; US Department of Agriculture: Washington, DC, USA, 1954.

29. Mehlich, A. *Determination of P, Ca, Mg, K, Na, and NH4*; North Carolina Soil Test Division: Raleigh, NC, USA, 1953; pp. 23–89.

30. Saxton, K.E.; Rawls, W.J.; Romberger, J.S.; Papendick, R.I. Estimating Generalized Soil-water Characteristics from Texture. *Soil Sci. Soc. Am. J.* **1986**, *50*, 1031–1036. [CrossRef]

31. Akbari, G.A.; Heshmati, S.; Soltani, E.; Dehaghi, M.A. Influence of Seed Priming on Seed Yield, Oil Content and Fatty Acid Composition of Safflower (*Carthamus tinctorius* L.) Grown Under Water Deficit. *Int. J. Plant Prod.* **2019**, 245–258. [CrossRef]

32. Estakhr, A.; Heidari, B.; Ahmadi, Z. Evaluation of kernel yield and agronomic traits of European maize hybrids in the temperate region of Iran. *Arch. Agron. Soil Sci.* **2015**, *61*, 475–491. [CrossRef]

33. Jahahngirlou, R.M. Investigation of the Morphophysiological Aspects of Yield Formation in Maize (*Zea Mays* L.) Cultivars and the Agro-Ecological Assessment of Their Performance in Khouzestan and Fars Provinces. Master's Thesis, Shahid Beheshti University, Tehran, Iran, 2015.

34. Soufizadeh, S. *GDD Calculator Program*; Shahid Beheshti University: Tehran, Iran, 2012.

35. Cutforth, H.; Shaykewich, C. A temperature response function for corn development. *Agric. For. Meteorol.* **1990**, *50*, 159–171. [CrossRef]

36. FAOSTAT. Statistical Databases and Data-Sets of the Food and Agriculture Organization of the United Nations. Available online: http://www.fao.org (accessed on 5 November 2020).

37. Hanway, J. How a corn plant develops. *Iowa State Univ. Coop. Ext. Serv. Spec. Rep.* **1966**, *48*, 1–17.

38. Megazyme, Amylose/Amylopectin. Assay Procedure for Measurement of Amylose and Amylopectin of Starch. Available online: http://www.megazyme.com (accessed on 5 November 2020).

39. AOAC. *Official Methods of Analysis of AOAC International*, 20th ed.; AOAC International: Rockville, MD, USA, 2016; Volume 2, pp. 20–25.

40. Minitab, L.L.C. *SAS, v.19*; Minitab. Inc.: State College, PA, USA, 2006.

41. SAS Institute. *The SAS System. v.9.4*; SAS Institute: Cary, NC, USA, 2003.

42. Hammac, W.A.; Maaz, T.M.; Koenig, R.T.; Burke, I.C.; Pan, W.L. Water and Temperature Stresses Impact Canola (*Brassica napus* L.) Fatty Acid, Protein, and Yield over Nitrogen and Sulfur. *J. Agric. Food Chem.* **2017**, *65*, 10429–10438. [CrossRef]

43. Singletary, G.; Banisadr, R.; Keeling, P. Heat Stress During Grain Filling in Maize: Effects on Carbohydrate Storage and Metabolism. *Funct. Plant Biol.* **1994**, *21*, 829–841. [CrossRef]

44. Pirasteh-Anosheh, H.; Ranjbar, G.; Pakniyat, H.; Emam, Y.; Azooz, M.; Ahmad, P. *Plant-Environment Interaction: Responses and Approaches to Mitigate Stress*; Wiley Blackwell: Hoboken, NJ, USA, 2016.

45. Sehgal, A.; Sita, K.; Siddique, K.H.M.; Kumar, R.; Bhogireddy, S.; Varshney, R.K.; Hanumantharao, B.; Nair, R.M.; Prasad, P.V.V.; Nayyar, H. Drought or/and Heat-Stress Effects on Seed Filling in Food Crops: Impacts on Functional Biochemistry, Seed Yields, and Nutritional Quality. *Front. Plant Sci.* **2018**, *9*, 1705. [CrossRef] [PubMed]

46. Saastamoinen, M.; Kumpulainen, J.; Nummela, S.; Häkkinen, U. Effect of Temperature on Oil Content and Fatty Acid Composition of Oat Grains. *Acta Agric. Scand.* **1990**, *40*, 349–356. [CrossRef]

47. Elferjani, R.; Soolanayakanahally, R.Y. Canola Responses to Drought, Heat, and Combined Stress: Shared and Specific Effects on Carbon Assimilation, Seed Yield, and Oil Composition. *Front. Plant Sci.* **2018**, *9*, 1224. [CrossRef] [PubMed]

48. Riccardi, F. Protein Changes in Response to Progressive Water Deficit in Maize. Quantitative Variation and Polypeptide Identification. *Plant Physiol.* **1998**, *117*, 1253–1263. [CrossRef] [PubMed]

49. Castro, M.; Peterson, C.J.; Rizza, M.D.; Dellavalle, P.D.; Vázquez, D.; Ibanez, V.; Ross, A. Influence of Heat Stress on Wheat Grain Characteristics and Protein Molecular Weight Distribution. In *Molecular Breeding of Forage and Turf*; Hopkins, A., Wang, Z.Y., Mian, R., Sledge, M., Barker, R.E., Eds.; Springer: Berlin/Heidelberg, Germany, 2007; Volume 12, pp. 365–371.

50. Butts-Wilmsmeyer, C.J.; Seebauer, J.R.; Singleton, L.; Below, F.E. Weather during Key Growth Stages Explains Grain Quality and Yield of Maize. *Agronomy* **2019**, *9*, 16. [CrossRef]

51. Kaya, Y.; Ii, I.; Akcura, M. Effects of genotype and environment on grain yield and quality traits in bread wheat (*T. aestivum* L.). *Food Sci. Technol.* **2014**, *34*, 386–393. [CrossRef]

52. Nielsen, D.C.; Calderón, F.J.; Hatfield, J.L.; Sauer, T.J. Fallow Effects on Soil. In *Soil Management: Building a Stable Base for Agriculture*; Hatfield, J.L., Saue, T.J., Eds.; Soil Science Society of America: Madison, WI, USA, 2015; pp. 287–300.

53. Rahimi, J.M.; Kambouzia, J.; Zand, E.; Rezayi, M. Investigation of grain yield and some related traits in different maize cultivars (*Zea mays* L.). *J. Plant Physiol.* **2019**, *10*, 150–166.

54. Hurkman, W.J.; McCue, K.F.; Altenbach, S.B.; Korn, A.; Tanaka, C.K.; Kothari, K.M.; Johnson, E.L.; Bechtel, D.B.; Wilson, J.D.; Anderson, O.D.; et al. Effect of temperature on expression of genes encoding enzymes for starch biosynthesis in developing wheat endosperm. *Plant Sci.* **2003**, *164*, 873–881. [CrossRef]

55. Chen, G.; Zhen, S.; Liu, Y.; Yan, X.; Zhang, M.; Yan, Y. In vivo phosphoproteome characterization reveals key starch granule-binding phosphoproteins involved in wheat water-deficit response. *BMC Plant Biol.* **2017**, *17*, 168. [CrossRef]

56. Yao, S.; Wu, D.; Yang, W.; Yang, X.; Huang, L. Sprinkler irrigation enhancing accumulation and quality properties of starch in wheat grain. *Trans. Chin. Soc. Agric. Eng.* **2015**, *31*, 97–102.

57. Uribelarrea, M.; Below, F.E.; Moose, S.P. Grain Composition and Productivity of Maize Hybrids Derived from the Illinois Protein Strains in Response to Variable Nitrogen Supply. *Crop. Sci.* **2004**, *44*, 1593–1600. [CrossRef]

58. Jiyun, J.; Ping, H.; Hailong, L.; Wenjuan, L.; Shaowen, H.; Xiufang, W.; Lichun, W.; Jiagui, X.; Guogang, Z. Comparison of nitrogen absorption, yield and quality between high-starch and common corn as affected by nitrogen application. *J. Plant Nutr. Soil Sci.* **2004**, *10*, 568–573.

59. Kaplan, M.; Karaman, K.; Kardes, Y.M.; Kale, H. Phytic acid content and starch properties of maize (*Zea mays* L.): Effects of irrigation process and nitrogen fertilizer. *Food Chem.* **2019**, *283*, 375–380. [CrossRef]

60. Kaufman, R.C.; Wilson, J.; Bean, S.R.; Presley, D.R.; Blanco-Canqui, H.; Mikha, M. Effect of Nitrogen Fertilization and Cover Cropping Systems on Sorghum Grain Characteristics. *J. Agric. Food Chem.* **2013**, *61*, 5715–5719. [CrossRef]

61. Zhu, J.; Zhang, S.; Zhang, B.; Qiao, D.; Pu, H.; Liu, S.; Li, L. Structural features and thermal property of propionylated starches with different amylose/amylopectin ratio. *Int. J. Biol. Macromol.* **2017**, *97*, 123–130. [CrossRef]

62. Jane, J.-L. Current Understanding on Starch Granule Structures. *J. Appl. Glycosci.* **2006**, *53*, 205–213. [CrossRef]

63. Li, L.; Blanco, M.; Jane, J.-L. Physicochemical properties of endosperm and pericarp starches during maize development. *Carbohydr. Polym.* **2007**, *67*, 630–639. [CrossRef]

64. Pierre, C.S.; Peterson, C.J.; Ross, A.; Ohm, J.-B.; Verhoeven, M.C.; Larson, M.; Hoefer, B. White Wheat Grain Quality Changes with Genotype, Nitrogen Fertilization, and Water Stress. *Agron. J.* **2008**, *100*, 414–420. [CrossRef]

65. Yang, Y.; Zhang, M.; Zheng, L.; Cheng, D.-D.; Liu, M.; Geng, Y.-Q. Controlled Release Urea Improved Nitrogen Use Efficiency, Yield, and Quality of Wheat. *Agron. J.* **2011**, *103*, 479–485. [CrossRef]

66. Gous, P.W.; Warren, F.J.; Mo, O.W.; Gilbert, R.G.; Fox, G.P. The effects of variable nitrogen application on barley starch structure under drought stress. *J. Inst. Brew.* **2015**, *121*, 502–509. [CrossRef]

67. Zhang, L.; Liang, Z.-Y.; He, X.-M.; Meng, Q.-F.; Hu, Y.; Schmidhalter, U.; Zhang, W.; Zou, C.-Q.; Chen, X. Improving grain yield and protein concentration of maize (*Zea mays* L.) simultaneously by appropriate hybrid selection and nitrogen management. *Field Crop. Res.* **2020**, *249*, 107754. [CrossRef]

68. Zhong, Y.; Xu, D.; Hebelstrup, K.H.; Yang, D.-L.; Cai, J.; Wang, X.; Zhou, Q.; Cao, W.; Dai, T.; Jiang, D. Nitrogen topdressing timing modifies free amino acids profiles and storage protein gene expression in wheat grain. *BMC Plant Biol.* **2018**, *18*, 353. [CrossRef]

69. Harrigan, G.G.; Stork, L.G.; Riordan, S.G.; Reynolds, T.L.; Taylor, J.P.; Masucci, J.D.; Cao, Y.; LeDeaux, J.R.; Pandravada, A.; Glenn, K.C. Impact of environmental and genetic factors on expression of maize gene classes: Relevance to grain composition. *J. Food Compos. Anal.* **2009**, *22*, 158–164. [CrossRef]

70. Huang, M.; Zhang, H.; Zhao, C.; Chen, G.; Zou, Y. Amino acid content in rice grains is affected by high temperature during the early grain-filling period. *Sci. Rep.* **2019**, *9*, 2700. [CrossRef] [PubMed]

71. Aguirrezabal, L.A.N.; Martre, P.; Pereyrairujo, G.; Echarte, M.M.; Izquierdo, N.G. Improving grain quality: Ecophysiological and modeling tools to develop management and breeding strategies. In *Crop Physiology*; Sadras, V.O., Calderini, D.F., Eds.; Academic Press: Cambridge, MA, USA, 2015; pp. 423–465.

72. Geesing, D.; Diacono, M.; Schmidhalter, U. Site-specific effects of variable water supply and nitrogen fertilisation on winter wheat. *J. Plant Nutr. Soil Sci.* **2014**, *177*, 509–523. [CrossRef]

73. Singer, S.D.; Zou, J.; Weselake, R.J. Abiotic factors influence plant storage lipid accumulation and composition. *Plant Sci.* **2016**, *243*, 1–9. [CrossRef] [PubMed]

74. Alves, L.S.; Stark, E.M.L.M.; Zonta, E.; Fernandes, M.S.; Dos Santos, A.M.; De Souza, S.R. Different nitrogen and boron levels influence the grain production and oil content of a sunflower cultivar. *Acta Sci. Agron.* **2017**, *39*, 59. [CrossRef]

75. Obeng, E.; Obour, A.K.; Nelson, N.O.; Moreno, J.A.; Ciampitti, I.A.; Wang, D.; Durrett, T.P. Seed yield and oil quality as affected by Camelina cultivar and planting date. *J. Crop. Improv.* **2019**, *33*, 202–222. [CrossRef]

76. Van Der Merwe, R.; Labuschagne, M.; Herselman, L.; Hugo, A. Effect of heat stress on seed yield components and oil composition in high- and mid-oleic sunflower hybrids. *S. Afr. J. Plant Soil* **2015**, *32*, 121–128. [CrossRef]

77. Schulte, L.R.; Ballard, T.C.; Samarakoon, T.B.; Yao, L.; Vadlani, P.V.; Staggenborg, S.; Rezac, M. Increased growing temperature reduces content of polyunsaturated fatty acids in four oilseed crops. *Ind. Crop. Prod.* **2013**, *51*, 212–219. [CrossRef]

78. Martínez-Rivas, J.M.; Sánchez-García, A.; Sicardo, M.D.; García-Díaz, M.T.; Mancha, M. Oxygen-independent temperature regulation of the microsomal oleate desaturase (FAD2) activity in developing sunflower (Helianthus annuus) seeds. *Physiol. Plant.* **2003**, *117*, 179–185. [CrossRef]

79. Parsons, D.; Cherney, J.; Gauch, H.G. Alfalfa Fiber Estimation in Mixed Stands and Its Relationship to Plant Morphology. *Crop. Sci.* **2006**, *46*, 2446–2452. [CrossRef]

Article

Effects of Camu-Camu (*Myrciaria dubia*) Powder on the Physicochemical and Kinetic Parameters of Deteriorating Microorganisms and *Salmonella enterica* Subsp. *enterica* Serovar Typhimurium in Refrigerated Vacuum-Packed Ground Beef

Jorge Luiz da Silva [1,2,3], Vasco Cadavez [1], José M. Lorenzo [4], Eduardo Eustáquio de Souza Figueiredo [2,*] and Ursula Gonzales-Barron [1]

1   Centro de Investigação de Montanha (CIMO), Instituto Politécnico de Bragança (IPB), Campus of Santa Apolónia, 5300-253 Braganza, Portugal; jorge.silva@svc.ifmt.edu.br (J.L.d.S.); vcadavez@ipb.pt (V.C.); ubarron@ipb.pt (U.G.-B.)
2   Federal University of Mato Grosso (UFMT), Cuiabá, MT 78060-900, Brazil
3   Federal Institute of Education, Science and Technology of Mato Grosso (IFMT), Campo Verde, MT 78106-970, Brazil
4   Meat Technology Centre Foundation of Galicia, Technologic Park of the Galicia, San Cibrao das Viñas, 32900 Ourense, Spain; jmlorenzo@ceteca.net
*   Correspondence: figueiredoeduardo@ufmt.br; Tel.: +55-65-3615-8589

Citation: da Silva, J.L.; Cadavez, V.; Lorenzo, J.M.; Figueiredo, E.E.d.S.; Gonzales-Barron, U. Effects of Camu-Camu (*Myrciaria dubia*) Powder on the Physicochemical and Kinetic Parameters of Deteriorating Microorganisms and *Salmonella enterica* Subsp. *enterica* Serovar Typhimurium in Refrigerated Vacuum-Packed Ground Beef. *Agriculture* **2021**, *11*, 252. https://doi.org/10.3390/agriculture11030252

Academic Editor: Alessandra Durazzo

Received: 13 February 2021
Accepted: 4 March 2021
Published: 17 March 2021

Publisher's Note: MDPI stays neutral with regard to jurisdictional claims in published maps and institutional affiliations.

Abstract: This study aims to evaluate the effects of camu-camu powder (CCP), Amazonian berry fruit with documented bioactive properties, physicochemical meat parameters, and the growth kinetics parameters of *S. enterica* ser. Typhimurium, psychrotrophic bacteria (PSY), and lactic acid bacteria (LAB) in vacuum-packed ground beef. Batches of ground beef were mixed with 0.0%, 2.0%, 3.5%, and 5.0% CCP (w/w), vacuum-packed as 10 g portions, and stored at 5 °C for 16 days. Centesimal composition analyses (only on the initial day), pH, TBARS, and color were quantified on storage days 1, 7, and 15, while PSY and LAB were counted on days 0, 3, 6, 9, 13, and 16. Another experiment was conducted with the same camu-camu doses by inoculating *S. enterica* ser. Typhimurium microbial kinetic curves were modeled by the Huang growth and Weibull decay models. CCP decreased TBARS in beef from 0.477 to 0.189 mg MDA·kg^{-1}. No significant differences in meat pH between treated and control samples were observed on day 15. CCP addition caused color changes, with color a* value decreases (from 14.45 to 13.44) and color b* value increases (from 17.41 to 21.25), while color L* was not affected. Higher CCP doses caused progressive LAB growth inhibition from 0.596 to 0.349 log CFU·day^{-1} at 2.0% and 5.0% CCP, respectively. Similarly, PSY growth rates in the treated group were lower (0.79–0.91 log CFU·day^{-1}) compared to the control (1.21 log CFU·day^{-1}). CCP addition at any of the investigated doses produced a steeper *S. enterica* ser. Typhimurium inactivation during the first cold storage day, represented by Weibull's concavity α shape parameter, ranged from 0.37 to 0.51, in contrast to 1.24 for the control. At the end of the experiment, however, *S. enterica* ser. Typhimurium counts in beef containing CCP were not significantly different ($p < 0.05$) from the control. Although CCP affects bacterial kinetics, it does not protect ground beef against spoilage bacteria and *Salmonella* to the same degree it does against lipid peroxidation.

Keywords: camu-camu powder; meat; pH; *S. enterica* ser. Typhimurium; spoilage bacteria; TBARS

1. Introduction

The camu-camu (*Myrciaria dubia*) is a reddish fruit color found in trees of typical Amazon forest wild species [1–4]. This fruit exhibits a high phytochemical profile, conferring nutritional and functional values to this fruit and economic importance. Camu-camu is essential for the food, pharmaceutical, and cosmetics industries due to its bioactive, antioxidant, anti-inflammatory, and antimicrobial properties while also containing high vitamin C, carotenoid, and phenolic compound contents [3–10].

The beef industry has attempted to use natural compounds displaying functional and antimicrobial properties to formulate products to promote health and food safety [11]. The presence of *Salmonella* in beef is a relevant concern among biological risks, as this pathogen is involved with the most food-borne disease outbreaks [12]. Thus, alternatives such as natural compounds to control *Salmonella* in foodstuffs are increasingly required.

Previous assessments concerning camu-camu powder (CCP) have indicated phenolic compounds such as flavonoids, anthocyanins, ellagic acid derivatives, ellagitannins, gallic acid derivatives, and proanthocyanidins in the pulp, seed, and fruit peel [4,13]. Among phenolic compounds, CCP contains flavones, quercetin, and naringenin, which play a role in microorganism inhibition [14]. Both fresh and dry camu-camu exhibit the highest antioxidant capacity and polyphenol contents among 18 native non-traditional tropical Brazilian fruits [15].

Alcoholic camu-camu seed and pulp extracts exhibit high anti-bactericidal activity against *Streptococcus mutans* and *Streptococcus sanguinis* [16], while aqueous camu-camu extract fractions, rich in phenolic compounds, display inhibitory action against *Staphylococcus aureus* [3]. In another study, CCP methanol/water extracts (70:30 v/v) showed inhibitory effects against *S. aureus*. However, no effects were observed against other microorganisms, such as *Escherichia coli*, *Enterobacter aerogenes*, *Listeria monocytogenes*, *S. enterica* ser. Typhimurium or *Salmonella enterica ser.* Enteritidis [6]. Some studies have evaluated the antimicrobial properties of camu-camu extracts, although CCP effects in meat have not yet been evaluated.

In this context, the present study aimed to evaluate CCP effects on physicochemical meat properties and the behavior of psychrotrophic bacteria, lactic acid bacteria, and *S. enterica* ser. Typhimurium in vacuum-packed ground beef during cold storage. So, these results can be used in meat industries to use one natural compound in the processes.

2. Material and Methods

This study comprised two experiments. In the first experiment, physicochemical characteristics and spoilage microorganisms were quantified along time in non-inoculated ground beef samples to evaluate if CCP impacts the shelf-life and quality attributes of ground meat. The second experiment consisted of a challenge study where *S. enterica* ser. Typhimurium was inoculated in ground beef containing CCP to characterize its kinetic parameters of behavior. Beef samples were obtained from a slaughterhouse located in Braganza, Portugal, while raw whole camu-camu powder (CCP) was purchased from Bio-Aurora Industry (Lima, Peru). The proximate CCP composition, informed by the industry, was as follows: moisture 6.08%, fat 0.06%, protein 1.64%, ashes 1.52%, carbohydrates 90.73%, fiber 0.81% with 15.02% titratable acidity (citric acid), and 7397 mg/100 g vitamin C.

2.1. Sample Preparation

A total of four batches (420 g each) of ground beef were mixed with CCP at 0.0%, 2.0%, 3.5%, and 5.0% (w/w), respectively, using a semi-professional mixer (model Artisan 5KSM125, KitchenAid, Benton Harbor, MI, USA) for 5 min at speed 4 (~100 rpm). Ten-gram portions were then vacuum packed in transparent gas-tight polyamide and polyethylene bags (grades vacuum bags, Orved®, Musile di Piave, Venice, Italy, with the permeability of 84 $\pm$ 4.20 cc/m^2/24 h/atm for O$_2$, 361 $\pm$ 18.05 cc/m^2/24 h/atm for CO$_2$, 22 $\pm$ 1.10 cc/ m^2/24 h/atm for N$_2$ and 9.0 $\pm$ 0.45 cc/m^2/24 h/atm for H$_2$O and a density of $\pm$100 μm) using a vacuum sealer (Silvercrest SFS 110B2, Bochum, Germany), and stored at 5 °C in a cooling chamber (Froztec, Miramar, FL, USA).

2.1.1. Physicochemical Analyses

Three samples (10 g portions) per treatment were analyzed to determine the physicochemical parameters (pH, color, and lipid oxidation) on days 1, 7, and 15. The proximate composition (moisture, lipids, protein, and ash) was determined on day 0, per treatment, according to AOAC [17] and ISO [18] in duplicate subsamples, while physicochemical parameters were determined in five subsamples. The pH values were determined using a potentiometer (HI 9913, Hanna Instruments, Eibar, Spain) and color (L*, a*, b*) by a Minolta colorimeter (Konica Minolta CM-600d, Osaka, Japan). Lipid oxidation was determined through the thiobarbituric acid reaction (TBARS) according to Vyncke [19]. Briefly, a 10 g portion was added to 97.5 mL of distilled water containing 2.5 mL of 4 N HCl, homogenized and distilled. Then, 5 mL of the distillate was added to 5 mL of a 0.02 M TBA solution and heated in a water bath at 100 °C for 35 min. Measurements were carried out using a Specord 200 spectrophotometer (Analytik Jena AG, Jena, Germany) at 528 nm, and the results were expressed as mg of malondialdehyde (MDA)/kg.

2.1.2. Microbiological Analyses (PSY and LAB)

Total psychrotrophic and lactic acid bacteria counts were performed on storage days 0, 3, 6, 9, 13, and 16. Sample (10 g) was added to 90 mL of buffered peptone water 2% (Oxoid, Cheshire, UK) and homogenized for 60 s (Stomacher 400, Seward, West Sussex, UK), followed by serial dilutions preparations.

According to APHA [20], psychrotrophic plate counts were performed where 1 mL of appropriate dilutions, in duplicate, were pour-plated into PCA agar (Liofilchem® s.r.l., Roseto Degli Abruzzi, Italy) and incubated at 7 °C for 10 days. For lactic acid bacteria, 1 mL of appropriate dilutions were pour-plated in duplicate in MRS agar (Liofilchem® s.r.l., Roseto Degli Abruzzi, Italy) supplemented with Tween 80 (Liofilchem® s.r.l., Roseto Degli Abruzzi, Italy), and overlaid with MRS agar. The plates were then incubated in anaerobic conditions at 30 °C for 48 h. Typical colonies were counted with a colony counter (Digital S®, J.P. Selecta S.A., Barcelona, Spain), and the data were converted to $\log_{10}$ CFU·g^{-1}.

2.2. *S. enterica ser. Typhimurium Behaviour in Vacuum-Packed Ground Beef Containing CCP*

One loop of *S. enterica* ser. Typhimurium ATCC 14028 strain stored in a cryo-vial at −80 °C was transferred to TSB broth (Liofilchem® s.r.l., Roseto Degli Abruzzi, Italy) and incubated 35 °C for 24 h. Subsequently, a loopful was streaked on XLD agar (Liofilchem® s.r.l., Roseto Degli Abruzzi, Italy) and incubated at 35 °C for 24 h to obtain isolated colonies. One isolated colony was then transferred to a new TSB broth tube and further incubated at 35 °C for 24 h. A loopful was finally transferred to TSB broth (Liofilchem® s.r.l., Roseto Degli Abruzzi, Italy) and incubated at 8 °C (slow growth condition) until reaching the early stationary phase, determined with the aid of a previously constructed calibration curve determined using a Specord 200 spectrophotometer (Analytik Jena AG, Jena, Germany) at 600 nm.

Batches (200 g) of ground beef were mixed with CCP at 0.0%, 2.0%, 3.5%, or 5.0% (w/w) using a semi-professional mixer (model Artisan 5KSM125, KitchenAid, Benton Harbor, MI, USA) for 5 min at speed 4 (~100 rpm). An appropriate amount of the refrigerated inoculum was diluted in physiological water (5 mL) added to the bulk ground beef to target an *S. enterica* ser. Typhimurium concentration of 5 log CFU/g^{-1}. The ground beef was further mixed for 7 min. Three portions of ten-gram each were then vacuum packed in transparent gas-tight polyamide and polyethylene bags (grades vacuum bags, Orved®, Spain, with the permeability of 84 ± 4.20 cc/m^2/24 h/atm for O$_2$, 361 ± 18.05 cc/m^2/24 h/atm for CO$_2$, 22 ± 1.10 cc/m^2/24 h/atm for N$_2$ and 9.0 ± 0.45 cc/m^2/24 h/atm for H$_2$O and density of ± 100μm) using a vacuum sealer (Silvercrest SFS 110B2, Bochum, Germany), and stored at 5 °C.

S. enterica ser. Typhimurium was enumerated on storage days 0, 3, 6, 9, 12, and 15 in three samples analyzed per treatment and time point. Ten-gram samples were then mixed with 90 mL of buffered peptone water 2% (Oxoid, Cheshire, UK) and homogenized for 60 s (Stomacher 400, Seward, West Sussex, UK). Serial dilutions were performed, and 0.1 mL of the appropriate dilutions were inoculated, in duplicate, onto plates containing Hektoen Enteric Agar (Liofilchem® s.r.l., Roseto Degli Abruzzi, Italy) and incubated at 35 °C for

24 h. Typical colonies were counted with a colony counter (Digital S®, J.P. Selecta S.A., Barcelona, Spain), and the data were converted to $\log_{10}$ CFU·g^{-1}.

2.3. Statistical Analyses

The physicochemical data, namely pH, L*, a*, b*, TBARS, and proximate analysis, underwent an analysis of variance and Dunnett's comparison of means tests to determine potential differences CCP levels at a significance probability level of 0.05. All statistical analyses were carried out using the R software [21].

Each of the experimental psychrotrophic bacteria and LAB growth curves was modeled by adjusting the integrated form of the Huang primary model (Equation (1)) proposed for a constant environmental condition [22], as follows:

$$Y(t) = Y_0 + Y_{max} - ln\{\exp(Y_0) + (\exp(Y_{max}) - \exp(Y_0)) \times \exp(-\mu_{max}B(t))\}$$
$$B(t) = t + \frac{1}{\alpha}ln\frac{1+\exp(-\alpha(t-\lambda))}{1+\exp(\alpha\lambda)} \tag{1}$$

where: Y_0, Y_{max}, and Y represent the natural logarithms of microbial concentrations at an initial time point ($t = 0$), maximum population and actual time t, respectively; μ_{max} accounts for a maximum specific growth rate (day^{-1}); λ is the delay interval (or lag time) of a curve depicting microbial behavior through time (day); α is the coefficient that accounts for the lag phase shift (set to 4, as recommended by Huang [23]); and t is the time interval. Since LAB curves did not exhibit any lag phase, λ was set to zero in Equation (1) when adjusting the Huang model.

S. enterica ser. Typhimurium decay behavior was modeled by a three-parameter modified Weibull equation (Equation (2)), defined as

$$Y(t) = Y_0 - \left(\frac{t}{\chi}\right)^{\beta} \tag{2}$$

where the scale and shape parameters of the underlying Weibull distribution are χ and β, respectively. In shape parameter $\beta > 1$, convex curves are obtained, and for $\beta < 1$, concave curves are represented. Although the Weibull model is basically empirical, van Boekel [24] suggested that $\beta < 1$ presumes that the surviving microorganisms at any point in the inactivation curve display the capacity to adapt to the applied stress, whereas $\beta > 1$ indicates that the remaining cells become increasingly susceptible to heat. The parameter χ is called scale parameter (a characteristic time). Y is defined as above, and the parameter Y_0 represents the natural logarithm of the initial microbial concentration, and t represents the time.

3. Results

3.1. Physicochemical Characterization of Vacuum-Packed and Refrigerated Beef Containing Camu-Camu Powder

The centesimal composition of the ground beef samples presented mean values of 68.13%, 3.79%, 23.71%, and 1.31% for moisture, lipids, proteins, and ashes, respectively. Table 1 presents the proximate composition of the meat samples subjected to CCP treatments.

Table 1. Proximate composition (% wb) of ground beef control and treatments with the addition of camu-camu powder as determined on the initial day of the experiment.

CCP (%)	n	Moisture (%)	Lipids (%)	Protein (%)	Ashes (%)
0.0%	5	71.2	3.4	24.1	1.3
2.0%	5	68.7	4.3	23.7	1.3
3.5%	5	67.5	4.3	23.7	1.3
5.0%	5	66.8	3.3	23.6	1.4

CCP = camu-camu powder; n = number of samples analyzed.

Concerning physicochemical parameters, significant effects were observed as a function of CCP addition, storage time (days), and the interactions of these factors. pH values varied ($p < 0.05$) as a function of CCP addition and time. However, higher CCP concentrations resulted in increases in pH values of 5.4 (0.0%) to 5.6 (5.0%). Considering storage time, pH values decreased from 5.9 to 5.3 between days 1 and 15, respectively. Regarding color changes, significant effects on L*, a*, b* were observed with CCP addition, while significant effects during the storage time were observed only for a* and b* (Table 2).

Table 2. Physicochemical parameters determined in vacuum-packed ground beef containing different levels of camu-camu powder and stored at 5 °C.

Main Effects	Parameters				
	pH	L*	a*	b*	TBARs [mg MDA/kg]
CCP (%)	***	***	***	***	***
0.0	5.476 [a] ± 0.051	49.11 [c] ± 0.544	14.45 [b] ± 0.401	17.41 [a] ± 0.354	0.573 [c] ± 0.065
2.0	5.576 [b] ± 0.069	44.34 [b] ± 0.260	14.44 [b] ± 0.306	20.08 [b] ± 0.180	0.464 [bc] ± 0.069
3.5	5.640 [c] ± 0.085	43.03 [ab] ± 0.333	13.96 [ab] ± 0.229	21.23 [c] ± 0.189	0.289 [ab] ± 0.048
5.0	5.662 [c] ± 0.069	41.92 [a] ± 0.222	13.44 [a] ± 0.212	21.25 [c] ± 0.215	0.207 [a] ± 0.048
Time (days)	***	NS	***	***	***
1	5.933 [c] ± 0.029	44.24 [a] ± 0.645	14.28 [b] ± 0.281	19.8 [a] ± 0.541	0.066 [a] ± 0.017
7	5.488 [b] ± 0.026	44.82 [a] ± 0.831	14.48 [b] ± 0.222	20.6 [b] ± 0.296	0.403 [b] ± 0.044
15	5.344 [a] ± 0.010	44.75 [a] ± 0.560	13.46 [a] ± 0.251	19.6 [a] ± 0.334	0.501 [c] ± 0.050
CCP × Time	***	NS	***	**	**

[a,b,c] Means ± standard error followed by the same letter in the columns do not differ according to Dunnett's test at 5% significant: NS = non-significant. *** and ** = 0.001% and 0.01% of significance, respectively, by ANOVA test.

Camu-camu powder addition promoted decreases ($p < 0.05$) in meat lipid oxidation with values from 0.573 mg malondialdehyde (MA)/Kg (control group) to 0.207 mg malondialdehyde (MA)/kg (5.0% CCP) (Table 2). Interactions between CCP addition and storage time were also observed for TBARS values ($p < 0.05$) (Table 2). Higher decreases in lipid oxidation were noted for the 5.0% CCP concentration during nine storage days.

Considering the interaction between CCP concentration and time, pH decreases ($p < 0.05$) were observed from the first day until the end of the experiment for all treatments, while the CCP addition caused lipid oxidation declines, resulting in the lowest TBARS concentrations in samples containing 5.0% CCP. A progressive lipid oxidation inhibition as a function of increasing CCP concentrations was also observed (Figure 1).

Concerning color characteristics, interactions between CCP addition and time resulted in meat color changes, with decreases in L* and a*, and increases in b* values (Table 2). The control treatment presented higher L* values (49.11) compared to the CCP samples (44.34–41.92) ($p < 0.05$). Decreased ($p < 0.05$) a* values from 14.45 (control) to 13.44 (5.0% CCP) were observed, causing a red color reduction in the meat samples. Concerning b* (yellow hue), CCP addition caused increases from 17.4 (control) to 20.08 up to 21.25 in the CCP-containing samples (Figure 2).

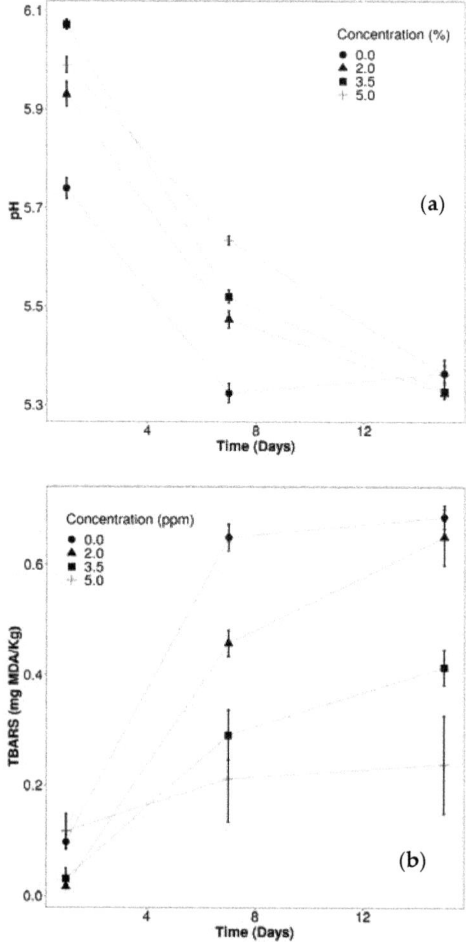

Figure 1. pH (**a**) and TBARS (**b**) of vacuum-packed ground beef containing different camu-camu powder levels stored at 5 °C.

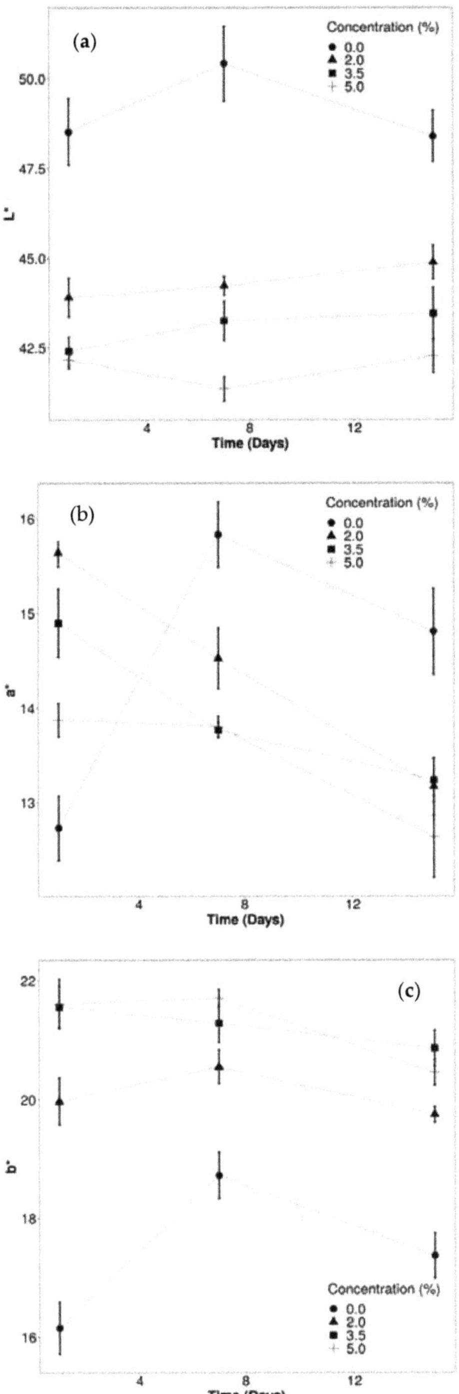

Figure 2. Colors, L* (**a**), a* (**b**), b* (**c**) of vacuum-packed ground beef containing different camu-camu powder levels stored at 5 °C.

3.2. Behavior of Deteriorating Microorganisms and S. enterica ser. Typhimurium in Vacuum-Packed Ground Beef

Kinetic parameter effects on both spoilage bacteria and *Salmonella* were caused by CCP addition. Although CCP additions decreased LAB kinetic parameters, such as μ_{max}, microbial growth did not show changes, with decreased initial concentrations and increased final concentrations (Table 3).

Table 3. Kinetic parameters of the Huang growth model describing lactic acid bacteria concentrations in vacuum-packed ground beef containing different camu-camu powder levels stored at 5 °C.

Model Parameters [1]	Camu-Camu Powder Concentration			
	0.0%	2.0%	3.5%	5.0%
Y_0	3.696 ± 0.227 *[2]	3.655 ± 0.277 *	3.584 ± 0.106 *	3.851 ± 0.375 *
Y_{max}	6.941 ± 0.175 *	7.019 ± 0.186 *	6.985 ± 0.091 *	7.601 ± 0.532 *
μ_{max}	0.490 ± 0.079 *	0.596 ± 0.112 *	0.451 ± 0.033 *	0.349 ± 0.092 *

[1] Y_0: initial counts (log CFU/g); Y_{max}: final counts (log CFU/g); μ_{max}: maximum growth rate (day^{-1}) (these parameters were expressed as means and standard error). [2] Asterisks represent the significance of the estimated parameter at $p < 0.05$.

CCP addition caused a psychrotrophic bacteria lag phase increase from 7.955 to 8.156 days in the controls and 5.0% in the treated groups, respectively (Table 4). However, no significant initial and final concentration effects were observed.

Table 4. Kinetic parameters of the Huang growth model describing psychrotrophic bacteria concentrations in vacuum-packed ground beef containing different camu-camu powder levels stored at 5 °C.

Model Parameters [1]	Camu-Camu Powder Concentration			
	0.0%	2.0%	3.5%	5.0%
Y_0	5.092 ± 0.016 *[2]	5.060 ± 0.022 *	4.910 ± 0.080 *	4.845 ± 0.158 *
λ	7.955 ± 0.308 *	7.438 ± 0.270 *	7.599 ± 0.917 *	8.156 ± 0.804 *
Y_{max}	7.871 ± 0.030 *	7.470 ± 0.046 *	7.309 ± 0.166 *	7.579 ± 0.315 *
μ_{max}	1.210 ± 0.350 *	0.788 ± 0.133 *	0.828 ± 0.527 *	0.912 ± 0.694 *

[1] Y_0: initial counts (log CFU/g); λ: lag phase duration (day); Y_{max}: final counts (log CFU/g); μ_{max}: maximum growth rate (day^{-1}) (these parameters were expressed as means and standard error). [2] Asterisks represent the significance of the estimated parameter at $p < 0.05$.

The results for other kinetic parameters did not present a statistical difference with initial concentration, and the maximum concentration at final storage time was between 7.30 to 7.87 log CFU/g, indicating the start of deterioration. Although no statistical effect, it should be highlighted the μ_{max} difference between the control group and treated groups showing kinetic action.

Significant effects ($p < 0.05$) were observed concerning *Salmonella* concentrations and kinetics in vacuum-packed ground beef stored at 5 °C (Table 5). Increasing in χ parameters, and the decrease in β parameters indicate higher declines in pathogen concentrations on the first days of cold storage.

Table 5. Kinetic parameters of the Weibull decay model describing *S. enterica* ser. Typhimurium behavior in vacuum-packed ground beef containing different camu-camu powder levels stored at 5 °C.

Model Parameters [1]	Camu-Camu Powder Concentration			
	0.0%	2.0%	3.5%	5.0%
Y_0	5.121 ± 0.048 *[2]	5.046 ± 0.072 *	5.091 ± 0.028 *	5.089 ± 0.040 *
χ	13.65 ± 1.710 *	13.61 ± 5.192 *	26.90 ± 7.319 *	40.32 ± 20.34 [ns]
β	1.247 ± 0.442 *	0.519 ± 0.240 *	0.433 ± 0.110 *	0.359 ± 0.103 *

[1] Y_0: initial counts (log CFU/g); χ: scale parameter (day^{-1}); β: shape parameter (dimensionless) (these parameters were expressed as means and standard error). [2] Asterisks represents the significance of the estimated parameter at $p < 0.05$; ns: non-significant.

No variations were observed in final pathogen concentrations, with values ranging from 4.61 to 4.79 log CFU/g. Therefore, the decreased bacterial concentrations observed in the present study were due to the low storage temperature (Figure 3).

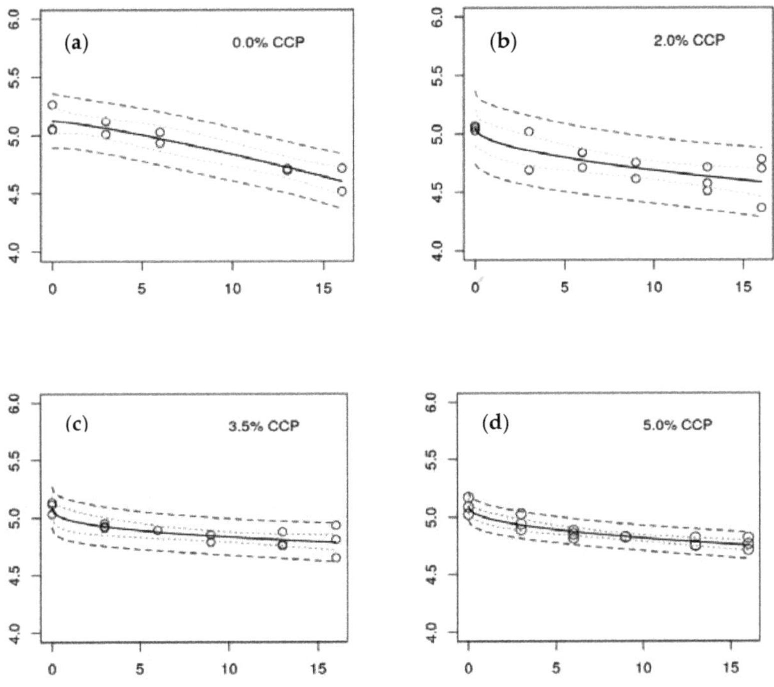

Figure 3. Weibull model fitting to experimental survival curves of *S. enterica* ser. Typhimurium in vacuum-packed ground beef containing different camu-camu powder (CCP) levels stored at 5 °C: 0.0% (**a**), 2.0% (**b**), 3.5% (**c**) and 5.0% (**d**). The 95% confidence intervals are displayed as green lines and the prediction intervals as blue lines. X-axis: Time (day); y-axis: Counts (log CFU/g).

4. Discussion

The addition of CCP caused moisture variations between 71.2% (control) and 66.78% (5.0%), while for lipids these values were between 3.34% (5.0% CCP) and 4.28% (2.0% CCP). Protein and ashes presented values between 23.70% and 24.13% and 1.27% and 1.37% respectively. Proximate composition alterations are not unexpected since camu-camu berries, apart from water, are mainly composed of carbohydrates (4.84% wb), crude fiber (0.56% wb), and several minerals such as potassium (87.0 mg/100 g, phosphorous (18.2 mg/100 g), sulphate (14.7 mg/100 g), calcium (9.1 mg/100 g), magnesium (7.4 mg/100 g), cobalt (1.2 mg/100 g), and iron (0.42 mg/100 g), among others [9].

In one experiment, the addition of 100 ppm of camu-camu peel and seed extracts led to increases in pH values from 6.01 to 6.16 in ground-lamb meat stored at 4 °C for nine days under normal atmosphere [25], which was also observed herein (Table 2). Another study reported that pH values increased in lamb meat after 10 days of refrigeration due to medium alkalinization by bacteria and endogenous enzymatic protein degradation [26]. In the present study, bacterial spoilage growth (PSY and LAB) was observed during refrigerated storage of the CCP containing- vacuum-packed beef.

Lipid oxidation decreases were observed herein with CCP addition. In another experiment, camu-camu extracts affected ($p < 0.05$) lipid oxidation values in chilled ground lamb meat treated with camu-camu peel and seed extracts and stored at 4 °C for 9 days under a normal atmosphere, where the control group presented 0.108 mg MDA/kg while treat groups presented 0.079 and 0.068 mg MDA/kg for peel and seed extracts respectively [25]. The camu-camu extracts promoted food stability, which may be associated with phenolic compounds present in camu-camu fruits. In the present survey, TBARS decreases were also observed in the meat containing CCP, indicating that phenolic compounds are available in both the extract and powdered forms of the fruit.

Lipid peroxidation decreases between 24% and 86% ($p \leq 0.05$) have been reported in egg homogenates due to the use of aqueous and ethyl alcohol camu-camu seed extracts in studies using DPPH (2,2-diphenyl-1-picrylhydrazyl) radical activity and FCRC (Follin-Ciocalteu reducing capacity) tests [27]. Lipid oxidation inhibition was correlated with total phenolic contents ($r = 0.78$), total flavonoid contents ($r = 0.67$), and condensed tannin contents ($r = 0.68$), and may be an effect of reaction mechanisms, resulting in greater antioxidant activity [27]. Another study reported camu-camu flour (CCF) and camu-camu pulp powder (CCPP) antioxidant capacities, as both the products contain both vitamin C and phenolic compounds [13]. In a DPPH assay, CCF and CCPP contained 1036.4 and 510.5 μmol Trolox/g dry material, respectively, while other polyphenol-rich powders exhibit values between 300 and 450 μmol Trolox/g dry material. Therefore, the antioxidant activity was higher with CCP addition in the present study, probably due to higher phenolic compound concentrations.

Furthermore, the highest antioxidant capacity among five ripe exotic fresh fruit was observed for camu-camu in another survey [28]. Camu-camu residue powders are a relevant source of vitamin C (8.2 mg/g of dried weight), and its extracts contain considerable phenolic content, even accounting for dehydration losses [3].

In the present study L* values were significantly different ($p < 0.05$) as a function of CCP concentrations, with values ranging from 49.11 (control) to 41.92 (5.0% CCP). However, no effects were observed concerning storage time or interactions with CCP concentrations. Similar to this study, another experiment noted no significant effects on L* values in refrigerated ground-lamb meat treated with camu-camu peel and seed extracts, with values ranging between 43.30 and 45.67 [25].

Fresh camu-camu residues presented an L* value of 51.8, increasing to 58.9 and 70.3, respectively, after dehydration by heat or freeze-drying [3], while L* values of 60.45 and 36.60 to 40.84 were observed for freeze-dried and dried camu-camu pulps, respectively [6].

During the storage period, and concerning CCP concentrations versus time, L* did not differ between the initial and final storage days, with a mean value of 44.75. Another survey observed that the luminosity value of vacuum-packed ground beef changed from 44.6 to 48.4 during 20 days of refrigerated storage [29].

Concerning the other investigated color characteristics, significant effects were observed, with decreases in a* values and increases in b* values as a function of CCP concentrations and decreases and increases in a* and b* values, respectively, as a function of time storage.

One survey demonstrated that peel and seed camu-camu extracts resulted in a* reductions from 11.13 to 6.51 and 15.60 to 6.71, respectively, in vacuum-packed ground lamb meat stored at 4 °C for nine days. Decreases in b* values were also observed, with values ranging from 16.29 to 11.25 and 13.33 to 10.19 for peel and seed camu-camu extracts, respectively [25]. These findings corroborate with the present study, where decreased red hues (a*) in CCP-containing meat samples were observed. In addition, the same aforementioned study indicated that camu-camu peel and seed extracts led to a* value decrease, from 16.29 and 13.33 to 10.19 in vacuum-packed ground lamb meat stored at 4 °C for nine days [25]. It probably occurred due to color pigments oxidation by the action of radical species from lipid oxidation [25].

Concerning the microbial analyses, LAB did not present lag phase (λ), exhibiting growth since the beginning of the experiment. CCP addition did not interfere with the initial microorganism concentrations in the investigated meat samples. The control and CCP treatment values were 3.69 log CFU/g and between 3.58 and 3.85 log CFU/g respectively. Camu-camu powder addition did not inhibit LAB growth in meat, determined as 7.01, 6.98, and 7.60 log CFU/g in the treated group and 6.94 log CFU/g (control group).

On the other hand, CCP interfered in the kinetic behavior of LAB, resulting in decreased maximum growth rates, ranging from 0.45 to 0.35 log CFU/g/day for samples containing 3.5% and 5.0% CCP, respectively, and 0.49 log CFU/g/day in the control samples.

Some authors have reported the high antibacterial activity of alcoholic camu-camu pulp and seed extracts against *Streptococcus mutans* and *Streptococcus sanguinis*. These bacteria exhibit a high prevalence in the human oral cavity. Antibacterial activity was observed only for pulp extracts, with a minimum inhibitory concentration (MIC) of 62.5 µg/mL for the tested microorganisms [16]. It should be noted that *Streptococcus* belongs to the LAB group.

Another survey reported that camu-camu peel and seed extracts presented great antimicrobial activity against Gram-positive bacteria [4]. Other authors [3] reported MIC values ranging from 0.312 to 0.62 mg/mL for polyphenol-rich aqueous camu-camu extracts against *Staphylococcus aureus*. However, another assessment reported a lack of antibacterial effects of camu-camu seed and pulp extracts against *Staphylococcus aureus*, *Escherichia coli*, and *Saccharomyces cerevisiae* strains [30].

In the present study, slight decreases in initial PSY concentrations were observed in meat containing CCP with the value of 4.84 log CFU/g (5.0% CCP sample) and 5.092 log CFU/g in the control sample. CCP concentrations decreased the maximum PSY growth rate (μ_{max}) to 0.788–0.912 log CFU/g/day in the CCP groups and 1.21 log CFU/g/day in the controls.

Another study found that dried and freeze-dried raw CCP extracts resulted in effects only against *Staphylococcus aureus*. In addition, the extracts displayed no antimicrobial activity against *Enterobacter aerogenes*, *Escherichia coli*, *Listeria monocytogenes*, *S. enterica* ser. Enteritidis, and *S. enterica* ser. Typhimurium [6]. These results corroborate the present study, where CCP did not inactivate *S. enterica* ser. Typhimurium concentrations in beef samples.

Strong antimicrobial activity of n-hexane and seed CCP extracts containing acylphloroglucinols and rhodomyrtone compounds against Gram-positive bacteria has been reported [4], with significant action against *Bacillus cereus*, *Micrococcus luteus*, *Staphylococcus epidermidis*, but no effects against Gram-negative bacteria such as *Pseudomonas aeruginosa* and *S. enterica* ser. Typhimurium. No CCP action against *Salmonella* and psychrotrophic bacteria was observed in the present study, probably due to the cell wall differences between Gram-positive and Gram-negative bacteria. However, effects against *Pseudomonas aeruginosa* were observed in a culture medium applying freeze-dried optimized camu-camu seed extracts [27]. Most studies have tested the antimicrobial effects of camu-camu extracts than CCP, and a few researches have analyzed the antimicrobial effects of CCP addition in meat, mainly in beef. Thus, we aimed to study only CCP effects to simplify camu-camu processing and use.

Regarding the pathogen behavior analyzed herein, *Salmonella* initial concentrations (Y_0) were significant ($p < 0.05$), with values ranging between 5.121 log CFU/g (control sample) and 5.046, 5.091, and 5.089 log CFU/g (treated samples). A fast decline in *Salmonella* concentrations with increasing CCP concentrations was observed, and it can be observed due to χ values of 26.9 and 40.32 in the 3.5% and 5.0% camu-camu addition treatments, respectively (Table 5), indicating more time for pathogen concentration decrease.

According to the shape parameter of the Weibull model, the 5.0% CCP addition ($\beta = 0.359$) caused more damage and stress to the evaluated pathogens than in the control group ($\beta = 1.247$) ($p < 0.05$). These shape parameters were 0.519 and 0.433 for the 2.0% and 3.5% CCP additions, respectively. Consequently, *Salmonella* exhibited higher adaptation in control group samples (0.0% CCP), demonstrating that camu-camu powder affected pathogen adaptability in ground beef.

One study using the micro-dilution method did not present any effects of dried and freeze-dried CCP extracts against *S. enterica* ser. Typhimurium and *Salmonella* Enteritidis [6], while another report indicated no antimicrobial activity of acylphloroglucinol and rhodomyrtone isolated from camu-camu peels and seeds against *S. enterica* ser. Typhimurium [4]. On the other hand, other authors [27] using the diffusion method on plate-cavity agar technic observed antimicrobial activity of freeze-dried optimized camu-camu seed extracts against *Salmonella* Enteritidis and *S. enterica* ser. Typhimurium, with 6.82 mm and 6.42 mm inhibition halos, respectively. *S. enterica* ser. Typhimurium and *S. enterica* ser. Enteritidis growth both were not inhibited by the action of freeze-dried and spray-dried CCP extracts in another survey [31]. Another assessment concluded that camu-camu seed and peel extracts had no effect against *S. enterica* ser. Typhimurium, concluding that these extracts exhibit strong action against Gram-positive bacteria, but little or no effect against Gram-negative bacteria such as *Salmonella* [4].

The results finding shows CCP addition did not decrease *Salmonella* concentrations in vacuum-packed ground beef, although it caused significant effects on some kinetic parameters.

5. Conclusions

CCP addition decreases lipid oxidation of vacuum-packed ground beef, although it affects color aspects, leading to a decreased red hue in the CCP-containing meat. Concerning antimicrobial activity, CCP does not interfere in *S. enterica* ser. Typhimurium behavior does not extend the *shelf-life* of vacuum-packed ground beef based on the concentration of certain spoilage microorganisms, acting only on the kinetic bacterial behavior parameters.

Author Contributions: Conceptualization, V.C., U.G.-B.; Data curation, V.C., U.G.-B.; Formal analysis, V.C., U.G.-B.; Funding acquisition, E.E.d.S.F., V.C., U.G.-B.; Investigation, J.L.d.S., J.M.L., V.C., U.G.-B.; Methodology, V.C., U.G.-B.; Project administration, V.C., U.G.-B.; Resources, V.C., J.M.L., U.G.-B.; Supervision, V.C., U.G.-B.; Validation, V.C., U.G.-B.; Visualization, J.L.d.S., V.C., E.E.d.S.F., U.G.-B.; Writing—original draft, J.L.d.S., E.E.d.S.F., V.C., U.G.-B.; Writing—review & editing, J.L.d.S., E.E.d.S.F. All authors have read and agreed to the published version of the manuscript.

Funding: This research was funded by the Foundation for Science and Technology (FCT, Portugal), Coordination for the Improvement of Higher Education Personnel (CAPES, Brazil) and Federal University of Mato Grosso, Brazil (UFMT).

Institutional Review Board Statement: Not applicable.

Informed Consent Statement: Not applicable.

Acknowledgments: The authors would like to thank CAPES (Coordination for the Improvement of Higher Education Personnel)—Brazil, for supporting the first author with a scholarship from the international Sandwich Exchange Program—PDSE 047/2017/Process no. 88881.189927/2018-01 and the National Council for Scientific and Technological Development—CNPq (Process: 310462/2018-5), and to PROPeq/PROPG-UFMT, Brazil. U. Gonzales-Barron and V. Cadavez are grateful to the Foundation for Science and Technology (FCT, Portugal) for financial support through national funds

FCT/MCTES to CIMO (UIDB/00690/2020). U. Gonzales-Barron acknowledges FCT, P.I., for the institutional scientific employment program contract.

Conflicts of Interest: The authors declare no conflict of interest.

References

1. Chirinos, R.; Galarza, J.; Betalleluz-Pallardel, I.; Pedreschi, R.; Campos, D. Antioxidant compounds and antioxidant capacity of Peruvian camu-camu (*Myrciaria dubia* (H.B.K.) McVaugh) fruit at different maturity stages. *Food Chem.* **2010**, *120*, 1019–1024. [CrossRef]
2. Yuyama, K.; Mendes, N.B.; Valente, J.P. Longevity camu-camu seeds submitted to different storage environments and forms of conservation. *Rev. Bras. Frutic.* **2011**, *33*, 601–607. [CrossRef]
3. Azevêdo, J.C.S.; Fujita, A.; Oliveira, E.L.; Genovese, M.I.; Correia, R.T.P. Dried camu-camu (*Myrciaria dubia* H.B.K. McVaugh) industrial residue: A bioactive-rich Amazonian powder with functional attributes. *Food Res. Int.* **2014**, *62*, 934–940. [CrossRef]
4. Kaneshima, T.; Myoda, T.; Toeda, K.; Fujimori, T.; Nishizawa, M. Antimicrobial constituents of peel and seeds of camu-camu (*Myrciaria dubia*). *Biosci. Biotechnol. Biochem.* **2017**, *81*, 1461–1465. [CrossRef] [PubMed]
5. Zanatta, C.F.; Cuevas, E.; Bobbio, F.O.; Winterhalter, P.; Mercadante, A.Z. Determination of anthocyanins from camu-camu (*Myrciaria dubia*) by HPLC-PDA, HPLC-MS and NMR. *J. Agric. Food Chem.* **2005**, *53*, 9531–9535. [CrossRef]
6. Fujita, A.; Borges, K.; Correia, R.; Franco, B.D.G.M.; Genovese, M.I. Impact of spouted bed drying on bioactive compounds, antimicrobial and antioxidant activities of commercial frozen pulp of camu-camu (*Myrciaria dubia* Mc. Vaugh). *Food Res. Int.* **2013**, *54*, 495–500. [CrossRef]
7. Kaneshima, T.; Myoda, T.; Toeda, K.; Fujimori, T.; Nishizawa, M. Antioxidative constituents in camu-camu fruit juice residue. *Food Sci. Technol. Res.* **2013**, *19*, 223–228. [CrossRef]
8. Borges, L.; Conceição, E.; Silveira, D. Active compounds and medicinal properties of *Myrciaria genus*. *Food Chem.* **2014**, *15*, 224–233. [CrossRef] [PubMed]
9. Castro, J.C.; Maddox, J.D.; Imán, S.A. Camu-camu—*Myrciaria dubia* (Kunth) McVaugh. In *Exotic Fruits Reference Guide*, 1st ed.; Rodrigues, S., Brito, E.S., Eds.; Embrapa Agroindústria Tropical: Fortaleza, Brazil, 2018; Volume 1, pp. 97–105. [CrossRef]
10. Anhê, F.F.; Nachbar, R.T.; Varin, T.V.; Trottier, J.; Dudonné, S.; Barz, M.L.; Feutry, P.; Pilon, G.; Barbier, O.; Desjardins, Y.; et al. Treatment with camu-camu (*Myrciaria dubia*) prevents obesity by altering the gut microbiota and increasing energy expenditure in diet-induced obese mice. *Microbiome* **2019**, *68*, 453–464. [CrossRef] [PubMed]
11. Tiwari, B.K.; Valdramidis, V.P.; O'donnell, P.O.; Muthukumarappan, K.; Bourke, P.; Cullen, P.J. Application of natural antimicrobials for food preservation. *J. Agric. Food Chem.* **2009**, *57*, 5987–6000. [CrossRef] [PubMed]
12. Ferrari, R.G.; Rosario, D.K.A.; Cunha-Neto, A.; Mano, S.B.; Figueiredo, E.E.S.; Conte-Junior, C.A. Worldwide epidemiology of *Salmonella* serovars in animal-based foods: A meta-analysis. *Appl. Environ. Microbiol.* **2019**, *85*, 1–21. [CrossRef]
13. Fracassetti, D.; Costa, C.; Moulay, L.; Tomás-Barberán, F.A. Ellagic acid derivatives, ellagitannins, proanthocyanidins and other phenolics, vitamin C and antioxidant capacity of two powder products from camu-camu fruit (*Myrciaria dubia*). *Food Chem.* **2013**, *139*, 578–588. [CrossRef]
14. Rauha, J.P.; Remes, S.; Heinonem, M.; Hopia, A.; Kähkönen, M.; Kujala, T.; Pihlaja, K.; Vuorela, H.; Vuorela, P. Antimicrobial effects of finish plant extracts containing flavonoids and other phenolic compounds. *Int. J. Food Microbiol.* **2000**, *56*, 3–12. [CrossRef]
15. Rufino, M.S.M.; Alves, R.E.; Brito, E.S.; Péere-Jiménez, J.; Saura-Calixto, F.; Mancini-Filho, J. Bioactive compounds and antioxidant capacities of 18 non-traditional tropical fruits from Brazil. *Food Chem.* **2010**, *121*, 996–1002. [CrossRef]
16. Camere-Colarossi, R.; Ulloa-Urizar, G.; Medina-Flores, D.; Caballero-García, S.; Mayta-Tovalino, F.; Valle-Mendoza, J. Antibacterial activity of *Myrciaria dubia* (camu-camu) against *Streptococcus mutans* and *Streptococcus sanguinis*. *Asian Pac. J. Trop. Biomed.* **2016**, *6*, 740–744. [CrossRef]
17. AOAC. *Official Methods of Analysis*, 19th ed.; Association of Official Analytical Chemists: Washington, DC, USA, 2012.
18. ISO 937. *International Standards Meat and Meat Products—Determination of Nitrogen Content*; International Organization for Standarization: Geneva, Switzerland, 1978.
19. Vyncke, W. Evaluation of the direct thiobarbituric acid extraction method for determining oxidative rancidity in mackerel (*Scomber scombrus* L.). *Fette Seifen Anstrichm.* **1975**, 239–240. [CrossRef]
20. American Public Health Association—APHA. *Compendium of Methods for the Microbiological Examination of Foods*, 4th ed.; APHA: Washington, DC, USA, 2001.
21. R Development Core Team. *R: A Language and Environment for Statistical Computing*; R Foundation for Statistical Computing: Vienna, Austria, 2018. Available online: https://www.r-project.org (accessed on 11 November 2019).
22. Huang, L. Growth kinetics of *Listeria monocytogenes* in broth and beef frankfurters—Determination of lag phase duration and exponential growth rate under isothermal conditions. *J. Food Sci.* **2008**, *75*, 235–242. [CrossRef] [PubMed]
23. Huang, L. Optimization of a new mathematical model for bacterial growth. *Food Control.* **2013**, *32*, 283–288. [CrossRef]
24. Van-Boekel, M.A.J.S. On the use of the Weibull model to describe thermal inactivation of microbial vegetative cells. *Int. J. Food Microbiol.* **2002**, *74*, 139–159. [CrossRef]
25. Guedes-Oliveira, J.M.; da Costa-Lima, B.R.C.; Cunha, L.C.M.; Salim, A.P.A.A.; Baltar, J.D.; Fortunato, A.R.; Conte-Junior, C.A. Impact of *Myrciaria dubia* peel and seed extracts on oxidation process and color stability of ground lamb. *CYTA J. Food.* **2018**, *16*, 931–937. [CrossRef]

26. Muela, E.; Sañudo, C.; Campo, M.M.; Mendel, I.; Beltrán, J.A. Effect of freezing method and frozen storage duration on instrumental quality of lamb throughout display. *Meat Sci.* **2010**, *84*, 662–669. [CrossRef] [PubMed]
27. Fidelis, M.; do Carmo, M.A.V.; da Cruz, T.M.; Azevedo, L.; Myoda, T.; Furtado, M.M.; Marques, M.B.; Sant'Ana, A.S.; Genovese, M.I.; Oh, W.Y.; et al. Camu-camu seed (*Myrciaria dúbia*)—From side stream to an antioxidant, antihyperglycemic, antiproliferative, antimicrobial, antihemolytic, anti-inflammatory, and antihypertensive ingredient. *Food Chem.* **2020**, *310*, 125909. [CrossRef] [PubMed]
28. Genovese, M.I.; Da Silva Pinto, M.; De Souza, S.G.A.; Lajolo, F.M. Bioactive compounds and antioxidant capacity of exotic fruits and commercial frozen pulps from Brazil. *Food Sci. Technol. Int.* **2008**, *14*, 207–214. [CrossRef]
29. Rogers, H.B.; Brooks, J.C.; Martin, J.N.; Tittor, A.; Miller, M.F.; Brashears, M.M. The impact of packaging system and temperature abuse on the shelf-life characteristic of ground beef. *Meat Sci.* **2014**, *97*, 1–10. [CrossRef] [PubMed]
30. Myoda, T.; Fujimura, S.; Park, B.; Nagashima, T.; Nakagawa, J.; Nishizawa, M. Antioxidative and antimicrobial potential of residues of camu-camu juice production. *J. Food Agric. Environt.* **2010**, *8*, 304–307. [CrossRef]
31. Fujita, A.; Sarkar, D.; Wu, S.; Kennelly, E.; Shetty, K.; Genovese, M.I. Evaluation of phenolic-linked bioactives of camu-camu (*Myrciaria dubia* Mc. Vaugh) for antihyperglycemia, antihypertension, antimicrobial properties and cellular rejuvenation. *Food Res. Int.* **2015**, *77*, 194–203. [CrossRef]

Article

Reasonable Nitrogen Fertilizer Management Improves Rice Yield and Quality under a Rapeseed/Wheat–Rice Rotation System

Peng Ma [1], Yan Lan [2], Xu Lv [3], Ping Fan [3], Zhiyuan Yang [3], Yongjian Sun [3], Rongping Zhang [1,*] and Jun Ma [3,*]

[1] School of Life Science and Engineering, Southwest University of Science and Technology, Mianyang 610000, China; Mapeng1@stu.sicau.edu.cn

[2] College of Agronomy, Sichuan Agricultural University, Wenjiang, Chengdu 611130, China; lanyan@stu.sicau.edu.cn

[3] Rice Research Institute, Sichuan Agricultural University, Wenjiang, Chengdu 611130, China; lvxu@stu.sicau.edu.cn (X.L.); 2020211006@stu.sicau.edu.cn (P.F.); 14236@sicau.edu.cn (Z.Y.); yongjians1980@sicau.edu.cn (Y.S.)

* Correspondence: zhzhrrpp@163.com (R.Z.); majun@sicau.edu.cn (J.M.); Tel.: +86-159-8340-0681 (R.Z.); +86-136-0822-2603 (J.M.)

Citation: Ma, P.; Lan, Y.; Lv, X.; Fan, P.; Yang, Z.; Sun, Y.; Zhang, R.; Ma, J. Reasonable Nitrogen Fertilizer Management Improves Rice Yield and Quality under a Rapeseed/Wheat–Rice Rotation System. *Agriculture* **2021**, *11*, 490. https://doi.org/10.3390/agriculture11060490

Academic Editor: Alessandra Durazzo

Received: 2 May 2021
Accepted: 21 May 2021
Published: 25 May 2021

Publisher's Note: MDPI stays neutral with regard to jurisdictional claims in published maps and institutional affiliations.

Abstract: To determine the influence of N fertilizer management on rice yield and rice quality under diversified rotations and establish a high-yield, high-quality, and environmentally friendly diversified planting technology, a rapeseed/wheat–rice rotation system for 2 successive years was implemented. In those rotation systems, a conventional N rate (Nc; 180 kg/hm^2 N in rape season, 150 kg/hm^2 N in wheat season) and a reduced N rate (Nr; 150 kg/hm^2 N in rape season, 120 kg/hm^2 N in wheat season) were applied. Based on an application rate of 150 kg/hm^2 N in the rice season, three N management models were applied, in which the application ratio of base:tiller:panicle fertilizer was 20%:20%:60% in treatment M1, 30%:30%:40% in treatment M2, and 40%:40%:20% in treatment M3. Zero N was used as the control (M0). The results showed that, under Nc and Nr in the rape season, M3 management produced an increase in rice yield. The average rice yields in 2018 and 2019 were 9.41 t/hm^2 and 9.54 t/hm^2, respectively. An increase in rice peak viscosity, hot viscosity, break disintegration, and chalkiness was achieved. Under Nc and Nr in the wheat season, the panicle fertilizer of 40%:40%:20% in rice season produced a higher rice yield. The average yield was 9.45 t/hm^2 and 9.19 t/hm^2, respectively, and an increase in rice peak viscosity, hot viscosity, and break disintegration was produced. Reduced N for rapeseed and the panicle fertilizer of 40%:40%:20% in rice season under a rapeseed–rice rotation system can be recommended to stabilize yield and ensure high-quality rice production and environmentally friendly rapeseed–rice rotation systems in southern China.

Keywords: rapeseed/wheat–rice rotation system; nitrogen management; rice yield; rice quality

1. Introduction

It is known that rice plays an important role in the world. With the improvement of living standards, consumers pay more attention to rice quality, particularly the eating/cooking quality. Studies have found that rice varieties with high amylose content have poor eating/cooking quality [1], while rice varieties with low amylose content generally have a higher eating/cooking quality. Quality is an important consideration in rice production. Rice quality is not only controlled by genetic factors but also affected by temperature, water, and nitrogen management. Nitrogen is an important element in fertilizer that can significantly affect the grain yield and quality of rice [2,3]. Reasonable nitrogen fertilizer management not only increases the yield of rice but is also an important cultivation measure to regulate the quality of production [4,5]. There is a large nitrogen fertilizer input in China's nitrogen-fertilized farmland, and the nitrogen utilization rate is only about 30%. The high nitrogen fertilizer input not only reduces the use of nitrogen but also causes environmental pollution. The nitrogen fertilizer enters the water body, soil, and air, causing water and soil pollution [6,7]. Diversified crop rotation models, such as wheat–rice crop rotations and rapeseed–rice crop rotations, are widely distributed and produce a large amount of straw every year in China. Rapeseed and wheat straw contains abundant nutrients, and its incorporation has become one of the important methods used to decrease the application of N and other chemical fertilizers. Straw returned to the field can effectively increase the rice yield, roughness rate, polished and whole rice rate, reduce chalky grain rate, chalkiness and amylose content, increase the aspect ratio and gel consistency, and improve the rice processing quality and appearance quality, taste quality, and nutritional quality [8,9]. Yan et al. [10] showed that returning wheat straw to the field can increase rice yield. At the same time, it can reduce the chalky grain rate and chalkiness and improve the rice quality under a rice-wheat rotation system. The optimization of straw return to the field and nitrogen fertilizer under different rotation modes can not only realize the efficient use of resources but also effectively improve economic benefits [11]. Under the wheat–rice and rapeseed–rice rotation model, straw returned to the field and nitrogen fertilizer management have a significant effect on the nitrogen use efficiency of hybrid rice [12]. Returning straw to the field combined with on-site nitrogen fertilizer management can increase yield and improve the appearance and taste quality of rice [13]. Under the condition of a nitrogen application rate of 276 kg·hm^{-2}, if only high-quality rice is required, a nitrogen fertilizer operation with a ratio of base tiller to panicle fertilizer of 10:0 should be used. For high-quality rice, nitrogen fertilizer management with a 7:3 ratio between the base tiller fertilizer and spike fertilizer should be used [14]. Under conditions of a nitrogen application rate of 2.25 t/hm^2 and 4.50 t/hm^2, and a nitrogen fertilizer operation with a ratio of the base tiller to panicle fertilizer of 6:4–8:2, the chalkiness and amylose content of rice are reduced, while the gel consistency and protein content are reduced, which can improve the cooking and eating quality and nutritional quality of rice [15]. The above research mainly focuses on the effects of the straw return to the field, nitrogen fertilizer management, and the supporting nitrogen fertilizer management under a single rotation mode on rice yield and rice quality, but there are few comparative studies between straw return to the field and nitrogen fertilizer management under different rotation modes. Therefore, in this study, under a rapeseed–rice and wheat–rice rotation system, the straw of rape and wheat were returned to the field, and different nitrogen fertilizer management treatments were used. The objective was to determine the effects of optimized nitrogen fertilizer application on the yield and quality of hybrid indica rice under a rapeseed/wheat–rice rotation system. In so doing, the regulation and control methods for the quality and yield improvement of hybrid indica rice under the diversified rotation system can be identified, with a view to improving the quality of rice under different rotation models in production.

2. Materials and Methods

2.1. Experimental Site Information

The experiments were conducted at the farm of the Rice Research Institute, Sichuan Agricultural University, Wenjiang, Sichuan Province, China (30.70° N, 103.83° E) from October 2017 to early September 2019. Immediately before the field experiment (2017), soil samples from the top 0.20 m of surface soil contained 1.52 g/kg total N (Kjeldahl method, UDK-169, ITA), 23.89 mg/kg of available phosphorus (Mo–Sb colorimetry after digestion with H_2SO_4 and $HClO_4$), 2.421% organic matter ($K_2Cr_2O_7$-volumetric method), and 52.61 mg/kg available K (flame spectrometry after NH_4OAc extraction) and had a pH of 6.19 (tested in a sample containing a 1:2.5 ratio of soil to water). The average air temperature and precipitation during the previous crop and rice-growing season, measured at the weather station close to the experimental site, are detailed in Figure 1.

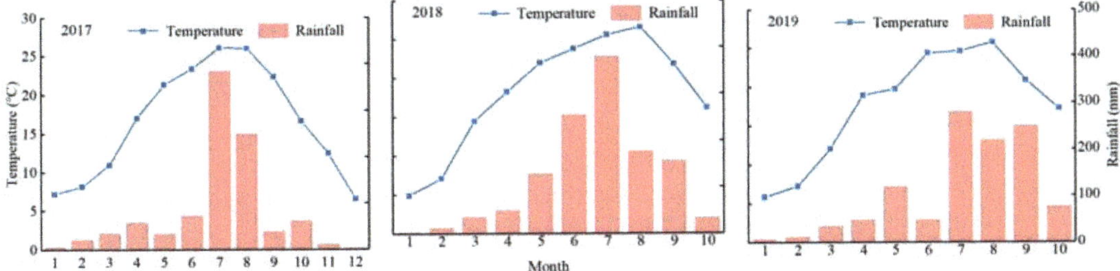

Figure 1. The meteorological data of the experimental area, including temperature and rain full in 2017–2019.

2.2. Experiment Design

The experiment adopted a three-factor design. The first factor (two levels) was the previous crop, which was rape and wheat represented by Pr and Pw, respectively. The second factor (two levels) was the N application rate in the previous season, with a conventional N application (Nc; 180 kg/hm^2 N in rape season, 150 kg/hm^2 N in wheat season), reduced N (Nr; 150 kg/hm^2 N in rape season, 120 kg/hm^2 N in wheat season), and N fertilizer as basal manure and top dressing at a 5:5 ratio. The third factor (three levels) was N fertilizer management, with common urea as the N source. Based on an application rate of 150 kg/hm^2 N in the rice season, three N management models were used, where the ratio of the application of base fertilizer, tiller fertilizer, and panicle fertilizer was 20%:20%:60% in treatment M1, 30%:30%:40% in treatment M2, and 40%:40%:20% in treatment M3. M0 was defined as the zero-N control. A total of 16 treatments were performed with three repetitions. Each experimental plot was 15.75 m^2 in area with a 30 cm-wide ridge covered with a plastic film to prevent water and nutrient penetration from the contiguous plots.

The rape variety used was 'Mianyou No. 15' (Mianyang Academy of Agricultural Sciences, Sichuan Province), and the wheat variety used was 'Shumai 969' (Wheat Research Institute of Sichuan Agricultural University). Rape seedlings were transplanted on 12 October 2017 and 2018 and spaced at 0.5 × 0.35 m (57,000 plants/hm^2) in both 2017 and 2018. The rapeseed was harvested on 1 May 2018 and 2019, and the wheat was harvested on 8 May 2018 and 2019. Straw was cut into 5 cm pieces and returned to the corresponding plots after the rape and wheat harvest. The N contribution to the plot by rape straw was 16.08–27.81 kg/hm^2, 13.08–19.88 kg/hm^2 by wheat straw. Urea (N, 46.4%) was used as the N source, phosphorus (P$_2$O$_5$, 12.0%) was used as the phosphorus source, and potassium chloride (K$_2$O, 60.0%) was used as the potassium source. Phosphorus and potassium were applied to the soil as a base fertilizer 1 day before sowing or transplanting. Nitrogen, phosphorus, and potassium fertilizers were applied in the rape season at a ratio of 2:1:2 and in the wheat season at a ratio of 2:1:1.

'Fyou 498' a commonly planted, high-yield, indica hybrid rice cultivar, was sown in a seedbed on 17 April 2018 and 2019, and seedlings were transplanted to the field on 23 May 2018 and 2019. The rice seedlings were transplanted and spaced at 0.333 × 0.167 m, in both 2018 and 2019, with one plant per hill. Ordinary urea was applied during the rice season. Nitrogen, phosphorus, and potassium fertilizers were applied in the rape season at a ratio of 2:1:2. Phosphate fertilizer (P$_2$O$_5$; 75 kg/hm^2) and potash fertilizer (K$_2$O; 150 kg/hm^2) were used as base fertilizers. The base fertilizer was applied 1 day before transplanting. Tiller fertilizer was applied 7 days after transplanting. Spike fertilizer, divided into flower-promoting fertilizer and flower-keeping fertilizer at a ratio of 5:5, was applied twice at the four-leaf and two-leaf stages. After the rice was harvested, the entire amount of straw was chopped and returned to the corresponding plot, and the N contribution to the plot by rape straw was 16.08–27.81 kg/hm^2. The weeds were controlled in the rape and rice plots with pretilachlor (1720 mL/hm^2) (Jiangsu Changlong Agrochemicals Co., Ltd. Taizhou, China). The herbicide was applied once at the seedling stage of rapeseed and the tillering stage of rice. Pests and diseases were controlled by imidacloprid (90 g/hm^2) (Hubei Xinhe Chemical Co., Ltd. Wuhan, China) and kasugamycin (1200 mL/hm^2) (Hubei Dibai Chemical Co., Ltd., Wuhan, China) to avoid yield loss.

2.3. Measurements and Methods

2.3.1. Plant Sampling and Measurements

At the maturity stage, all rice plants were selected from each plot to test the rice yield (GY) were calculated according to the actual number of harvested plants; the value was adjusted to 13.5% moisture to ensure safe storage.

2.3.2. Rice Quality Index Measurements

Rice grains were collected, dried, and stored for more than 3 months, according to NY/T83-1988 (1988). Grain samples of 120 g with 3 replications from each plot were collected for grain quality analysis according to GB/T 17891-1999 (1999). The brown rice, milled rice, and head rice rates were expressed as percentages of the total grain weights. Chalkiness was evaluated on 100 milled grains per plot. The number of grains containing over 20% white was considered as chalkiness rate. The chalkiness size was expressed as the percentage of the total area of the kernel. The amylose content was determined from the absorption at 620 nm by scanning the iodine absorption spectrum from 400 to 900 nm using a spectrophotometer (Ultrospec 6300 pro, Amersham Biosciences, Little Chalfont, UK). The values were converted to amylose content by reference to a standard curve prepared from rice. The protein content was measured with a grain analyzer (Infratec 1241, Foss, Denmark). Rice paste properties were determined using a Rapid Visco Analyser (RVA; Super3, Newport Scientific, Sydney, Australia), following the procedure of the American Association of Cereal Chemists. Three-gram samples of flour were sifted with a 0.15 mm sieve and mixed with 25 g of deionized water in an RVA sample tube. Peak viscosity, hot viscosity, cool viscosity in centipoise units (cp), and their derivative parameter breakdown (peak viscosity minus hot viscosity), setback (cool viscosity minus peak viscosity), and consistency (cool viscosity minus hot viscosity) were recorded with matching software, Thermal Cline for Windows (TCW). Cooking/eating quality was measured by Taste Analyzer RCTA11A (Satake Co., Hiroshima, Japan). The primary function of the taste analyzer was to convert various physicochemical parameters of rice into taste value.

2.4. Data Analysis

Data were analyzed using analysis of variance (ANOVA), and the means were compared based on the least significant difference (LSD) test at the 0.05 probability level using SPSS23 (Chinese version v22.0.0.0) (Statistical Product and Service Solutions Inc., Chicago, IL, USA). The Origin Pro 2017(OriginLab, Northampton, MA, USA) was used to draw the figures.

3. Results

3.1. Effects of N Application Rate in the Previous Season and N Management in Rice Season, on Rice Yield

The analysis of variance showed that the previous crop (P), nitrogen application rate (N), nitrogen fertilizer management (M), and their interaction effects reached significant levels, and there were also differences between treatments in the 2 years (Table 1).

Table 1. Analysis of the variance of rice yield by nitrogen fertilizer management under a rapeseed/wheat–rice rotation system.

Source of Variation	Degree of Freedom	Sum of Squares	Mean Square	Computed F	$F_{0.05}$	$F_{0.01}$
Replication	4	0.01	0.00	73.13 **	2.53	3.65
Treatment	30	169.82	5.66	116467.42 **	1.65	2.03
Year (Y)	1	4.31	4.31	88609.67 **	4.00	7.08
Previous crop (P)	1	1.19	1.19	24546.63 **	4.00	7.08
N rate (N)	1	0.33	0.33	6798.35 **	4.00	7.08
N management (M)	3	156.81	52.27	175441.77 **	2.76	4.13
Y × P	1	1.19	1.19	24453.13 **	4.00	7.08
Y × N	1	0.01	0.01	199.67 **	4.00	7.08
Y × M	3	4.25	1.42	29128.42 **	2.76	4.13
P × N	1	0.02	0.02	344.81 **	4.00	7.08
P × M	3	2.94	0.98	20151.89 **	2.76	4.13
N × M	3	0.11	0.04	774.41 **	2.76	4.13
Y × P × N	1	0.08	0.08	1715.07 **	4.00	7.08
Y × P × M	3	1.49	0.50	10198.31 **	2.76	4.13
Y × N × M	3	0.82	0.27	5635.96 **	2.76	4.13
P × N × M	3	0.35	0.12	2423.26 **	2.76	4.13
Y × P × N × M	3	0.23	0.08	1567.64 **	2.76	4.13
Error	60	0.00	0.00			
Total variation	95	174.14				

Y: year; P: previous crop; N: nitrogen rate; M: nitrogen management. * and ** mean significance at the 0.05 and 0.01 probability levels, respectively. F: Analysis of variance.

Further analysis shows that the change in rice output between the years is basically the same (Figure 2). The yield of rice under different previous crops was rapeseed (Pr) > wheat (Pw), and Pr increased by 2.67% relative to Pw. Under different nitrogen application rates, the performance was ranked as: conventional nitrogen application (Nc) > reduction (Nr). Under different nitrogen operations, the performance was: M3 > M2 > M1 > M0, and M3 was relative to M2, M1, and M0, and increased by 1.39%, 4.61%, and 55.67%, respectively. The interaction effect of Pr and M3 nitrogen fertilizer management on seed setting rate, thousand-grain weight, and yield was significantly higher than the interaction effect of other previous crops and different nitrogen fertilizer management treatments, and the interaction effect of Nc and M3 nitrogen fertilizer management treatment had a higher impact on yield. Interaction effects of other nitrogen application rates and different nitrogen fertilizer management strategies indicated that in the rape season, the rice yield under Nc and Nr was the highest under the M3 operation, and the 2-year average yields were 9.41 t/hm^2 and 9.54 t/hm^2, respectively. Compared with Nc, the rice yield under the M3 operation in the rice season increased by 1.38% on average in 2 years. In the wheat season, both Nc and Nr had the highest rice yield under the M2 operation, and the 2-year average yields were 9.45 t/hm^2 and 9.19 t/hm^2. Compared with Nc in wheat season, rice yield under M2 operation decreased by 2.75% on average in 2 years, and the difference was not significant. This indicates that reducing nitrogen by 20% in rapeseed season, with a ratio of the application of base fertilizer, tiller fertilizer, and panicle fertilizer of 40%:40%:20% (M3) in rice season, was more conducive to increasing rice yield.

3.2. Effects of N Application Rate in the Previous Season and N Management in Rice Season, on Rice Quality Characteristics

The analysis of variance shows that there are significant differences among the various indicators of rice quality depending on the year, the previous crop, the amount of nitrogen applied, the proportion of nitrogen fertilizer, and the interaction between them. The interaction of the three was not significant (Table 2). It can be seen that nitrogen fertilizer management under the rapeseed/wheat–rice rotation had a greater impact on various indicators of rice quality.

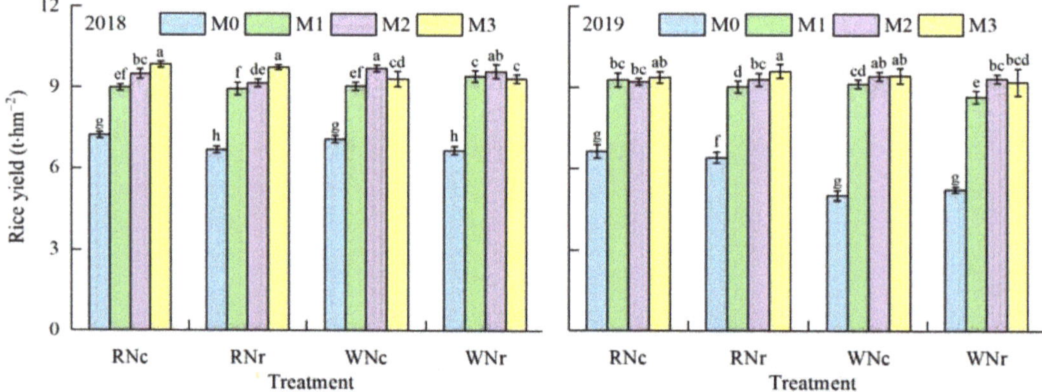

Figure 2. The effects of the N application rate in the previous season and N management in the rice season on the rice yield. RNc and RNr represent the conventional nitrogen application and reduced nitrogen application in the rape season, respectively. WNc and WNr represent the conventional nitrogen application and reduced nitrogen application in the wheat season, respectively. M0 represents zero N was used in rice season; M1, M2, and M3 represent based on an application rate of 150 kg/hm^2 N in the rice season, three N management models were applied, in which the application ratio of base:tiller:panicle fertilizer was 20%:20%:60%, 30%:30%:40%, and 40%:40%:20%, respectively. Lower case letters indicate that the yields of the hybrid rice are significantly different among the treatments ($p < 0.05$, LSD method).

3.2.1. Processing and Nutritional Quality

The brown rice rate, polished rice rate, and amylose content of rice are the highest in 2019 (Table 3). The brown rice rate, protein, and amylose are higher in the wheat season than in the rape season; the polished rice rate and the whole rice rate are higher in the rapeseed season than in the wheat season. Different previous crops have a significant impact on the processing and nutritional quality of rice. Brown rice rate, polished rice rate, and whole rice rate are the largest under different nitrogen application rates in rapeseed and wheat seasons under reduced nitrogen applications, while protein and amylose are the highest under conventional nitrogen application in rapeseed and wheat seasons. The brown rice rate, polished rice rate, and protein content under different nitrogen fertilizer management showed as M1 > M2 > M3 > M0; protein content increased by 4.24%, 6.97%, and 28.72% compared to M2, M3, and M0 under the treatment of M1; the content of amylose was M0 > M3 > M2 > M1 under the different nitrogen fertilizer strategies. The amylose content of rice, except for the control without nitrogen fertilizer (M0), was the highest, and all gradually decreased with the decrease of the ratio of basal tiller fertilizer. The change in protein content was opposite, with significant differences between treatments. Increasing the ratio of panicle fertilizer can improve the nutritional quality of rice.

Table 2. Significance of variance estimates related to years (Y), previous crop (P), nitrogen rate (N), nitrogen management (M), and their interactions on rice grain quality traits.

Source of Variation	Year (Y)	Previous Crop (P)	N Rate (N)	N Management (M)	Y × P	Y × N	Y × M	P × N	P × M	N × M	Y × P × N	Y × P × M	Y × N × M	P × N × M	Y × P × N × M
Degree of freedom	1	1	1	3	1	1	3	1	3	3	1	3	3	3	3
BR	5.02 *	51.21 **	0.16 ns	35.84 **	0.02 ns	1.38 ns	3.20 *	2.23 ns	2.48 ns	2.32 ns	1.20 ns	0.24 ns	0.69 ns	2.44 ns	0.42 ns
MR	2.20 ns	781.90 **	0.01 ns	15.80 **	0.22 ns	0.50 ns	3.71 *	11.20 **	5.31 **	1.84 ns	0.05 ns	4.06 *	1.93 ns	1.19 ns	2.18 ns
HMR	6.47 *	1204.87 **	15.72 **	109.73 **	14.91 **	28.00 **	45.54 **	19.50 **	111.92 **	21.43 **	4.80 *	36.51 **	9.16 **	4.12 *	4.56 **
CP	16.43 **	1218.23 **	1.02 ns	1202.13 **	90.00 **	196.23 **	24.70 **	2.73 ns	22.47 **	33.47 **	49.44 **	30.64 **	14.60**	1.38 ns	59.22 **
CD	2.21 ns	748.80 **	0.17 ns	392.30 **	49.82 **	87.85 **	6.84 **	8.82 **	1.45 ns	17.37 **	18.17 **	8.64 **	6.62 **	0.80 ns	27.13 **
AC	661.61 **	613.48 **	4.70 *	0.42 ns	261.03 **	1.47 ns	11.51 **	0.76 ns	3.62 *	4.65 **	0.76 ns	3.62 *	1.08 ns	1.62 ns	1.62 ns
PC	198.58 **	1473.91 **	4.23 *	1341.81 **	9.28 **	159.25 **	132.13 **	0.01 ns	0.01 ns	53.73 **	0.01 ns	0.01 ns	57.12 **	0.01 ns	0.02 ns
Mouthfeel	10.48 **	523.49 **	9.82 **	102.31 **	15.76 **	1.32 ns	1.05 ns	2.45 ns	0.41 ns	0.78 ns	0.39 ns	1.31 ns	3.12 *	10.48 **	8.90 **
Comprehensive	12.53 **	448.13 **	10.49 **	98.40 **	9.54 **	2.50 ns	2.24 ns	0.05 ns	7.90 **	1.23 ns	4.17 *	2.51 ns	2.92 *	6.28 **	8.55 **
PV	8.99 **	1396.57 **	69.67 **	92.61 **	0.38 ns	63.88 **	6.30 **	102.62 **	2.43 ns	97.07 **	34.60 **	46.77 **	44.78 **	76.85 **	44.31 **
HV	29.38 **	153.92 **	16.62 **	57.96 **	11.28 **	0.01 ns	31.43 **	25.64 **	7.29 **	4.04 *	3.12 ns	18.84 **	88.48 **	69.24 **	5.09 **
BD	1.03 ns	205.19 **	0.39 ns	8.10 **	33.87 **	1.01 ns	4.15 **	6.37 *	4.38 **	7.07 **	4.16 *	6.22 **	3.83 *	4.17 **	11.93 **
CV	5.56 *	291.82 **	24.30 **	66.79 **	0.06 ns	30.69 **	0.70 ns	40.89 **	7.14 **	25.54 **	3.12 ns	12.47 **	32.59 **	28.21 **	9.35 **
SB	1.47 ns	1126.67 **	7.46 **	50.35 **	1.26 ns	5.30 *	81.42 **	0.90 ns	7.20 **	5.59 **	1.84 ns	15.59 **	17.65 **	4.17 **	18.49 **
PT	1.97 ns	84.40 **	0.01 ns	0.11 ns	0.73 ns	0.40 ns	0.48 ns	0.82 ns	1.95 ns	2.16 ns	0.08 ns	0.24 ns	1.31 ns	0.87 ns	3.85 *
$F_{0.05}$	4.01	4.01	4.00	2.76	4.00	4.00	2.76	4.00	2.76	2.76	4.00	2.76	2.76	2.76	2.76
$F_{0.01}$	7.07	7.08	7.08	4.13	7.08	7.08	4.13	7.08	4.13	4.13	7.08	4.13	4.13	4.13	4.13

Y: year; P: previous crop; N: nitrogen rate; M: nitrogen management; BR: brown rice rate; MR: milled rice rate; HMR: head rice rate; CP: chalk grain rate; CD: chalkiness degree; AC: amylose content; PC: protein content; PV: peak viscosity; HV: hot viscosity; BD: break disintegration; CV: cool viscosity; SB: setback; PT: peak time. * and ** mean significance at the 0.05 and 0.01 probability levels, respectively.

Table 3. The effects of the N application rate in the previous season and N management in the rice season on rice quality characteristics.

Year			Treatment	BR (%)	MR (%)	HMR (%)	PC (%)	AC (%)
2018	Pr	Nc	M0	79.18 l	65.51 de	52.00 jkl	4.22 q	19.63 g
			M1	81.42 bc	67.90 abc	59.50 bcd	5.85 h	16.31 j
			M2	80.76 gh	68.48 ab	58.60 ef	5.63 ij	17.78 hi
			M3	81.14 cd	68.32 abc	58.44 ef	5.61 ij	18.57 h
			average	80.63	67.55	57.14	5.33	18.07
		Nr	M0	79.48 l	64.62 ef	51.02 mn	4.33 q	17.85 hi
			M1	81.18 cd	68.99 a	61.38 a	6.05 g	16.90 ij
			M2	80.84 fg	69.54 a	60.70 abc	5.87 h	17.48 i
			M3	81.50 bc	68.96 a	60.18 abc	5.75 hi	17.69 hi
			average	80.75	68.03	58.32	5.50	17.48
	Pw	Nc	M0	80.03 k	63.10 fgh	52.23 jkl	4.90 mn	22.78 cd
			M1	81.82 abc	63.02 fgh	44.94 o	6.53 bc	22.39 de
			M2	81.96 abc	63.15 fgh	54.883 h	6.31 def	22.55 de
			M3	81.52 bc	62.36 ghi	54.21 hi	6.29 ef	22.59 de
			average	81.33	62.91	51.57	6.01	22.58
		Nr	M0	80.71 gh	61.13 i	52.05 jkl	5.01 m	22.60 de
			M1	82.28 ab	62.68 ghi	51.08 lm	6.73 a	21.97 f
			M2	81.71 abc	61.78 hi	57.34 fg	6.55 b	22.47 de
			M3	81.48 bc	63.19 fgh	53.53 hij	6.43 bcd	22.51 de
			average	81.55	62.20	53.50	6.18	22.39
2019	Pr	Nc	M0	80.18 jk	66.62 cd	56.83 g	4.60 p	23.03 abc
			M1	81.61 abc	69.16 a	60.74 abc	5.64 ij	22.30 ef
			M2	81.35 bc	68.05 abc	61.00 ab	5.59 j	22.43 de
			M3	81.23 cd	67.16 bc	59.34 cd	5.58 j	22.83 bc
			average	81.09	67.75	59.48	5.35	22.65
		Nr	M0	80.29 ij	68.12 abc	54.52 h	4.68 op	23.04 abc
			M1	81.08 de	69.20 a	60.82 abc	5.83 h	21.99 f
			M2	80.57 hi	68.46 ab	58.00 efg	5.16 l	22.23 ef
			M3	80.98 ef	68.54 ab	59.22 de	4.78 no	22.89 bc
			average	80.73	68.58	58.14	5.11	22.54
	Pw	Nc	M0	80.71 gh	62.66 ghi	52.56 jkl	5.18 l	24.02 a
			M1	81.56 bc	62.46 ghi	51.79 kl	6.22 f	23.29 abc
			M2	82.17 ab	63.75 fg	52.96 ijk	6.17 fg	23.42 abc
			M3	81.95 abc	63.15 fgh	50.14 n	6.16 fg	23.82 abc
			average	81.60	63.01	51.86	5.93	23.64
		Nr	M0	81.36 bc	61.86 hi	51.62 kl	5.26 kl	24.03 a
			M1	82.60 a	63.58 fgh	54.32 hi	6.41 cde	22.98 abc
			M2	82.08 ab	62.42 ghi	52.44 jkl	5.74 hi	23.22 abc
			M3	81.13 cd	62.06 ghi	52.64 jkl	5.36 k	23.88 ab
			average	81.79	62.48	52.76	5.69	23.53

Pr represents rapeseed; Nc and Nr represent the conventional nitrogen application and reduced nitrogen application in the rape season, respectively; Pw represents wheat; Nc and Nr represent the conventional nitrogen application and reduced nitrogen application in the wheat season, respectively. M0 represents zero N was used in rice season; M1, M2, and M3 represent based on an application rate of 150 kg/hm2 N in the rice season, three N management models were applied, in which the application ratio of base:tiller:panicle fertilizer was 20%:20%:60%, 30%:30%:40%, and 40%:40%:20%, respectively. Lower case letters indicate that the rice quality characteristic are significantly different among the treatments ($p < 0.05$, LSD method). BR: brown rice rate; MR: milled rice rate; HMR: head rice rate; AC: amylose content; PC: protein content.

3.2.2. Appearance Quality

The rice chalkiness rate, chalkiness, aspect ratio, appearance, hardness, and eating quality (such as taste and eating value) were the highest in 2019 (Table 4). The hardness, taste, and taste values of the wheat season were the highest, followed by the rape season; the aspect ratio was the highest in the rape season, followed by the wheat season. The previous crop had a significant impact on the appearance and taste quality of rice. Appearance (chalkiness rate, aspect ratio, appearance, hardness) and taste quality (taste and eating value) were highest under different nitrogen application rates in the rapeseed season and wheat season. The chalkiness is based on the rapeseed or wheat season. Conventional nitrogen fertilization was the highest in the rape season and wheat season. Different nitrogen application rates have a great impact on the appearance and taste quality of rice. Different nitrogen fertilizer management had a great impact on the chalkiness rate, chalkiness size, and chalkiness of rice appearance. Chalkiness and chalkiness rate were M0 > M3 > M2 > M1 under different nitrogen fertilizer operations; aspect ratio was M0 > M3 > M2 > M1 under different nitrogen fertilizer operations. Appearance under different nitrogen fertilizer operations was M0 > M2 > M3 > M1; under the oil–rice rotation, the rice season M3 treatment increased the rice length-to-width ratio, but at the same time, increased the rice chalkiness and reduced the appearance quality of the rice. Under the wheat–rice rotation, the M2 treatment in the rice season increased the aspect ratio of the rice, and at the same time, increased the chalkiness of the rice and also reduced the appearance quality of the rice. The rice taste and mouthfeel under different nitrogen fertilizer managements were M0 > M3 > M2 > M1. Among them, rice taste M3 increased 0.35% and 2.43% compared with M2 and M1, respectively. Nitrogen fertilizer management under the oil/wheat–rice rotation had a great impact on the eating quality. The taste of rice under the oil–rice rotation increased with the increase in the ratio of the base tiller fertilizer to the total nitrogen application, and the M3 operation was the best treatment. Under the wheat–rice rotation, the taste of rice increased first and then later changed with the increase in the ratio of the base tiller fertilizer to the total nitrogen application. The M2 operation is the best. Therefore, it is more appropriate to reduce the application ratio of ear fertilizer, which is then conducive to improving the eating quality of rice.

3.2.3. RVA Profile Characteristic Value of Rice

The characteristic of the starch RVA profile is an important indicator of the taste of rice. Generally speaking, varieties with better eating quality generally have a larger disintegration value and lower cut-off value. The peak viscosity, hot paste viscosity, cold glue viscosity, and disintegration value were the largest in 2019, while the peak time, reduction value, and gelatinization temperature were the highest in 2018 (Table 5). The gum viscosity and disintegration values were highest in the wheat season, followed by the rape season, while the peak time, reduction value, and gelatinization temperature were the highest in the rape season, followed by the wheat season. Different previous crops will affect the RVA of rice starch. The effect of nitrogen application rate on the RVA spectral characteristics of starch is clear. Peak viscosity, hot paste viscosity, cold glue viscosity, disintegration value, and peak time were higher under conventional nitrogen application, while the reduction value and gelatinization temperature were the largest under reduced nitrogen application. Different nitrogen fertilizer operations have a great impact on the characteristic value of the RVA profile of rice starch. The peak viscosity and hot slurry viscosity under different nitrogen fertilizer operations was M0 > M1 > M2 > M3; the disintegration value under different nitrogen fertilizer operations was M3 > M2 > M1 > M0; M3 increased relative to M2, M1, and M0 by 2.54%, 2.88%, and 6.98%, respectively. The cold glue viscosity and gelatinization temperature under different nitrogen fertilizer operations was expressed as M0 > M1 > M2 > M3; the reduction value was M1 > M3 > M2 > M0; M1 relative to M3, M2 and M0 increased by 27.13%, 52.62%, and 71.04%, respectively. The peak viscosity, hot pulp viscosity, disintegration value, and cold gel viscosity of rice under different nitrogen fertilizer management treatments under the oil–rice rotation (except for the control treatment without nitrogen fertilizer; M0) were the highest and were the same as the base tiller fertilizer. The proportion of nitrogen increased gradually, while the change in the reduction value was opposite; under the wheat–rice rotation, the proportion of base tiller fertilizer in the total nitrogen application increased first and then decreased. The difference between the treatments was significant, and the disintegration value was the largest under the M2 treatment. The reduction value was the smallest at M2. It shows that reasonable nitrogen fertilizer management under the rapeseed/wheat–rice rotation is beneficial in improving the eating quality of rice.

Table 4. The effects of the N application rate in the previous season and N management in the rice season on rice appearance quality and eating quality.

Year	Treatment			CP (%)	CD (%)	L/W	Appearance	Comprehensive	Hardness	Mouthfeel
2018	Pr	Nc	M0	57.06 c	20.26 b	2.50 cd	8.43 cd	85.33 cd	3.36 h	7.56 def
			M1	35.92 o	11.68 klm	2.50 cd	7.80 ijk	74.33 g	3.96 bc	6.30 m
			M2	35.57 o	11.26 lm	2.50 cd	7.63 jk	77.66 f	4.13 abc	6.73 jkl
			M3	38.36 lmn	12.26 jkl	2.60 a	7.23 m	80.00 e	3.86 cd	6.90 ijk
			average	41.73	13.87	2.53	7.77	79.33	3.83	6.8725
		Nr	M0	53.80 d	17.21 cdef	2.50 cd	8.50 cd	85.33 cd	3.50 gh	7.60 def
			M1	33.23 p	10.63 m	2.56 ab	7.60 jk	76.33 fg	4.33 ab	6.50 lm
			M2	35.52 o	11.07 lm	2.52 cd	7.67 jk	77.33 f	4.10 abc	6.66 kl
			M3	37.03 mno	11.23 lm	2.61 a	7.30 lm	77.66 f	3.66 fg	6.70 jkl
			average	39.90	12.54	2.55	7.77	79.16	3.90	6.865
	Pw	Nc	M0	62.63 a	23.03 a	2.60 a	8.83 ab	88.33 ab	4.13 abc	8.33 b
			M1	55.76 cd	20.61 b	2.43 e	8.43 cd	84.00 d	3.50 gh	7.70 def
			M2	50.50 ef	17.60 cde	2.51 cd	8.36 de	84.66 cd	3.63 fg	7.76 de
			M3	45.96 hi	16.66 def	2.46 de	8.40 de	84.66 cd	4.16 abc	7.63 def
			average	53.71	19.48	2.50	8.51	85.41	3.86	7.855
		Nr	M0	60.43 b	22.36 a	2.61 a	9.01 a	89.33 a	3.30 h	8.70 a
			M1	43.26 k	14.66 hi	2.51 cd	8.56 bc	84.33 d	3.60 gh	7.63 def
			M2	43.70 jk	15.86 fgh	2.50 cd	8.36 de	86.33 bcd	4.36 a	7.90 cd
			M3	49.76 fg	18.05 cd	2.50 cd	8.53 bc	86.00 bcd	4.23 ab	7.86 cd
			average	49.29	17.73	2.53	8.62	86.50	3.87	8.0225
2019	Pr	Nc	M0	55.70 cd	18.54 c	2.61 a	8.66 bc	86.00 bcd	3.53 gh	7.86 cd
			M1	38.83 lm	12.74 jk	2.51 cd	7.76 ijk	77.00 f	4.16 abc	6.63 klm
			M2	38.56 lm	13.46 ij	2.50 cd	8.03 hi	76.66 f	3.83 de	6.71 jkl
			M3	40.15 l	12.87 jk	2.56 ab	7.53 kl	81.00 e	4.20 ab	7.16 ghi
			average	43.31	14.40	2.55	8.00	80.17	3.93	7.09
		Nr	M0	55.85 cd	18.26 c	2.60 a	8.26 fg	84.00 d	3.56 gh	7.41 efg
			M1	35.51 o	11.76 klm	2.50 cd	8.26 fg	80.00 e	3.80 ef	7.03 hij
			M2	38.76 lm	12.35 jkl	2.50 cd	8.03 hi	80.33 e	4.20 ab	7.10 ghi
			M3	49.23 fg	16.33 efg	2.56 ab	7.86 ij	83.66 d	3.53 gh	7.36 fgh
			average	44.84	14.68	2.54	8.10	82.00	3.77	7.225
	Pw	Nc	M0	63.43 a	23.03 a	2.62 a	8.83 ab	88.33 ab	4.16 abc	8.33 b
			M1	36.50 no	12.97 jk	2.50 cd	8.43 cd	83.66 d	3.80 ef	7.60 def
			M2	45.37 ij	15.16 gh	2.53 bc	8.20 gh	86.33 bcd	3.50 gh	7.90 cd
			M3	44.06 ijk	14.66 hi	2.50 cd	8.61 bc	84.33 d	3.40 h	7.40 efg
			average	47.34	16.46	2.54	8.52	85.66	3.72	7.8075
		Nr	M0	63.03 a	22.36 a	2.61 a	9.01 a	89.33 a	3.30 h	8.71 a
			M1	47.74 gh	17.24 cdef	2.50 cd	8.73 ab	84.33 d	3.56 gh	7.56 def
			M2	51.85 e	18.21 c	2.50 cd	8.33 ef	87.00 abc	3.66 fg	8.13 bc
			M3	49.57 fg	18.19 c	2.50 cd	8.50 cd	85.33 cd	4.33 ab	7.66 def
			average	53.05	19.00	2.53	8.64	86.50	3.71	8.015

Pr represents rapeseed; Nc and Nr represent the conventional nitrogen application and reduced nitrogen application in rape season, respectively; Pw represents wheat; Nc and Nr represent the conventional nitrogen application and reduced nitrogen application in the wheat season, respectively. M0 represents zero N was used in rice season; M1, M2 and M3 represent based on an application rate of 150 kg/hm2 N in the rice season, three N management models were applied, in which the application ratio of base:tiller:panicle fertilizer was 20%:20%:60%, 30%:30%:40%, and 40%:40%:20%, respectively. Lower case letters indicate that the rice appearance quality and eating quality are significantly different among the treatments ($p < 0.05$, LSD method). CP: chalk grain rate; CD: chalkiness degree; L/W: Grain length/Width ratio.

Table 5. The effects of the N application rate in the previous season and N management in the rice season on the RVA profile characteristic value of rice.

Year	Treatment		PV (cp)	HV (cp)	BD (cp)	CV (cp)	SB (cp)	Pt (cp)	PT (cp)
2018	Pr	Nc	M0	305.55 ij	200.72 cde	104.77 jkl	317.61 ef	16.10 cdef	6.13 abc	78.45 a
			M1	281.41 o	161.65 op	96.14 m	301.39 lm	19.50 bc	6.11 abc	78.13 ab
			M2	289.77 lmn	186.30 ij	96.11 m	308.36 ij	22.63 ab	6.18 ab	78.35 ab
			M3	289.97 lmn	193.91 efg	106.94 hi	312.66 gh	16.47 cdef	6.07 abc	77.28 bc
			average	291.68	185.65	100.96	310.01	18.68	6.12	78.05
		Nr	M0	300.35 jk	207.02 abc	104.30 klm	311.22 gh	11.66 hij	6.09 abc	78.15 ab
			M1	284.27 no	165.58 nop	94.53 n	309.66 hi	18.41 cd	6.15 ab	77.61 abc
			M2	283.95 no	180.13 kl	108.97 gh	296.61 m	22.50 ab	5.95 def	77.41 abc
			M3	285.67 no	192.85 efg	114.97 fg	303.43 kl	17.64 cde	6.02 abc	78.26 ab
			average	288.56	186.40	105.66	305.23	17.55	6.05	77.86

Table 5. *Cont.*

Year	Treatment			PV (cp)	HV (cp)	BD (cp)	CV (cp)	SB (cp)	Pt (cp)	PT (cp)
2019	Pw	Nc	M0	342.45 b	200.31 cde	129.43 abc	335.67 b	−2.25 n	6.02 abc	76.62 ghi
			M1	296.60 kl	172.44 mn	139.01 a	331.01 bc	4.39 lm	6.01 bcd	76.88 fg
			M2	328.03 de	196.83 def	124.69 cd	315.11 fg	3.36 lm	5.95 cde	76.81 fg
			M3	307.52 hi	192.59 fgh	121.89 cd	304.58 jkl	12.47 ghi	5.97 cde	76.68 ghi
			average	318.65	190.54	128.70	321.59	4.49	5.99	76.75
		Nr	M0	337.66 bc	202.43 bcd	126.45 bc	337.33 ab	−2.107 n	5.94 def	76.58 ghi
			M1	316.38 fg	197.58 def	122.14 cd	322.58 def	6.57 kl	6.03 bcd	76.49 ghi
			M2	321.30 ef	197.97 def	126.72 bc	321.53 def	6.32 kl	6.13 abc	76.95 ef
			M3	313.94 gh	189.64 ghi	117.47 def	319.57 def	10.25 ij	6.03 bcd	77.13 de
			average	322.32	196.91	123.15	325.25	5.26	6.03	76.79
	Pr	Nc	M0	298.65 jk	195.38 defg	108.08 ghi	313.14 gh	15.30 defg	6.09 abc	78.31 ab
			M1	271.52 p	162.86 op	106.52 ij	314.50 fg	24.27 a	6.08 abc	78.13 ab
			M2	285.47 no	172.19 mn	108.91 gh	294.28 n	25.08 a	5.95 def	77.63 abc
			M3	296.69 kl	188.61 hi	115.80 efg	296.69 m	15.39 defg	6.02 abc	78.06 abc
			average	288.08	179.76	109.80	304.65	20.01	6.04	78.03
		Nr	M0	295.60 klm	186.47 ij	115.05 fg	312.72 gh	13.01 fghi	6.06 abc	78.13 ab
			M1	284.15 no	167.52 mno	102.27 lm	306.77 ijk	26.08 a	6.09 abc	77.86 abc
			M2	284.83 no	180.68 kl	114.55 fg	304.25 jkl	16.72 cdef	6.02 abc	78.21 ab
			M3	288.80 mno	174.64 lm	116.94 def	303.28 kl	16.69 cdef	5.95 def	77.83 abc
			average	288.35	177.33	112.15	306.76	18.13	6.03	78.01
	Pw	Nc	M0	324.85 e	208.75 ab	131.17 abc	326.78 cd	3.39 lm	5.94 def	76.25 i
			M1	313.62 gh	159.25 p	108.27 ghi	324.22 cde	13.94 efgh	6.15 ab	77.18 cd
			M2	314.62 fg	194.98 defg	124.08 cd	284.33 o	1.56 m	5.79 f	77.45 abc
			M3	265.56 p	186.52 ij	116.19 def	314.89 fg	5.19 klm	6.01 bcd	76.93 ef
			average	304.66	187.38	119.85	312.56	6.02	5.97	76.95
		Nr	M0	359.51 a	210.89 a	104.96 jk	344.87 a	−9.117 o	6.22 a	77.31 abc
			M1	304.61 ij	183.34 jk	118.66 de	306.47 ijk	8.52 jk	5.99 bcd	77.45 abc
			M2	332.61 cd	196.66 def	135.05 ab	343.87 a	4.25 lm	5.91 ef	76.32 hi
			M3	325.36 e	195.40 defg	126.19 bcd	322.94 def	11.55 hij	6.09 abc	77.39 abc
			average	330.52	196.57	121.15	329.54	3.80	6.05	77.12

Pr represents rapeseed; Nc and Nr represent the conventional nitrogen application and reduced nitrogen application in rape season, respectively; Pw represents wheat; Nc and Nr represent the conventional nitrogen application and reduced nitrogen application in the wheat season, respectively. M0 represents zero N was used in rice season; M1, M2 and M3 represent based on an application rate of 150 kg/hm^2 N in the rice season, three N management models were applied, in which the application ratio of base:tiller:panicle fertilizer was 20%:20%:60%, 30%:30%:40%, and 40%:40%:20%, respectively. Lower case letters indicate that the RVA profile characteristic value of rice are significantly different among the treatments ($p < 0.05$, LSD method). PV: peak viscosity; HV: hot viscosity; BD: break disintegration; CV: cool viscosity; SB: setback; PT: peak time.

4. Discussion

4.1. Effects of Nitrogen Fertilizer Management on Yield in Different Rotation Modes

How to increase crop yields and reduce nitrogen fertilizer input to increase the efficient absorption and utilization of nitrogen fertilizer by crops is one of the current hot spots in the domain of agricultural research. Existing studies have shown that straw return to the field, nitrogen fertilizer management, and straw return to the field combined with nitrogen fertilizer have important regulatory effects on rice efficiency and yield, carbon and nitrogen metabolism, and high-efficiency utilization of nitrogen [16]. A large number of studies have shown that appropriate nitrogen fertilizer management, nitrogen application rate [17], and straw return to the field and nitrogen fertilizer management [18] can promote a significant increase in the cumulative amount of nitrogen uptake by rice at the mature stage, which can greatly reduce the amount of nitrogen fertilizer applied. Studies have shown that reduced fertilization has little effect on yield in crop rotation systems such as wheat–rice, rapeseed–rice, and corn-cole [19,20]. Zhang Weile et al. showed that under the condition of returning straw to the field, the nitrogen demand of crops could be met by the post-fertilization of nitrogen fertilizer, and high and stable crop yields can be ensured [21]. Yanfengjun et al. [22] reported that when the ratio of base tiller fertilizer to ear fertilizer was 6:4, high yields were ensured. This study believes that whether rice yield increases significantly under the rapeseed/wheat–rice rotation is closely related to the ratio of base tiller fertilizer to panicle fertilizer nitrogen fertilizer. The results of this study show that in the rapeseed–rice planting system, the yield of rice under different treatments depends on the reduction of nitrogen in the rape season, which is the largest when combined with the nitrogen fertilizer M3 in the rice season. The advantages of the huge root system of rapeseed with a large accumulation of nutrients, large biomass of straw returned to the field, and high nutrient release efficiency may be the main reasons for its significant role in promoting rice production in the rice season. In the wheat–rice rotation system, rice yield under different treatments is represented by conventional nitrogen application in the wheat season, which is highest when combined with the nitrogen fertilizer M2 operation in the rice season. It is possible that the reasonable management of nitrogen fertilizer in the wheat–rice rotation system reduces the nitrogen accumulation in the early stage of rice growth, inhibits the occurrence of ineffective tillers, and then meets the nitrogen demand during the grain development process through top dressing, ensuring the production of rice. In different crop rotation systems, reasonable nitrogen fertilizer management can better coordinate straw rot and rice growth to compete for nitrogen, ensure early and stable rice tillers, achieve the expected number of panicles, and coordinate the contradiction between foot panicles and large panicles to achieve high yields. Nitrogen fertilizer management under other crop rotation modes and the background value of soil nutrients will affect nitrogen fertilizer management. The effect of returning straw into the field on the formation of rice yield remains to be further studied in this respect.

4.2. Effects of Nitrogen Fertilizer Management on Rice Processing, Appearance, RVA and Nutritional Quality under Different Rotation Models

The types of straw and the amount of nitrogen applied have substantial effects on rice processing, appearance quality, and nutritional quality. After returning the straw to the field, it can reduce the chalkiness rate, the size and the degree of chalkiness, increase the brown rice rate, the polished rice rate, and the whole rice rate, and improve the processing quality and appearance quality of rice [23,24]. Previous research reveals that an appropriate amount of N fertilizer can decrease the chalky kernel rate, while the overuse of N can increase the chalky kernel rate and undesirable grain appearance [25,26]. The degree of chalkiness was significantly negatively related to eating/cooking quality. This was primarily due to the fact that high chalkiness implies a low density of starch granules, and therefore, the grains are more prone to breakage during cooking [27]. The amylose content was decreased with the increasing nitrogen level. According to a previous study, there are A- and B-types of starch granules in the endosperm [28]. This study shows that under different rotation modes, returning crop stalks to the field had significant or effects on the rice milling rate, chalkiness, hardness, and protein content of hybrid indica rice. We believe that the effect of straw return on rice chalkiness may be mainly related to the nitrogen and carbon supply of grain filling and the dynamic changes in grain filling [29]. The mechanism of returning all straws to the field to improve the appearance of rice needs to be further explored. The protein of rice is an ideal plant protein, which is easily absorbed by the human body and is the main indicator of the nutritional quality of rice. Reasonable nitrogen fertilizer management can substantially improve the quality of rice [30]. Wopereis-Pura et al. [31] researched that more panicle fertilizer application can significantly improve the processing quality of rice. The results of this study also show that as the percentage of panicle fertilizer in the total nitrogen application increases, the brown rice rate, polished rice rate, and whole rice rate increase. Increasing the ratio of panicle fertilizer to the total nitrogen application can significantly improve the processing quality of rice. Marwanto et al. [32] believed that an increase in the proportion of nitrogen fertilizer does not increase the chalkiness rate of rice, but the chalkiness becomes larger. The results of the current study are consistent with this. It shows that increasing the ratio of ear fertilizer to the total nitrogen application can significantly improve the nutritional quality of machine-grown, high-quality edible rice under a rapeseed–rice rotation system. The effect of returning the whole amount of straw to the field on the eating quality of rice is still lacking. Starch RVA profile characteristics are important indicators for evaluating rice quality and are closely related to cooking and eating quality. After the straw is returned to the field, the maximum viscosity and disintegration value both increase, but the reduction value becomes smaller [33]. In this study, the reduced amount of nitrogen fertilization in the rape season under the rapeseed–rice rotation and the eating quality of the rice under the treatment of the M3 operation in the rice season are the best, and the rice taste quality was best following the conventional nitrogen application in the wheat season combined with the M2 operation in the rice season under wheat–rice rotation. This shows that reasonable nitrogen fertilizer management under the rapeseed/wheat–rice rotation is beneficial in improving the eating quality of rice.

5. Conclusions

Optimizing nitrogen fertilizer management can increase rice yield and rice quality under rapeseed/wheat–rice rotation systems. Reduced N for rapeseed and the panicle fertilizer of 40%:40%:20% in rice season under a rapeseed–rice rotation system can be recommended to stabilize yield and high-quality rice production and can be used as an N-saving and environmentally friendly measure in rapeseed–rice rotation systems in southern China.

Author Contributions: P.M. and Y.L. are co-first authors. P.M.: Investigation, Methodology, Writing—original draft. Y.L.: Resources, Software, Writing—original draft. X.L.: Data curation. P.F.: Investigation, Methodology. Z.Y.: Methodology. Y.S.: Software. R.Z. and J.M.: Conceptualization, Funding acquisition, Supervision, Validation. All authors have read and agreed to the published version of the manuscript.

Funding: This study was supported by the National Key Research and Development Program of China (Nos. 2017YFD0301701; 2017YFD0301706) and the Scientific Research Fund of Sichuan Provincial Education Department (18ZA0390).

Institutional Review Board Statement: Not applicable.

Informed Consent Statement: Informed consent was obtained from all subjects involved in the study.

Data Availability Statement: The data presented in this study are available on request from the authors.

Acknowledgments: We would like to thank our teacher for carefully reading and correcting our manuscript and providing technical assistance and financial support for the study, as well as our scientific research team for their contribution to this paper.

Conflicts of Interest: The authors declare that they have no known competing financial interests or personal relationships that could appear to influence the work reported in this paper.

References

1. Xuan, Y.; Yi, Y.; Liang, H.; Wei, S.Q.; Chen, N.P.; Jiang, L.; Ali, I.; Ullah, S.; Wu, X.C.; Cao, T.Y.; et al. Amylose content and RVA profile characteristics of noodle rice under different conditions. *Agron. J.* **2020**, *112*, 117–129. [CrossRef]
2. Ding, C.Q.; You, J.; Chen, L.; Wang, S.H.; Ding, Y. Nitrogen fertilizer increases spikelet number per panicle by enhancing cytokinin synthesis in rice. *Plant Cell Rep.* **2014**, *33*, 363–371. [CrossRef] [PubMed]
3. Jiang, Q.; Du, Y.L.; Tian, X.Y.; Wang, Q.S.; Xiong, R.H.; Xu, G.H.; Yan, C.; Ding, Y.F. Effect of panicle nitrogen on grain filling characteristics of high-yielding rice cultivars. *Eur. J. Agron.* **2016**, *74*, 185–192. [CrossRef]
4. Wang, Y.; Cui, J.; Wang, X.B.; Zhao, M.; Zhu, C.J.; Shi, L.L.; Zhang, X. Effect of fertilization method on soil available nutrients and taste of Japanese and Chinese rice. *Chin. J. Eco-Agric.* **2010**, *18*, 286–289, (In Chinese with English abstract). [CrossRef]
5. Peng, J.; Feng, Y.; Wang, X. Effects of nitrogen application rate on the photosynthetic pigment, leaf fluorescence characteristics, and yield of indica hybrid rice and their interrelations. *Sci. Rep.* **2021**, *11*, 7485. [CrossRef]
6. Wang, J.; Wang, D.; Zhang, G.; Wang, Y.; Wang, C.; Teng, Y.; Christie, P. Nitrogen and phosphorus leaching losses from intensively managed paddy felds with straw retention. *Agric. Water Manag.* **2014**, *141*, 66–73. [CrossRef]
7. Wang, Y.; Chang, S.X.; Fang, S.; Tian, Y. Contrasting decomposition rates and nutrient release patterns in mixed vs singular species litter in agroforestry systems. *J. Soil Sediment.* **2014**, *14*, 1071–1081. [CrossRef]
8. Yuan, L.; Zhang, X.; Zhang, J.; Yang, C.L.; Cao, X.C.; Wu, L.H. Effects of different cultivation methods and straw incorporation on grain yield and nutrition quality of rice. *Acta. Agron. Sin.* **2013**, *39*, 350–359, (In Chinese with English abstract). [CrossRef]
9. Zhang, S.; Shi, Z.L.; Yang, S.J.; Gu, K.J.; Dai, T.B.; Wang, F. Effects of nitrogen application rates and straw returning on nutrient balance and grain yield of late sowing wheat in rice-wheat rotation. *Chin. J. Appl. Ecol.* **2015**, *26*, 2714. Available online: http://en.cnki.com.cn/Article_en/CJFDTOTAL-YYSB201509017.htm (accessed on 15 April 2021).

10. Yan, F.J.; Sun, Y.J.; Xu, H.; Ma, J. Effects of wheat straw mulch application and nitrogen management on rice root growth, dry matter accumulation and rice quality in soils of different fertility. *Paddy Water Environ.* **2018**, *16*, 507–518. [CrossRef]

11. Zhang, L.; Zhou, L.H.; Wei, J.B. Integrating cover crops with chicken grazing to improve soil nitrogen in rice fields and increase economic output. *Sci. Total Environ.* **2020**, *713*, 135–218. [CrossRef] [PubMed]

12. Peng, Z.Y.; Xiang, K.H.; Yang, Z.Y.; Sun, Y.J.; Ma, J. Effects of straw returning to paddy field and nitrogen fertilizer management on nitrogen utilization characteristics of direct seeded hybrid rice under wheat/rape-rice rotation. *Chin. J. Rice Sci.* **2020**, *34*, 57–68. [CrossRef]

13. Xu, G.W.; Tan, G.L.; Wang, Z.Q.; Liu, L.J.; Yang, J.C. Effect of wheat-residue application and site-specific nitrogen management on grain yield and quality and nitrogen use efficiency in diect-seeding rice. *Sci. Agric. Sin.* **2009**, *42*, 2736–2746. Available online: http://en.cnki.com.cn/Article_en/CJFDTOTAL-ZNYK200908016.htm (accessed on 15 April 2021). (In Chinese with English abstract).

14. Jia, D.; Jing, L.U.; Sun, Y. Effect of Different Nitrogen Fertilizer Application Strategies on Rice Growth and Yield. *Asian Agric. Res.* **2016**, *8*, 37–43. [CrossRef]

15. Liang, J.Y.; Wang, C.Q.; Bing, L.; Long, S.F.; Chen, L. Effects of combined application of pig manure with urea on grain yield and nitrogen utilization efficiency in rice-wheat rotation system. *J. Appl. Ecol.* **2019**, *30*, 1088–1096. [CrossRef]

16. Chen, S.; Liu, S.; Zheng, X.; Yin, M.; Chu, G.; Xu, C.; Yan, J.; Chen, L.; Zhang, X. Effect of various crop rotations on rice yield and nitrogen use efficiency in paddy-upland systems in southeastern China. *Crop J.* **2018**, *6*, 576–588. Available online: http://www.cqvip.com/QK/72059X/20186/7000960603.html (accessed on 15 April 2021). [CrossRef]

17. Russell, C.A.; Dunn, B.W.; Batten, G.D. Soil tests to predict optimum fertilizer nitrogen rate for rice. *Field Crops. Res.* **2006**, *97*, 286–301. [CrossRef]

18. Xu, G.W.; Yang, L.N.; Wang, Z.Q.; Liu, L.J.; Yang, J.C. Effects of Wheat-Residue Application and Site-Specific Nitrogen Management on Absorption and Utilization of Nitrogen, Phosphorus, and Potassium in Rice Plants. *Acta. Agron. Sin.* **2008**, *34*, 1424–1434.

19. Zhang, J.H.; Liu, J.L.; Zhang, J.B. Effects of Nitrogen Application Rates on Translocation of Dry Matter and Nitrogen Utilization in Rice and Wheat. *Acta Agronomica Sinica* **2010**, *36*, 1736–1742. [CrossRef]

20. Sheng, F.M.; Liu, X.J.; Feng, J.R.; Suo, Z.F.; Hua, L.U.S.; Zhong, Z.X. Effects of non-flooded mulching cultivation and n rates on productivity and N utilization in rice-wheat cropping systems. *Acta. Agron. Sin.* **2004**, *24*, 2591–2596. [CrossRef]

21. Zhang, W.L.; Dai, Z.G.; Ren, T.; Zhou, X.Z.; Wang, Z.L. Effects of Nitrogen Fertilization Managements with Residues Incorporation on Crops Yield and Nutrients Uptake Under Different Paddy-Upland Rotation Systems. *Sci. Agric. Sin.* **2016**, *49*, 1254–1266. Available online: http://en.cnki.com.cn/Article_en/CJFDTOTAL-ZNYK201607004.htm (accessed on 15 April 2021).

22. Yan, F.J.; Sun, Y.J.; Ma, J. Effects of straw mulch and nitrogen management on root growth and nitrogen utilization characteristics of hybrid rice. *J. Plant Nutr. Fertil.* **2015**, *21*, 23–35. Available online: http://en.cnki.com.cn/Article_en/CJFDTOTAL-ZWYF20150 1004 (accessed on 15 April 2021). (In Chinese with English abstract).

23. Liu, S.P.; Nie, X.T.; Dai, Q.G.; Huo, Z.Y.; Xu, K. Effects of interplanting with zero tillage and wheat straw manuring on rice growth an grain quality. *Chin. J. Rice Sci.* **2007**, *21*, 71–76, (In Chinese with English abstract). [CrossRef]

24. Ge, L.L.; Wang, K.J.; Fan, M.M.; Ma, Y.H.; Zhang, L.; Liu, L.J. Effect of straw returning on soil fertility, grain yield and quality of rice. *Chin. Agric. Sci. Bull.* **2012**, *28*, 1–6, (In Chinese with English abstract). [CrossRef]

25. Zhou, L.J.; Liang, S.S.; Ponce, K.; Marundon, S.; Ye, G.Y.; Zhao, X.Q. Factors affecting head rice yield and chalkiness in indica rice. *Field Crops Res.* **2015**, *172*, 1–10. [CrossRef]

26. Xi, M.; Lin, Z.M.; Zhao, Y.L.; Zhang, X.C.; Yang, X.Y.; Liu, Z.H.; Li, G.H.; Wang, S.H.; Ding, Y.F. Effects of nitrogen fertilizer application on the formation of white-belly and white-core as well as biochemical composition of japonica rice grains. *Chin. J. Rice Sci.* **2016**, *30*, 193–199. (In Chinese) [CrossRef]

27. Yang, L.X.; Wang, Y.L.; Dong, G.C.; Gu, H.; Huang, J.Y.; Zhu, J.G.; Yang, H.J.; Liu, G.; Han, Y. The impact of free-air CO_2 enrichment (FACE) and nitrogen supply on grain quality of rice. *Field Crops Res.* **2007**, *102*, 128–140. [CrossRef]

28. Kaufman, R.C.; Wilson, J.D.; Bean, S.R.; Presley, D.R.; Blanco, C.H.; Mikha, M. Effects of nitrogen fertilization and cover cropping systems on sorghum grain characteristics. *J. Agric. Food Chem.* **2013**, *61*, 5715–5719. [CrossRef]

29. Yan, F.J.; Effects of Straw Mulch and Water-Nitrogen Fertilizer Management on Yield of rice, Grain Quality and the Properties of Soil[D]. Ya'an: Sichuan Agricultural University. 2015. Available online: http://cdmd.cnki.com.cn/article/cdmd-10626-10160505 67.htm (accessed on 15 April 2021).

30. Pan, S.G.; Qu, J.; Cao, C.G.; Cai, M.L.; Wang, R.H.; Huang, S.Q.; Li, J.S. Effects of nitrogen management practices on nutrition uptake and grain qualities of rice. *Plant Nutr. Fert. Sci.* **2010**, *16*, 522–527, (In Chinese with English abstract). [CrossRef]

31. Wopereis-Pura, M.M.; Watanabe, H.; Moreira, J.; Wopereis, M.C.S. Effect of late nitrogen application on rice yield, grain quality and profitability in the Senegal River valley. *Eur. J. Agron.* **2002**, *17*, 191–198. [CrossRef]

32. Marwanto, M.; Nasiroh, N.; Mucitro, B.G. Effects of Combined Application of Cow Manure and Inorganic Nitrogen Fertilizer on Growth, Yield and Nitrogen Uptake of Black Rice. *Akta. Agrosia.* **2018**, *21*, 27–32. [CrossRef]

33. Liu, Q.H.; Zhou, X.B.; Yang, L.Q. Effects of Chalkiness on Cooking, Eating and Nutritional Qualities of Rice in Two indica Varieties. *Rice Sci.* **2009**, *16*, 161–164. [CrossRef]

Article

Bioactive Compounds of Tomato Fruit in Response to Salinity, Heat and Their Combination

María Ángeles Botella [1], Virginia Hernández [2], Teresa Mestre [3], Pilar Hellín [2,4], Manuel Francisco García-Legaz [5], Rosa María Rivero [3,4], Vicente Martínez [3,4], José Fenoll [2,4] and Pilar Flores [2,4,*]

[1] Departamento de Biología Aplicada, EPSO, Universidad Miguel Hernández, Orihuela, 03312 Alicante, Spain; mangeles.botella@umh.es
[2] Instituto Murciano de Investigación y Desarrollo Agrario (IMIDA), Santo Ángel, 30151 Murcia, Spain; virginia.hernandez5@carm.es (V.H.); mariap.hellin@carm.es (P.H.); jose.fenoll@carm.es (J.F.)
[3] Centro de Edafología y Biología Aplicada del Segura, CSIC, 30100 Murcia, Spain; tmestre@cebas.csic.es (T.M.); rmrivero@cebas.csic.es (R.M.R.); vicente@cebas.csic.es (V.M.)
[4] Unidad Asociada Grupo de Fertirriego y Calidad Hortofrutícola (IMIDA-CSIC), 30150 Murcia, Spain
[5] Departamento de Agroquímica y Medio Ambiente, EPSO, Universidad Miguel Hernández, Orihuela, 03312 Alicante, Spain; F.Garcia-Legaz@umh.es
* Correspondence: mpilar.flores@carm.es

Citation: Botella, M.Á.; Hernández, V.; Mestre, T.; Hellín, P.; García-Legaz, M.F.; Rivero, R.M.; Martínez, V.; Fenoll, J.; Flores, P. Bioactive Compounds of Tomato Fruit in Response to Salinity, Heat and Their Combination. *Agriculture* **2021**, *11*, 534. https://doi.org/10.3390/agriculture11060534

Academic Editor: Alessandra Durazzo and Isabel Lara Ayala

Received: 22 April 2021
Accepted: 7 June 2021
Published: 10 June 2021

Publisher's Note: MDPI stays neutral with regard to jurisdictional claims in published maps and institutional affiliations.

Abstract: In light of foreseen global climatic changes, we can expect crops to be subjected to several stresses that may occur at the same time, but information concerning the effect of long-term exposure to a combination of stresses on fruit yield and quality is scarce. This work looks at the effect of a long-term combination of salinity and high temperature stresses on tomato yield and fruit quality. Salinity decreased yield but had positive effects on fruit quality, increasing TSS, acidity, glucose, fructose and flavonols. High temperatures increased the vitamin C content but significantly decreased the concentration of some phenolic compounds (hydroxycinnamic acids and flavanones) and some carotenoids (phytoene, phytofluene and violaxanthin). An idiosyncrasy was observed in the effect of a combination of stresses on the content of homovanillic acid *O*-hexoside, lycopene and lutein, being different than the effect of salinity or high temperature when applied separately. The effect of a combination of stresses may differ from the effects of a single stress, underlining the importance of studying how stress interactions may affect the yield and quality of crops. The results show the viability of exploiting abiotic stresses and their combination to obtain tomatoes with increased levels of health-promoting compounds.

Keywords: sugars; carotenoids; phenolic; antioxidants; nutritional quality; high temperature; NaCl

1. Introduction

Tomato (*Solanum lycopersicum* L.) is an important horticultural crop worldwide and one of the most consumed vegetables in the world. Several abiotic stresses, such as water deficit, salinity and extreme temperatures, can affect crop production. The effects of one of these single stresses on plant production and physiological, biochemical and molecular changes have been widely studied in the literature. In particular, tomato plants are often cultivated in arid or semi-arid regions of the world, where salinity and high temperature threaten to become, or already are, a problem. The effect of irrigation with saline waters on tomato fruit has been well documented, indicating a decrease in yield and changes in fruit quality [1], usually leading to better tasting fruits [2,3]. In relation to high temperatures, several studies have shown a decrease in tomato fruit yield [4,5], and some authors have indicated that secondary metabolites were more affected than primary metabolites [6]. Moreover, different responses to heat conditions amongst tomato genotypes have been associated with the different effects of heat on some photosynthetic parameters [7].

Agricultural land in arid or semi-arid regions can be affected not only by a single stress, but by several stress conditions simultaneously. Moreover, considering the predicted global climatic changes, we can expect this situation to be exacerbated, with serious consequences [8]. Recently, several studies have focused on the effect of combinations of various stresses on plant physiological responses [9,10]. Some results have indicated that when plants are subjected to a combination of abiotic stresses, the response may be different from that under each stress applied separately [10].

Similarly, results have indicated that the combination of various stresses had a greater impact on plant growth and productivity than a single stress [8,11]. Nevertheless, some reports have shown that a combination of stresses (e.g., salinity and heat) may lead to better plant behavior than when each stress was applied individually [9,12,13]. However, it is important to note that most of these studies reported on the short-term physiological effects of stress combinations, while information about the effect of long-term exposure to a combination of stresses on fruit yield and quality is scarce.

The aspects of productivity and sensory quality have attracted most attention, but recently, there has been increasing interest in the nutritional value of fruits and vegetables [14], as consumers demand products with a high content of health-promoting constituents. In this respect, tomato is an important source of carotenoids such as β-carotene, a precursor of vitamin A; lycopene, which has been associate to a reduction in the risk of cancer, cardiovascular disease and macular degeneration [15]; lutein, which plays a fundamental role in the protection of vision [16] and in preventing age-related maculopathy [17]; others that have been less well studied, such as phytoene and phytofluene, which may contribute to inhibiting the progression of atherosclerosis [18]. Furthermore, tomato is also a source of phenolic compounds such as flavonoids and hydroxycinnamic acid derivatives and vitamins such as ascorbic acid. All of these compounds contribute to its antioxidant properties and beneficial health effects [19].

The above-mentioned compounds are important for the commercial quality of tomato and can be affected by factors such as variety and environmental, agricultural and post-harvest conditions [20]. Moreover, using controlled abiotic stress may be an interesting approach to improve the nutraceutical value of fruits and vegetables [21]. Taking all of this into consideration, the aim of this work was to study the effect of a combination of different stresses (salinity and high temperature) on tomato yield and fruit quality. Unlike most previous studies found in the literature, this study involves the long-term exposure to the combination of stresses, according to the current growing conditions, and allows for elucidating the effect of these stresses on the final bioactive composition of the fruit.

2. Materials and Methods

The study was carried out from April (mid-spring) to July (mid-summer) in two polycarbonate greenhouses. Tomato seedlings (*Solanum lycopersicum* L.) were transplanted to 120 L containers (1 plant per container) with aerated Hoagland nutrient solution (pH = 5.5–6.1) prepared with osmosis-generated water and 1 mM NaCl in order to reach an optimum conductivity value (2.2 dS m^{-1}) for tomato plant and fruit development [22]. The cultivar used was Boludo, provided by Monsanto, which is an indeterminate hybrid variety for fresh consumption with a high fruit-setting capacity at high temperature and rounded fruits of medium size and homogeneous color at maturity. Thirteen days after transplanting (DAT), the temperature treatments were started, maintaining one of the greenhouses (greenhouse 1) at a maximum of 25 °C during the day, while in the other (greenhouse 2), the maximum temperature was gradually increased over three days to reach a final maximum temperature of 35 °C during the day. These temperatures were reached naturally (without any heat source). To avoid exceeding these temperatures, the greenhouses were fitted with a control system that included shade nets, zenithal windows and a cooling system (Munters, Madrid, Spain). The shade nets were activated simultaneously in both greenhouses for twelve hours per day starting at 8 a.m. In addition, zenithal windows were opened from 6 a.m. until the temperature exceeded a temperature value of 20 °C.

Thereafter, the maximum temperature set in each greenhouse was maintained using the cooling system. Night temperatures ranged between 20 and 13 °C throughout the growing period in a similar way in both greenhouses. The saline treatment (60 mM NaCl) was started (16 DAT) in half of the containers in each greenhouse, through the application of 20 mM NaCl for three consecutive days in order to avoid osmotic shock. The salinity level (60 mM NaCl, 7.8 dS m^{-1}) was selected on the basis of previous results, which showed that this level increased tomato fruit quality and reduced yield without drastically affecting plant development [22,23]. These combinations provided a total of four treatments: control (C, 25 °C + 1 mM NaCl), saline (S, 25 °C + 60 mM NaCl), heat (H, 35 °C + 1 mM NaCl) and heat + salinity (S + H, 35 °C + 60 mM NaCl), distributed in a completely randomized design with 6 replications (plants) per treatment (Figure 1). The nutrient solutions were analyzed every week, and nutrients were added when they were 10% below the starting level. The pH was adjusted every two days and water was added twice a week to replace that lost by evapotranspiration.

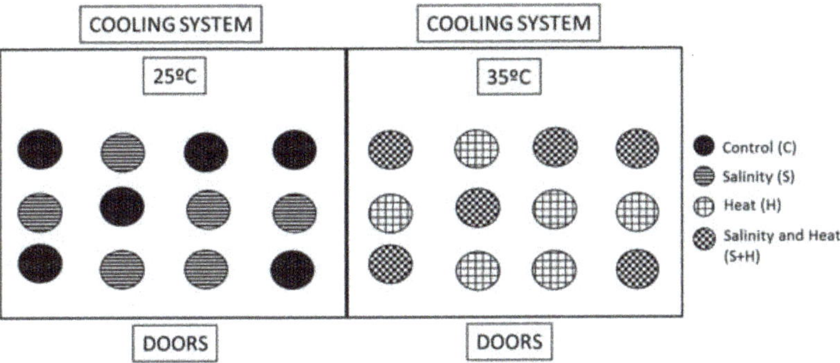

Figure 1. Experimental design layout of the greenhouse container experiment using a completely randomized design (CRD) with four treatments, control (C), saline (S) and heat conditions (H) and the combination of salinity and heat (S + H), and six replicates per treatment.

Plants were allowed to grow until they produced the ninth cluster, at which point the experiment terminated. Each tomato fruit was individually weighted to determine total and commercial production and mean fruit weight. Fruits under 70 g and/or affected by BER or cracking were classified as non-commercial. In order to analyze tomato quality, fruits at the full-red stage of ripening from trusses two and three were sampled during the period between 157 and 164 DAT. Fruit firmness of tomatoes with intact skin was determined with a texturometer (TA XT plus Texture Analyzer, Stable Micro System, Godalming, UK). Color was determined using a Minolta colorimeter CR200 model (Minolta Company, Limited, Ramsey, NJ, USA), taking three measurements for each fruit along the equatorial axis. Tomatoes taken from the same plant were cut into small pieces and mixed, constituting a sample (six samples per treatment). Later, the fruits were homogenized, and half of the homogenate was centrifuged to determine total soluble solids (TSS), pH and total acidity. The other half was kept at −80 °C for subsequent analysis of sugars, organic acids, vitamin C, phenolic compounds and carotenoids. Each sample was analyzed in triplicate.

Primary metabolites (soluble sugars and organic acids) and bioactive compounds (vitamin C, carotenoids and phenolic compound) were analyzed by high-performance liquid chromatography (HPLC) using a refraction index (IR) for sugars, a triple quadrupole mass spectrometer detector (MS/MS) for organic acids, vitamin C and phenolic compounds and a photodiode array UV-visible detector for carotenoids, following the methodologies described by Flores et al. [24], Fenoll et al. [25] and Flores et al. [26]. The IBM SPSS Statistic 21 software was used to statistically analyze the results with a one-way ANOVA and Duncan's test.

3. Results and Discussion

3.1. Yield Parameters

The total tomato yield obtained under control conditions was significantly reduced by the three different treatments ($p < 0.001$). The effect of salinity and heat individually was similar, and the combination of both stresses resulted in the highest yield reduction (Figure 2A). The reduction in commercial yield was even higher with all different treatments, which indicates a reduction in the percentage of commercial fruits (Figure 2B). Commercial yield was reduced from 91.8% under control conditions to 80.5, 73.5 and 65.4% with salinity, heat stress and the combination of both stresses, respectively. The above decrease in tomato yield with the different treatments was attributed to the significant reduction ($p < 0.001$) in fruit weight (Figure 2C) and not to a reduction in fruit number (Figure 2D). Several authors have described a reduction in tomato fruit size but no or little effect on fruit number under saline conditions [27–30]. In regard to the decrease in fruit weight under saline conditions, this effect has been attributed to a lower water uptake by the root, thus reducing water transport to the fruit [31–33]. Unlike salinity, heat stress may affect fruit set with negative consequences for the yield [34]. However, under our experimental conditions, heat stress alone or combined with salinity had no significant effect on fruit number.

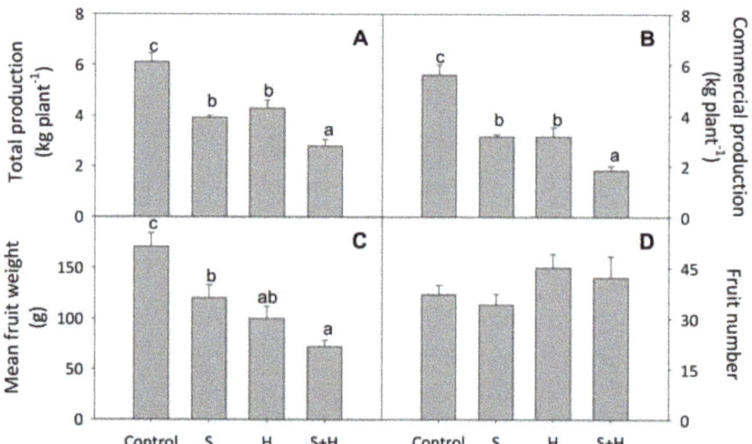

Figure 2. Total production (**A**), commercial production (**B**), fruit mean weight (**C**) and fruit number per plant (**D**) of tomato plants under control (C), saline (S) and heat conditions (H) or the combination of salinity and heat (S + H). Values are means ± SE ($n = 6$). Different letters indicate significant differences between means according to Duncan's test at the 5% level.

Different responses to combinations of stresses can be found: (1) additive, which is the addition of the single stress responses; (2) synergistic, which is the sum of each single stress; (3) idiosyncratic, when completely different from the individual stress responses; (4) dominant, if it is very close to one of the stresses [35]. Our results point to a higher negative effect of stress combinations than of each single stress on fruit yield (additive). Although Rivero et al. [9] reported that after 72 h, the heat treatment improved the salinity tolerance of tomato plants, long-term exposure to stress, such as in the present study, would be expected to have more pronounced effects on plant physiology and fruit yield. In agreement with our results, other authors studying drought, heat and their combination in tomato plants over a period of 6 days indicated that combined stress reinforced the negative effect of the individual stresses [36]. In addition, long-term studies have indicated that different stress interactions have a higher effect on yield than any of the stresses applied individually [8,11].

3.2. Fruit Organoleptic Properties

Total soluble solids (TSS) significantly increased by salinity, whether applied alone or, to a lesser extent, by the combination of salinity with high temperature, while temperature alone had no effect (Table 1). The combination of both stresses significantly decreased the pH in fruit, and the saline treatment applied as a single stress increased acidity. Other authors have reported similar results in relation to the effect of salt stress in tomato fruits, with both soluble solids and tritable acidity increasing [37,38]. Fruit firmness decreased with the combination of salinity and heat, but there were no differences between the other treatments. None of the treatments had any effect on L or hue values, while chroma increased only with the combination of stresses.

Table 1. Total soluble solids (TSS, °Brix), pH, acidity (g citric acid L^{-1}), firmness (N cm^2) and color parameters (L, hue and chroma) of tomato fruits under control (C), saline (S) and heat conditions (H) or the combination of salinity and heat (S + H). Values are means ($n = 6$).

Treatments	TSS	pH	Acidity	Firmness	L	Hue	Chroma
Control	4.7 a	4.3 bc	2.0 a	13.8 b	42.9	51.6	39.5 a
S	5.5 c	4.2 ab	2.7 b	11.6 ab	43.9	54.1	40.7 a,b
H	4.6 a	4.4 c	2.0 a	13.5 ab	43.3	52.6	39.7 a
S + H	5.2 b	4.1 a	2.3 a	10.8 a	44.0	51.2	42.0 b
	**	**	**	*	n.s.	n.s.	*

*,** Significant differences between means at the 5 or 1% level of probability, respectively; n.s., non-significant at $p = 5\%$. For each stage, different letters in the same column indicate significant differences between means according to Duncan's test at the 5% level.

The glucose and fructose contents significantly increased when salinity was applied as a single stress, but were not affected when heat was the only stress (Table 2). However, heat and salinity together had an additive effect, with the combination of both stresses resulting in the highest increase in both glucose and fructose. Many results can be found in the literature related to the increase in tomato fruit quality as a result of an increasing sugar content caused by salinity of the nutrient solution [3,27,39–41], which was attributed to the effect of saline stress on enzymes associated with sugar biosynthesis [42]. As for high temperature, no effect on the fruit's reducing sugar content has been described in tomato in spite of its impact on fruit mass production [43]. However, our findings indicated that under high temperature conditions, irrigation with saline water could increase the fruit sugar content and, therefore, lead to greater consumer preference because of the increase in sweetness and flavor.

Table 2. Concentration of soluble sugars and organic acids (mg g^{-1} fresh weight) in tomato fruits under control (C), saline (S) and heat conditions (H) or the combination of salinity and heat (S + H). Values are means ($n = 6$).

Treatments	Glucose	Fructose	Citric	Glutamic	Malic
Control	14.75 a	15.40 a	1.59	3.00 ab	0.42 ab
S	20.23 b	19.80 b	1.86	3.41 b	0.30 a
H	13.64 a	14.32 a	1.95	2.23 a	0.49 b
S + H	24.26 c	23.09 c	1.74	2.89 ab	0.41 ab
	***	***	n.s.	*	*

*,*** Significant differences between means at 5 or 0.1% level of probability, respectively; n.s., non-significant at $p = 5\%$. For each stage, different letters in the same column indicate significant differences between means according to Duncan's test at the 5% level.

Glutamic acid concentration was not affected by any treatment with regard to the control (Table 2), although significant differences were found between single stress applications ($p < 0.05$), being 1.5 times higher under saline than under heat stress. In the case of malic acid, the heat treatment led to a 1.6 times higher content than that obtained in saline conditions. Neither a single stress nor the combination of both significantly changed the citric acid concentration. Other authors have indicated that salinity increases organic acid as well as the sugar concentration [27,44], but this effect was closely dependent on the tomato variety [27]. The increased concentrations of both sugars and organic acids in tomato fruits by salinity and high temperature have been previously associated to a concentration effect as a result of a decreased sink/source ratio due to increased flower abortion [27,28]. However, the experimental conditions in the present study did not led to a decrease in the number of fruits as a consequence of any single stress or their combination. Therefore, the increased concentrations of sugars and organic acids could be attributed to an enhanced biosynthesis under these stress conditions.

3.3. Phenolic Compounds

The most abundant phenolic compound was homovanillic acid-*O*-hexoside, with an average concentration of 26.4 μg g^{-1}, followed by the flavonol derivative rutin (10.6 μg g^{-1}) and kaempferol-3-*O*-rutinoside (8.6 μg g^{-1}) and the flavanone naringenin (7.0 μg g^{-1}). Hydroxycinnamic acids were mainly represented by chlorogenic acid (5.9 μg g^{-1}). The dihydrochalcone phloretin-*C*-diglycoside was found at an average concentration of 3.8 μg g^{-1}. Other detected phenolic compounds were the flavonol derivatives rutin-*O*-hexoside (0.20 μg g^{-1}), rutin-*O*-pentoside (0.07 μg g^{-1}), quercetin (0.04 μg g^{-1}), the flavanone naringenin-*O*-hexoside (3.1 μg g^{-1}) and the hydroxycinnamic derivatives caffeic-acid-*O*-hexoside (2.4 μg g^{-1}), cryptochlorogenic acid (1.4 μg g^{-1}), ferulic acid-*O*-hexoside (1.3 μg g^{-1}) and *p*-coumaroyl quinic acid (0.19 μg g^{-1}), dicaffeoylquinic (0.15 μg g^{-1}), ferulic (0.14 μg g^{-1}), caffeic (0.12 μg g^{-1}) and *p*-coumaric acids (0.03 μg g^{-1}). Table S1 shows the values of each individual phenolic compound in the different treatments.

The content of hydroxycinnamic acids was significantly reduced by the high temperature, while the other treatments had no effect on these compounds (Figure 3A). Interestingly, salinity inhibited the detrimental effect of heat on this parameter. Flavanones were not affected by salinity but decreased significantly and in a similar manner with high temperature and the combination of salinity and heat (Figure 3B), which indicates that heat dominated the combined stress response. In contrast, flavonols were significantly increased by salinity and the combination of both stresses, and no significant differences were found between both conditions, indicating the dominant effect of salinity on this parameter. No effects of heat as a single stress were observed on the flavonol content (Figure 3C). Homovanillic acid-*O*-hexoside was significantly higher with both stresses applied together (Figure 4A), and phloretin was significantly reduced with the saline treatment (Figure 4B).

Phenolic compounds are important for the detoxification of free radicals [45] and environmental stress can increase the levels of these scavenger molecules [21]. Regarding salinity, contradictory reports of the effects on phenolic compounds in tomato fruits can be found in the literature, increasing [46,47], decreasing [48] or even remaining unchanged [49]. The same is true of flavonoids, with some authors finding an increase in the total flavonoid content of tomato fruits under saline conditions [48] and others reporting a reduction [50]. In the case of heat stress, some authors have pointed to an increase in specific phenolic compounds [6,51] under high temperature conditions. Martínez et al. [12] described a differential accumulation of phenolic compounds that was dependent on the type of abiotic stress, concluding that the accumulation of flavonols over hydroxycinnamic acids favored oxidative damage protection under abiotic stress. In agreement with these authors, our results indicated an increase in flavonols/hydroxycinnamic acids ratio under all stress conditions, with the highest values obtained when both stresses were combined.

The different results found in the literature could be due to the influence of several factors, such as stage of ripeness and tissue, growth conditions, genotype or the detection method [52,53]. Our results showed the specific effects of individual and combined stresses on each phenolic compound family, which may be underestimated when the total contents are analyzed with non-selective methods. Moreover, the effect of salinity on phenolic compounds may be influenced by other factors, as mentioned by Incerti et al. [54], who reported that their level decreased or increased, depending on the season (spring or autumn). These findings highlight the need to study the interaction between different factors that are expected to coexist when evaluating the impact of abiotic stress on fruit composition.

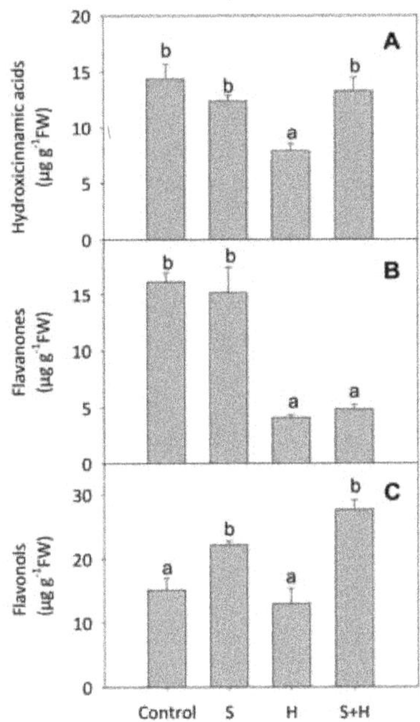

Figure 3. Concentration of hydroxycinnamic acids (**A**) flavanones (**B**) and flavonols (**C**) (μg g^{-1} fresh weight) in tomato fruits under control (C), saline (S) and heat conditions (H) or the combination of salinity and heat (S + H). Values are means $\pm$ SE (n = 6). Different letters indicate significant differences between means according to Duncan's test at the 5% level.

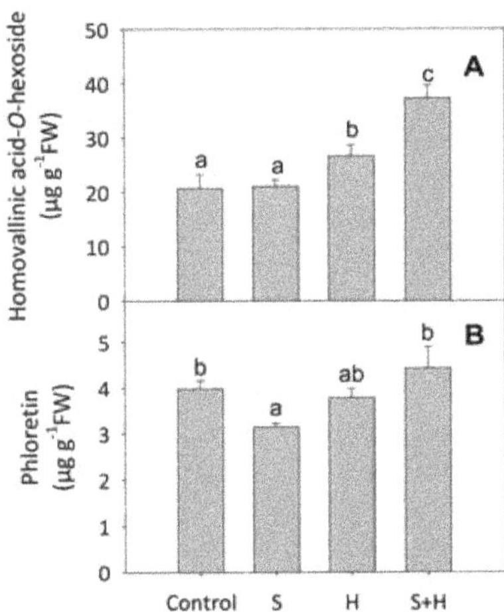

Figure 4. Concentration of homovanillic acid-*O*-hexoside (**A**) and phloretin (**B**) (μg g^{-1} fresh weight) in tomato fruits under control (C), saline (S) and heat conditions (H) or the combination of salinity and heat (S + H). Values are means $\pm$ SE (n = 6). Different letters indicate significant differences between means according to Duncan's test at the 5% level.

3.4. Vitamin C

Vitamin C concentrations increased ($p < 0.01$) similarly with high temperature and when both stresses were applied, indicating the dominant effect of heat stress and no effect of salinity (Figure 5). Gautier et al. [6] found that ascorbate levels decreased when temperature increased to 32 °C, and Rosales et al. [50] observed an increase in ascorbic acid under stress conditions due to high temperature. After an initial fall, an increase in vitamin C was found by Hernández et al. [55] after a long exposure to high temperatures, suggesting that plant metabolism adapted to a high temperature and/or when the temperature decreased during the night, a restoration of the ascorbate synthesis took place. Ehret et al. [30] suggested that vitamin C concentration increased as a response to abiotic stress through de novo synthesis or due to its regeneration from dihydrolipoic acid. In spite of the increase in vitamin C caused by salinity in tomato fruits reported by other authors [27,30,46,56], our results found that salinity had no effect when applied alone and no synergistic effect when applied at the same time as a high temperature.

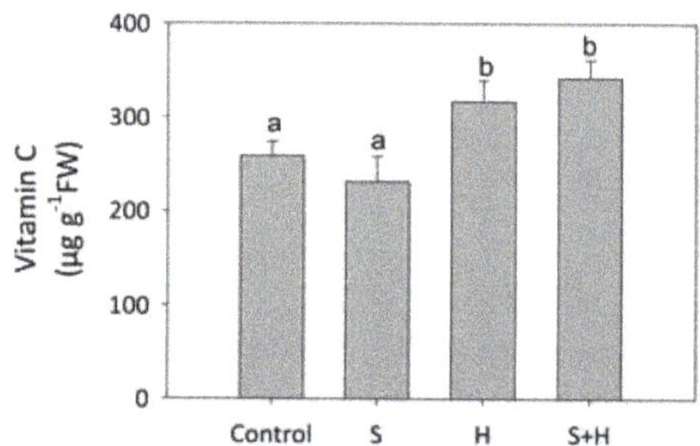

Figure 5. Concentration of vitamin C ($\mu g\ g^{-1}$ fresh weight) in tomato fruits under control (C), saline (S) and heat conditions (H) or the combination of salinity and heat (S + H). Values are means ± SE ($n = 6$). Different letters indicate significant differences between means according to Duncan's test at the 5% level.

3.5. Carotenoids

Salinity applied as a single stress did not significantly affect any of the precursors or carotenoids (Figures 6 and 7). High temperature did not increase carotenoids concentrations while the concentration of phytoene (Figure 6A), phytofluene (Figure 6B) and violaxanthin (Figure 7D) decreased with heat, whether applied as a single stress or combined with salinity. In spite of what occurred with the individual stresses, lycopene and lutein increased as a response to the combination of both stresses (Figure 7A,C). As for β-carotene, no significant differences were observed between any of the single stress treatments or their combination and the control treatment (Figure 7B).

Carotenoids can contribute to the fluidity and permeability of membranes in response to changes in temperature [57,58]. Although heat stress (32 °C) caused a decrease in lycopene levels, under certain conditions the fruits could recover or even increase the initial concentrations [55]. High temperature seems to have no effect on β-carotene and lutein [6,55]. However, some authors have reported a beneficial effect of salinity on the carotenoid content, [27,30,40,59], while Serio et al. [60] reported that salinity did not affect the lycopene content, in agreement with our results. Comparative studies have indicated that the response of carotenoids in tomato to salinity was genotype dependent [50,59].

Unlike the response to salinity or high temperature when applied separately, a specific and different response to the combination of both stresses was the increase in lycopene and lutein concentrations. Under such stress conditions, our results suggested a degradation of the precursors phytoene and phytofluene towards the accumulation of lycopene and lutein and the maintenance of β-carotene levels at the expense of a decreased accumulation of violaxanthin. These results of increased lycopene and lutein concentrations are especially important, considering the role of these metabolites in human health [61] and with lycopene being the principal carotenoid, which confers the red pigmentation to the fruit.

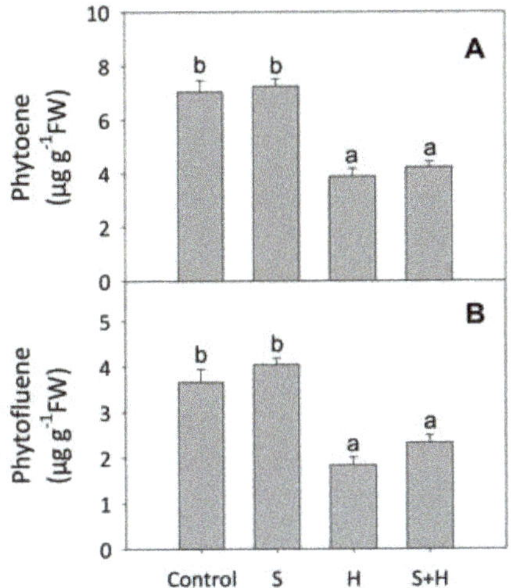

Figure 6. Concentration of phytoene (**A**) and phytofluene (**B**) (μg g^{-1} fresh weight) in tomato fruits under control (C), saline (S) and heat conditions (H) or the combination of salinity and heat (S + H). Values are means $\pm$ SE (n = 6). Different letters indicate significant differences between means according to Duncan's test at the 5% level.

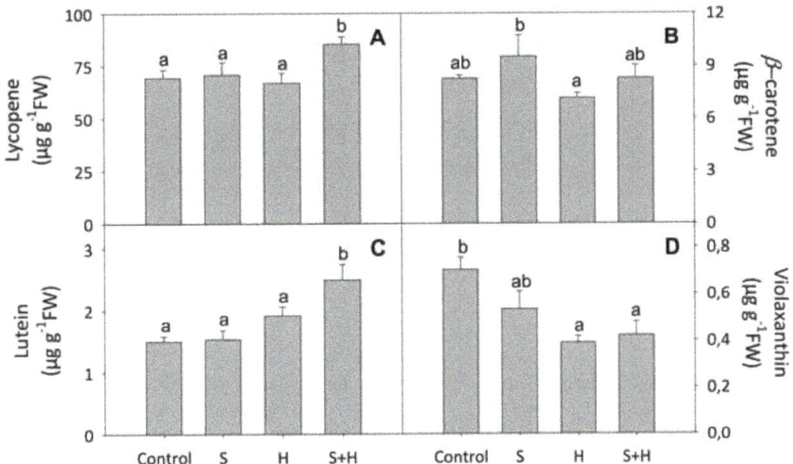

Figure 7. Concentration of lycopene (**A**), β-carotene (**B**), lutein (**C**) and violaxanthin (**D**) (μg g^{-1} fresh weight) in tomato fruits under control (C), saline (S) and heat conditions (H) or the combination of salinity and heat (S + H). Values are means $\pm$ SE (n = 6). Different letters indicate significant differences between means according to Duncan's test at the 5% level.

The effect of a combination of stresses may differ from those of single stresses, highlighting the importance of studying the effect of stress interactions on the yield and quality of crops. To summarize our findings, salinity applied as a single stress decreased the yield of tomato but had a positive effect on fruit quality by increasing sugars and flavonols. High temperatures increased the vitamin C content, but had a negative effect on yield and the content of various phenolic compounds (hydroxycinnamic acids and flavanones) and some carotenoids. Interestingly, an idiosyncrasy was found in the effect of the combination of stresses on the contents of homovallinic acid *O*-hexoside, lycopene and lutein. In addition, the combination of stresses inhibited the detrimental effect of high temperature on hydroxycinnamic acid content. The results from this preliminary study point to the viability of exploiting abiotic stresses and their combination to obtain tomatoes with increased levels of health-promoting compounds. However, it is to be expected that environmental, crop management and even varietal factors may affect the results obtained. Therefore, further studies are needed considering these factors and other abiotic stresses. Moreover, since abiotic stress combinations due to climate change are expected to severely restrict crop yield and fruit quality in the coming years, more studies that combine good crop management with new breeding tools and gene editing technologies will be needed in order to improve plant resilience and cope with the food, fiber and livestock feed demand.

Supplementary Materials: The following are available online at https://www.mdpi.com/article/10.3390/agriculture11060534/s1, Table S1: Concentration of individual phenolic compounds (μg g^{-1} fresh weight) in tomato fruits under control, salinity, heat or the combination of salinity and heat. Values are means $\pm$ SE (n = 6). Different letters indicate significant differences between means according to Duncan's test at the 5% level.

Author Contributions: Conceptualization, P.F., V.M., R.M.R. and P.H.; methodology, V.H., T.M., P.H. and P.F.; software V.H., T.M. and M.F.G.-L.; validation V.M., P.H., M.Á.B. and P.F.; formal analysis, V.H., T.M. and M.F.G.-L.; investigation, V.H., M.Á.B., V.M., P.H. and P.F.; resources, V.M., R.M.R., P.H. and P.F.; data curation, V.H., M.Á.B. and M.F.G.-L.; writing—original draft preparation, M.Á.B. and P.F.; writing—review and editing, M.Á.B., P.F. and V.H.; visualization, M.Á.B., P.H. and J.F.; project administration, V.M. and P.F.; funding acquisition, P.F., P.H., J.F. and V.M. All authors have read and agreed to the published version of the manuscript.

Funding: This research was funded by Fundación Séneca-Agencia de Ciencia y Tecnología de la Región de Murcia.

Institutional Review Board Statement: Not applicable.

Informed Consent Statement: Not applicable.

Data Availability Statement: Data are available upon reasonable request from the authors.

Acknowledgments: The authors are grateful to Inmaculada Garrido González, María V. Molina Menor, Elia Molina Menor and Juana Cava Artero for technical assistance.

Conflicts of Interest: The authors declare no conflict of interest. The funders had no role in the design of the study; in the collection, analyses or interpretation of data; in the writing of the manuscript or in the decision to publish the results.

References

1. Botella, M.Á.; Del Amor, F.; Amorós, A.; Serrano, M.; Martinez, V.; Cerdá, A. Polyamine, ethylene and other physico-chemical parameters in tomato (*Lycopersicon esculentum*) fruits as affected by salinity. *Physiol. Plant.* **2000**, *109*, 428–434. [CrossRef]
2. Petersen, K.K.; Willumsen, J.; Kaack, K. Composition and taste of tomatoes as affected by increased salinity and different salinity sources. *J. Hortic. Sci. Biotechnol.* **1998**, *73*, 205–215. [CrossRef]
3. Restuccia, G.; Marchese, M.; Mauromicale, G.; Restuccia, A.; Battaglia, M. Yield and Fruit Quality of Tomato Grown in Greenhouse with Saline Irrigation Water. *Acta Hortic.* **2003**, *614*, 699–704. [CrossRef]
4. Adams, S.R.; Cockshull, K.E.; Cave, C.R.J. Effect of Temperature on the Growth and Development of Tomato Fruits. *Ann. Bot.* **2001**, *88*, 869–877. [CrossRef]
5. Islam, M.T. Effect of Temperature on Photosynthesis, Yield Attributes and Yield of Tomato Genotypes. *J. Exp. Agric. Int.* **2011**, *2*, 8–11.
6. Gautier, H.; Diakou-Verdin, V.; Bénard, C.; Reich, M.; Buret, M.; Bourgaud, F.; Poëssel, J.L.; Caris-Veyrat, C.; Génard, M. How Does Tomato Quality (Sugar, Acid, and Nutritional Quality) Vary with Ripening Stage, Temperature, and Irradiance? *J. Agric. Food Chem.* **2008**, *56*, 1241–1250. [CrossRef]
7. Arena, C.; Conti, S.; Francesca, S.; Melchionna, G.; Hájek, J.; Barták, M.; Barone, A.; Rigano, M.M. Eco-Physiological Screening of Different Tomato Genotypes in Response to High Temperatures: A Combined Field-to-Laboratory Approach. *Plants* **2020**, *9*, 508. [CrossRef]

8. Mittler, R.; Blumwald, E. Genetic Engineering for Modern Agriculture: Challenges and Perspectives. *Annu. Rev. Plant Biol.* **2010**, *61*, 443–462. [CrossRef]

9. Rivero, R.M.; Mestre, T.C.; Mittler, R.; Rubio, F.; Garcia-Sanchez, F.; Martinez, V. The combined effect of salinity and heat reveals a specific physiological, biochemical and molecular response in tomato plants. *Plant Cell Environ.* **2013**, *37*, 1059–1073. [CrossRef] [PubMed]

10. Zandalinas, S.I.; Mittler, R.; Balfagón, D.; Arbona, V.; Gómez-Cadenas, A. Plant adaptations to the combination of drought and high temperatures. *Physiol. Plant.* **2018**, *162*, 2–12. [CrossRef] [PubMed]

11. Suzuki, N.; Rivero, R.M.; Shulaev, V.; Blumwald, E.; Mittler, R. Abiotic and biotic stress combinations. *New Phytol.* **2014**, *203*, 32–43. [CrossRef] [PubMed]

12. Martinez, V.; Mestre, T.C.; Rubio, F.; Girones-Vilaplana, A.; Moreno, D.A.; Mittler, R.; Rivero, R.M. Accumulation of Flavonols over Hydroxycinnamic Acids Favors Oxidative Damage Protection under Abiotic Stress. *Front. Plant Sci.* **2016**, *7*, 838. [CrossRef] [PubMed]

13. García-Martí, M.; Piñero, M.C.; García-Sanchez, F.; Mestre, T.C.; López-Delacalle, M.; Martínez, V.; Rivero, R.M. Amelioration of the Oxidative Stress Generated by Simple or Combined Abiotic Stress through the K^+ and Ca2+ Supplementation in Tomato Plants. *Antioxidants* **2019**, *8*, 81. [CrossRef] [PubMed]

14. Scarano, A.; Olivieri, F.; Gerardi, C.; Liso, M.; Chiesa, M.; Chieppa, M.; Frusciante, L.; Barone, A.; Santino, A.; Rigano, M.M. Selection of tomato landraces with high fruit yield and nutritional quality under elevated temperatures. *J. Sci. Food Agric.* **2020**, *100*, 2791–2799. [CrossRef] [PubMed]

15. Tan, H.-L.; Thomas-Ahner, J.M.; Grainger, E.M.; Wan, L.; Francis, D.M.; Schwartz, S.J.; Erdman, J.W.; Clinton, S.K. Tomato-based food products for prostate cancer prevention: What have we learned? *Cancer Metastasis Rev.* **2010**, *29*, 553–568. [CrossRef]

16. Sommerburg, O.; Keunen, J.E.; Bird, A.C.; Van Kuijk, F.J.G.M. Fruits and vegetables that are sources for lutein and zeaxanthin: The macular pigment in human eyes. *Br. J. Ophthalmol.* **1998**, *82*, 907–910. [CrossRef] [PubMed]

17. Giorio, G.; Yildirim, A.; Stigliani, A.L.; D'Ambrosio, C. Elevation of lutein content in tomato: A biochemical tug-of-war between lycopene cyclases. *Metab. Eng.* **2013**, *20*, 167–176. [CrossRef]

18. Riso, P.; Visioli, F.; Grande, S.; Guarnieri, S.; Gardana, C.; Simonetti, P.; Porrini, M. Effect of a Tomato-Based Drink on Markers of Inflammation, Immunomodulation, and Oxidative Stress. *J. Agric. Food Chem.* **2006**, *54*, 2563–2566. [CrossRef]

19. Slimestad, R.; Verheul, M. Review of flavonoids and other phenolics from fruits of different tomato (*Lycopersicon esculentum Mill.*) cultivars. *J. Sci. Food Agric.* **2009**, *89*, 1255–1270. [CrossRef]

20. Dumas, Y.; Dadomo, M.; Di Lucca, G.; Grolier, P. Effects of environmental factors and agricultural techniques on antioxidantcontent of tomatoes. *J. Sci. Food Agric.* **2003**, *83*, 369–382. [CrossRef]

21. Toscano, S.; Trivellini, A.; Cocetta, G.; Bulgari, R.; Francini, A.; Romano, D.; Ferrante, A. Effect of Preharvest Abiotic Stresses on the Accumulation of Bioactive Compounds in Horticultural Produce. *Front. Plant Sci.* **2019**, *10*, 1212. [CrossRef]

22. Dorai, M.; Papadopoulos, A.P. Influence of electric conductivity management on greenhouse tomato yield and fruit quality. *Agronomy* **2001**, *21*, 367–383. [CrossRef]

23. Navarro, J.M.; Flores, P.; Carvajal, M.; Martinez, V. Changes in quality and yield of tomato fruit with ammonium, bicarbonate and calcium fertilisation under saline conditions. *J. Hortic. Sci. Biotechnol.* **2005**, *80*, 351–357. [CrossRef]

24. Flores, P.; Hellín, P.; Fenoll, J. Determination of organic acids in fruits and vegetables by liquid chromatography with tandem-mass spectrometry. *Food Chem.* **2012**, *132*, 1049–1054. [CrossRef]

25. Fenoll, J.; Martínez, A.; Hellín, P.; Flores, P. Simultaneous determination of ascorbic and dehydroascorbic acids in vegetables and fruits by liquid chromatography with tandem-mass spectrometry. *Food Chem.* **2011**, *127*, 340–344. [CrossRef]

26. Flores, P.; Hernández, V.; Hellín, P.; Fenoll, J.; Cava, J.; Mestre, T.; Martinez, V. Metabolite profile of the tomato dwarf cultivar Micro-Tom and comparative response to saline and nutritional stresses with regard to a commercial cultivar. *J. Sci. Food Agric.* **2015**, *96*, 1562–1570. [CrossRef] [PubMed]

27. Magán, J.; Gallardo, M.; Thompson, R.; Lorenzo, P. Effects of salinity on fruit yield and quality of tomato grown in soil-less culture in greenhouses in Mediterranean climatic conditions. *Agric. Water Manag.* **2008**, *95*, 1041–1055. [CrossRef]

28. Li, Y.L.; Stanghellini, C.; Challa, H. Effect of electrical conductivity and transpiration on production of greenhouse tomato (*Lycopersicon esculentum L.*). *Sci. Hortic.* **2001**, *88*, 11–29. [CrossRef]

29. Eltez, R.; Tüzel, Y.; Gül, A.; Tuzel, I.; Duyar, H. Effects of different EC levels of nutrient solution on greenhouse tomato growing. *Acta Hortic.* **2002**, *573*, 443–448. [CrossRef]

30. Ehret, D.L.; Usher, K.; Helmer, T.; Block, G.; Steinke, D.; Frey, B.; Kuang, T.; Diarra, M. Tomato Fruit Antioxidants in Relation to Salinity and Greenhouse Climate. *J. Agric. Food Chem.* **2013**, *61*, 1138–1145. [CrossRef]

31. Mitchell, J.P.; Shennan, C.; Grattan, S.R. Developmental changes in tomato fruit composition in response to water deficit and salinity. *Physiol. Plant.* **1991**, *83*, 177–185. [CrossRef]

32. Mitchell, J.; Shennan, C.; Grattan, S.; May, D. Tomato Fruit Yields and Quality under Water Deficit and Salinity. *J. Am. Soc. Hortic. Sci.* **1991**, *116*, 215–221. [CrossRef]

33. Sakamoto, Y.; Watanabe, S.; Nakashima, T.; Okano, K. Effects of salinity at two ripening stages on the fruit quality of single-truss tomato grown in hydroponics. *J. Hortic. Sci. Biotechnol.* **1999**, *74*, 690–693. [CrossRef]

34. Gonzalo, M.J.; Li, Y.-C.; Chen, K.-Y.; Gil, D.; Montoro, T.; Nájera, I.; Baixauli, C.; Granell, A.; Monforte, A.J. Genetic Control of Reproductive Traits in Tomatoes Under High Temperature. *Front. Plant Sci.* **2020**, *11*, 326. [CrossRef] [PubMed]

35. Prasch, C.M.; Sonnewald, U. Signaling events in plants: Stress factors in combination change the picture. *Environ. Exp. Bot.* **2015**, *114*, 4–14. [CrossRef]

36. Zhou, R.; Kong, L.; Wu, Z.; Rosenqvist, E.; Wang, Y.; Zhao, L.; Zhao, T.; Ottosen, C.-O. Physiological response of tomatoes at drought, heat and their combination followed by recovery. *Physiol. Plant.* **2018**, *165*, 144–154. [CrossRef] [PubMed]

37. Balibrea, M.E.; Cuartero, J.; Bolarín, M.C.; Pérez-Alfocea, F. Sucrolytic activities during fruit development of Lycopersicon genotypes differing in tolerance to salinity. *Physiol. Plant.* **2003**, *118*, 38–46. [CrossRef] [PubMed]

38. Zushi, K.; Matsuzoe, N. Metabolic profile of organoleptic and health-promoting qualities in two tomato cultivars subjected to salt stress and their interactions using correlation network analysis. *Sci. Hortic.* **2015**, *184*, 8–17. [CrossRef]

39. Flores, P.; Navarro, J.M.; Carvajal, M.; Cerdá, A.; Martinez, V. Tomato yield and quality as affected by nitrogen source and salinity. *Agronomy* **2003**, *23*, 249–256. [CrossRef]

40. Fanasca, S.; Martino, A.; Heuvelink, E.; Stanghellini, C. Effect of electrical conductivity, fruit pruning, and truss position on quality in greenhouse tomato fruit. *J. Hortic. Sci. Biotechnol.* **2007**, *82*, 488–494. [CrossRef]

41. Zhang, P.; Senge, M.; Dai, Y. Effects of Salinity Stress on Growth, Yield, Fruit Quality and Water Use Efficiency of Tomato Under Hydroponics System. *Rev. Agric. Sci.* **2016**, *4*, 46–55. [CrossRef]

42. Saito, T.; Fukuda, N.; Matsukura, C.; Nishimura, S. Effects of Salinity on Distribution of Photosynthates and Carbohydrate Metabolism in Tomato Grown using Nutrient Film Technique. *J. Jpn. Soc. Hortic. Sci.* **2009**, *78*, 90–96. [CrossRef]

43. Kläring, H.-P.; Klopotek, Y.; Krumbein, A.; Schwarz, D. The effect of reducing the heating set point on the photosynthesis, growth, yield and fruit quality in greenhouse tomato production. *Agric. For. Meteorol.* **2015**, *214–215*, 178–188. [CrossRef]

44. Zhang, P.; Senge, M.; Yoshiyama, K.; Ito, K.; Dai, Y.; Zhang, F. Effects of Low Salinity Stress on Growth, Yield and Water Use Efficiency of Tomato Under Soilless Cultivation. *IDRE J.* **2017**, *85*, 15–21. [CrossRef]

45. Ksouri, R.; Megdiche, W.; Debez, A.; Falleh, H.; Grignon, C.; Abdelly, C. Salinity effects on polyphenol content and antioxidant activities in leaves of the halophyte *Cakile maritima*. *Plant Physiol. Biochem.* **2007**, *45*, 244–249. [CrossRef] [PubMed]

46. Sgherri, C.; Navari-Izzo, F.; Pardossi, A.; Soressi, G.P.; Izzo, R. The Influence of Diluted Seawater and Ripening Stage on the Content of Antioxidants in Fruits of Different Tomato Genotypes. *J. Agric. Food Chem.* **2007**, *55*, 2452–2458. [CrossRef]

47. Krauss, S.; Schnitzler, W.H.; Grassmann, A.J.; Woitke, M. The Influence of Different Electrical Conductivity Values in a Simplified Recirculating Soilless System on Inner and Outer Fruit Quality Characteristics of Tomato. *J. Agric. Food Chem.* **2006**, *54*, 441–448. [CrossRef]

48. Hernández-Fuentes, A.D.; López-Vargas, E.R.; Pinedo-Espinoza, J.M.; Campos-Montiel, R.G.; Valdés-Reyna, J.; Juárez-Maldonado, A. Postharvest Behavior of Bioactive Compounds in Tomato Fruits Treated with Cu Nanoparticles and NaCl Stress. *Appl. Sci.* **2017**, *7*, 980. [CrossRef]

49. Kim, H.-J.; Fonseca, J.M.; Kubota, C.; Kroggel, M.; Choi, J.-H. Quality of fresh-cut tomatoes as affected by salt content in irrigation water and post-processing ultraviolet-C treatment. *J. Sci. Food Agric.* **2008**, *88*, 1969–1974. [CrossRef]

50. Moles, T.M.; Francisco, R.D.B.; Mariotti, L.; Pompeiano, A.; Lupini, A.; Incrocci, L.; Carmassi, G.; Scartazza, A.; Pistelli, L.; Guglielminetti, L.; et al. Salinity in Autumn-Winter Season and Fruit Quality of Tomato Landraces. *Front. Plant Sci.* **2019**, *10*, 1078. [CrossRef]

51. Rosales, M.A.; Cervilla, L.M.; Sánchez-Rodríguez, E.; Rubio-Wilhelmi, M.D.M.; Blasco, B.; Rios, J.J.; Soriano, T.; Castilla, N.; Romero, L.; Ruiz, J.M. The effect of environmental conditions on nutritional quality of cherry tomato fruits: Evaluation of two experimental Mediterranean greenhouses. *J. Sci. Food Agric.* **2010**, *91*, 152–162. [CrossRef] [PubMed]

52. Slimestad, R.; Verheul, M.J. Seasonal Variations in the Level of Plant Constituents in Greenhouse Production of Cherry Tomatoes. *J. Agric. Food Chem.* **2005**, *53*, 3114–3119. [CrossRef] [PubMed]

53. Slimestad, R.; Fossen, T.; Verheul, M.J. The Flavonoids of Tomatoes. *J. Agric. Food Chem.* **2008**, *56*, 2436–2441. [CrossRef]

54. Incerti, A.; Navari-Izzo, F.; Pardossi, A.; Izzo, R. Seasonal variations in polyphenols and lipoic acid in fruits of tomato irrigated with sea water. *J. Sci. Food Agric.* **2009**, *89*, 1326–1331. [CrossRef]

55. Hernández, V.; Hellín, P.; Fenoll, J.; Flores, P. Increased Temperature Produces Changes in the Bioactive Composition of Tomato, Depending on Its Developmental Stage. *J. Agric. Food Chem.* **2015**, *63*, 2378–2382. [CrossRef]

56. Zushi, K.; Matsuzoe, N. Effect of Soil Water Deficit on Vitamin C, Sugar, Organic Acid, Amino Acid and Carotene Contents of Large-fruited Tomatoes. *J. Jpn. Soc. Hortic. Sci.* **1998**, *67*, 927–933. [CrossRef]

57. Nisar, N.; Li, L.; Lu, S.; Khin, N.C.; Pogson, B.J. Carotenoid Metabolism in Plants. *Mol. Plant* **2015**, *8*, 68–82. [CrossRef] [PubMed]

58. Spicher, L.; Glauser, G.; Kessler, F. Lipid Antioxidant and Galactolipid Remodeling under Temperature Stress in Tomato Plants. *Front. Plant Sci.* **2016**, *7*, 167. [CrossRef] [PubMed]

59. Borghesi, E.; González-Miret, M.L.; Escudero-Gilete, M.L.; Malorgio, F.; Heredia, F.J.; Melendez-Martinez, A.J. Effects of Salinity Stress on Carotenoids, Anthocyanins, and Color of Diverse Tomato Genotypes. *J. Agric. Food Chem.* **2011**, *59*, 11676–11682. [CrossRef] [PubMed]

60. Serio, F.; De Gara, L.; Caretto, S.; Leo, L.; Santamaria, P. Influence of an increased NaCl concentration on yield and quality of cherry tomato grown in Posidonia (*Posidonia oceanica* (L.) *Delile*). *J. Sci. Food Agric.* **2004**, *84*, 1885–1890. [CrossRef]

61. Przybylska, S. Lycopene—A bioactive carotenoid offering multiple health benefits: A review. *Int. J. Food Sci. Technol.* **2020**, *55*, 11–32. [CrossRef]

Article

Comparison of the Grain Quality and Starch Physicochemical Properties between *Japonica* Rice Cultivars with Different Contents of Amylose, as Affected by Nitrogen Fertilization

Yajie Hu, Shumin Cong and Hongcheng Zhang *

Jiangsu Co-Innovation Center for Modern Production Technology of Grain Crops, Jiangsu Key Laboratory of Crop Genetics and Physiology, Yangzhou University, Yangzhou 225009, China; huyajie@yzu.edu.cn (Y.H.); cong202105@163.com (S.C.)
* Correspondence: hczhang@yzu.edu.cn

Abstract: In order to determine the effects of nitrogen fertilizer on the grain quality and starch physicochemical properties of *japonica* rice cultivars with different contents of amylose, normal amylose content (NAC) and low amylose content (LAC) cultivars were grown in a field, with or without nitrogen fertilizer (WN). The relationships between the amylose content, starch physicochemical properties and eating quality were also examined. Compared with WN, nitrogen fertilizer (NF) significantly increased the grain yield but markedly decreased the grain weight. In addition, the processing quality tended to improve, but the appearance quality and eating quality deteriorated under NF application. The grain yield was similar between NAC and LAC cultivars. However, the grain quality and starch physicochemical properties were significantly different between NAC and LAC cultivars. The palatability of the cooked rice was significantly higher in the LAC than in NAC cultivar, which was due to its lower amylose content, protein content, hardness, and retrogradation enthalpy and degree, and its higher stickiness, peak viscosity, breakdown, relative crystallinity and peak intensity. The amylose content and protein content were significantly negatively correlated with the palatability. The amylose content was significantly positively correlated with the final viscosity and setback, and was significantly negatively correlated with the relative crystallinity, peak intensity, gelatinization enthalpy and breakdown. Palatability was significantly positively correlated with peak viscosity, breakdown and peak intensity, and was significantly negatively correlated with the final viscosity, setback, and retrogradation enthalpy and degree. Therefore, the selection of a low amylose content *japonica* rice cultivar grown without nitrogen fertilizer can reduce the amylose and protein contents, as well as improving the pasting properties, starch retrogradation properties and eating quality of the cooked rice.

Keywords: nitrogen fertilizer; *japonica* rice cultivars; grain quality; starch physicochemical properties

Citation: Hu, Y.; Cong, S.; Zhang, H. Comparison of the Grain Quality and Starch Physicochemical Properties between *Japonica* Rice Cultivars with Different Contents of Amylose, as Affected by Nitrogen Fertilization. *Agriculture* **2021**, *11*, 616. https://doi.org/10.3390/agriculture11070616

Academic Editor: Alessandra Durazzo

Received: 22 May 2021
Accepted: 28 June 2021
Published: 30 June 2021

1. Introduction

Rice (*Oryza sativa* L.) is one of the most staple food grains for more than half of the world's people, especially in China, where rice provides nutrients for two-thirds of the population [1]. With the increase in the population and the improvement of living standards in China, the demand for a high-yield, high-quality rice grain is rapidly increasing. Recently, the structural reform of the cereal grain supply was proposed because of the relative surplus of national cereal grain in China [2,3]. As a result, the focus has shifted from high yield to high quality in rice breeding and cultivation. In order to improve the quality of *japonica* rice, some cultivars with a high eating quality and low amylose content (AC: 10% to 15%) have been released in China, such as Nangeng 46, Nangeng 5055, Nangeng 9108, Suxianggeng 100, Zaoxianggeng 1, Huruan 1212 and Songxianggeng 1018, which have the characteristics of being soft but not sticky after cooking, and are defined as soft *japonica* rice [4,5].

Agriculture **2021**, *11*, 616. https://doi.org/10.3390/agriculture11070616

Rice quality is a very important factor affecting the choice of producers and consumers, including milling, appearance, nutrition and eating quality [6]. The four main quality traits are related to the content and composition of starch and protein in rice endosperm. The appearance quality, indicated by the opaqueness or chalkiness, is affected by the shape, size, and packing of starch granules, and by the starch composition [6]. The protein content stored in the seed determines the nutritional quality [7]. Of the four quality traits, eating quality is the most important, which is affected by multiple factors related to the amylose content, amylopectin content, gel consistency, gelatinization temperature and protein content [8]. A low amylose content is the main factor determining the high taste quality of cooked rice [9]. The protein content is also involved in the cooking quality, by impeding starch gelatinization (Chen et al. 2012; Ma et al. 2017). The pasting properties reflect the variation in the viscosity of the starch when rice is cooked, and in previous studies, a high taste quality was associated with a high peak viscosity and breakdown, and a low setback [10,11].

Starch is the dominant component in rice grains, accounting for approximately 90% of the endosperm's weight, and it is essentially composed of linear amylose and moderately branched amylopectin [12]. Previous studies indicate that the starch composition and structure influence the starch functional properties [13,14]. Cai et al. [15] found that the amylose and amylopectin contents, as well as the crystalline structure (relative crystallinity, lamellar peak intensity and lamellar distance), determine different functional properties; starch with a higher amylose content always shows faster retrogradation, leading to the higher setback viscosity of the starch. Wang et al. [16] found that rice starch with a lower amylose content had a higher relative crystallinity and short-range molecular order, which contributed to its higher swelling power, peak viscosity and breakdown viscosity. However, little information is available on the correlations between the starch structure and physicochemical properties and the rice grain quality, especially the eating quality.

The difference in the grain quality and starch physicochemical properties of rice cultivars with normal amylose content (NAC) and low amylose content (LAC) has been reported previously [17,18]. Nevertheless, the amylose content of rice cultivars with low AC is generally more susceptible to a varying environment than those with high AC. Cheng et al. [17] suggested that the effects of high temperatures on the AC of rice are cultivar-dependent, with high temperatures reducing the AC in a low-AC cultivar but increasing the AC in a high-AC cultivar.

Nitrogen (N) is a crucial factor that affects the AC, grain quality and the starch physicochemical properties. In previous studies, the amylose content decreased with an increase in the N level in NAC [19] and LAC [20] rice cultivars. How, the way in which the susceptibility of AC to N deficiency is affected by NAC and LAC cultivars is unclear. Several studies reported that N application improved the processing quality; in particular, topdressing N fertilizers increased the head rice percentage [21,22]. Zhu et al. [20] reported that increased N application decreased the appearance quality and improved the nutritional quality. Gao et al. [19] suggested that rice's taste value increases with a decrease in the N level, with the maximum value reached under an N deficiency. Therefore, we hypothesized that the rice eating quality would be improved by selecting an LAC rice cultivar grown under an N deficiency.

The objectives of this study were: (1) to compare the grain quality and starch physicochemical properties between NAC and LAC *japonica* rice cultivars with or without N fertilizer, and (2) to determine the relationships between the rice eating quality and the amylose content, protein content and starch properties. The traits of the rice quality (milling, appearance, nutrition, and eating quality), starch structure (starch granule size, relative crystallinity and lamellar structure) and physicochemical properties (pasting and gelatinization properties) were determined in order to understand their relationships.

2. Materials and Methods

2.1. Plant Materials and Growth Conditions

Field experiments were conducted in a paddy field in Shengao Town, Jiangyan County, Jiangsu Province, China (120.13° E, 32.58° N) in 2016 and 2017. In the paddy field, the clay soil had the following nutrient contents in 2016 and 2017, respectively: 27.1 and 26.9 g kg^{-1} organic matter; 1.8 and 1.7 g kg^{-1} total N; 18.9 and 18.3 g kg^{-1} Olsen-P; and 95.2 and 95.9 mg kg^{-1} available K. The soil nutrients were determined by the standard procedures [23]. The precipitation, sunshine hours and temperature during the rice growing season across the two years are shown in Figure 1.

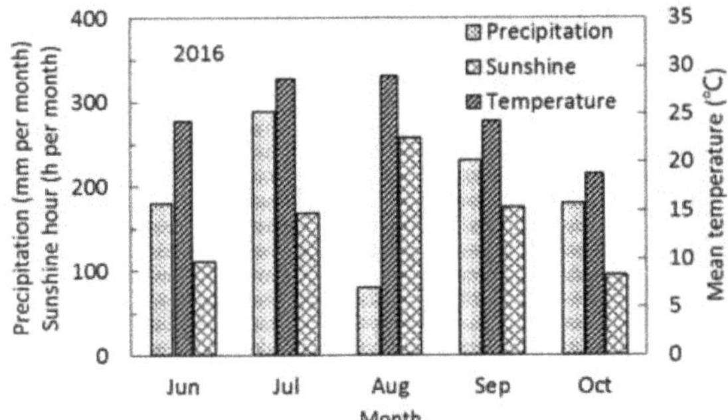
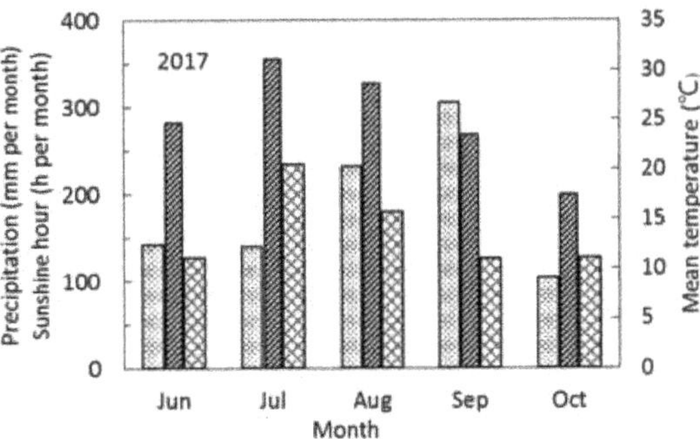

Figure 1. Precipitation, sunshine hours and temperature during the rice growing season of 2016 and 2017 in Jiangyan, Southeast China. The precipitation and sunshine hours are monthly totals. The temperatures are the monthly averages.

Eight conventional *japonica* rice cultivars or lines with different amylose contents were investigated in this study. We included four NAC cultivars with amylose contents ranging from 15% to 17.8% and four LAC cultivars with amylose contents ranging from 9.2% to 14.2%. The specific information on the rice cultivars or lines is shown in Table 1.

Table 1. The information on the *japonica* rice cultivars or lines with different contents of amylose.

Cultivar Type	Cultivar/Line	Amylose Content of Official Release (%)	Year of Official Release	Cross Information	Breeding Organization
NAC	Jingeng 818	17.8	2014	Jindao 9618 × Jindao 1007	Tianjing Academy of Agricultural Science
	Sugeng 815	15.0	2014	Zhengdao 99 × Wuyungeng 11 × Yandao 1229	Jiangsu Zhongjiang Seed Industry Company
	Liangeng 7	16.2	2014	Zhendao 88/Zhonggeng 8415 × Zhongjingchuan 2/Wuyugeng 3	Lianyungang Academy of Agricultural Science
	Wuyungeng 27	17.2	2012	Jia 45/9520 × Wuyungeng 21	Wujing Rice Research Institute
LAC	Songzaoxiang 1	13.2	2014	Zaoxiangruanfan 2 × Zaoxiangchangligeng	Shanghai Songjiang District Agricultural Technology Extension Center
	Zaoxianggeng 1	9.2	2019	Nanjing 46 × Wuyugeng 21	Changshu Agricultural Research Institute
	Yanggeng 239	12.9 *	Not released	/	Yangzhou Academy of Agricultural Science
	Ning 4725	14.2 *	Not released	/	Nanjing Agricultural University

NAC, *japonica* rice with normal amylose content; LAC, *japonica* rice with low amylose content. * The amylose content was determined in this experiment.

2.2. Treatments

The experiment was conducted in a rice–wheat rotation and was a split-plot design, with the N rate as the main plot and the cultivar amylose content as the split plot. The two N levels were 'no N application' (WN, 0 kg ha^{-1}) and 'N fertilizer' (NF, 225 kg ha^{-1}). The seeds were sown on 13 June 2016 and 15 June 2017. The seeding rate was 60 kg ha^{-1}. The size of each plot was 12 m^2 (4 m × 3 m). Three replicates were included in the experiment. Phosphorus (80 kg ha^{-1} as single superphosphate) and potassium (120 kg ha^{-1} as potassium chloride) were applied and incorporated before the direct sowing in both nitrogen treatments. In the NF treatment, basal N (90 kg ha^{-1} as urea) was applied and incorporated before the direct sowing, and top dressing N was also applied at the tillering (90 kg ha^{-1} as urea) and panicle branch differentiation (45 kg ha^{-1} as urea). Water, weeds, insects and diseases were controlled as required to avoid yield loss.

2.3. Grain Quality and Starch Property Measures

Once the rice had matured, the grain yield was determined from a harvest area of 6.0 m^2 in the middle of each plot. The yield components, i.e., the number of panicles per square meter, the number of spikelets per panicle, the percentage of filled grains, and the grain weight, were determined. The harvested grains were air-dried to 14% grain moisture content and then stored at 4 °C for three months. The rice quality analyses were performed according to GB/T 17891-2017.

2.3.1. Processing Quality

The rice grains were passed through a dehusker to obtain brown rice. The brown rice was polished to obtain milled rice. In order to obtain head-milled rice, grain with a length longer than 3/4 of its total length was separated from the milled rice. The brown rice, milled rice, and head-milled rice are expressed as percentages of the total grain weight.

2.3.2. Appearance Quality

The chalkiness was evaluated on 100 milled grains per plot. Chalky kernel rate (%) = the number of chalky kernels/100 milled grains × 100. Chalky area (%) = the chalky area/the total area of the kernel. Chalkiness degree (%) = the chalky kernel rate × the chalky area.

2.3.3. Nutritional Quality

The protein content and amylose content were measured in a 30 g sample of milled rice with a grain analyzer (Infratec 1241, Foss, Denmark). The protein content indicates the nutritional quality of rice. The amylose content is related to the eating quality.

2.3.4. Eating Quality

The sensory properties (palatability, hardness, stickiness) of the cooked rice were measured using an STA-1A rice sensory analyzer (Satake, Japan), which reflects the eating quality. Milled rice, 30 g, was washed in a stainless steel container and then transferred into a 50 mL aluminum box containing 40 mL water. The milled rice was cooked in an electric rice cooker (Z06YA3-S2, Supor, China). After the cooking, the sensory properties of the cooked rice were determined.

2.3.5. Pasting Properties

The milled rice was ground into flour. The rice flour pasting properties were determined using a rapid viscosity analyzer (RVA, Super 3, Newport Scientific, Australia). A total of 3 g flour sifted through 0.15-mm sieves was mixed with 25 g pure water in the RVA sample can. The peak viscosity, trough viscosity, final viscosity in cP (centipoise) and their derivative parameters, i.e., breakdown (peak viscosity-trough viscosity), setback (final viscosity-peak viscosity), and consistency (final viscosity-trough viscosity) were recorded with matching Software of Thermal Cline for Windows (TCW).

2.3.6. Flour and Starch Isolation

Polished rice was ground into flour in a mill (FOSS 1093 Cyclotec Sample Mill, Sweden) in order to pass through a 0.5 mm screen. The starch was isolated from the rice grains according to the method of Zhang et al. [13]. Polished rice grains (10 g) were soaked in an NaOH solution (0.2%) for 24 h. After soaking at room temperature, the swollen grains were rinsed and homogenized in a blender (IKA-T RCT-Basic, Germany) at 1500 rpm for 3 min. Then, 5 mg protease K was added to the slurry and mixed for 24 h at 37 °C. The slurry was sieved through a 150 μm mesh. The residue was mixed with the original volume of the 0.2% NaOH solution and passed through a 150 μm mesh. The combined starch filtrates were centrifuged at $3000 \times g$ for 15 min, and the supernatant was decanted. The process was repeated six times until the yellow tailings were removed. The starch sample was washed with ethanol, dried at 30 °C and stored in a container at 4 °C.

2.3.7. X-ray Diffraction Analysis of the Starch

X-ray diffraction (XRD) patterns were obtained using an X-ray diffractometer (D8 Advance, Bruker, Germany). The starch samples were analyzed by the diffractometer at 200 mA and 40 kV, with a diffraction angle (2θ) that ranged from 3 to 35° (2θ) at a scanning speed of 0.02°. The relative crystallinity (%) was quantified as the ratio of the crystalline area to the total area.

2.3.8. Small-Angle X-ray Scattering Analysis of the Starch

The lamellar structure of a starch sample was determined by using a Bruker NanoStar SAXS instrument equipped with a Vantec 2000 detector and pin-hole collimation for point focus geometry according to the method of Cai et al. [15]. The variables of the small-angle X-ray scattering (SAXS) were analysed according to the simple graphical method.

2.3.9. Gelatinization Properties

The starch gelatinization properties were measured using differential scanning calorimetry (DSC, 200-F3, Netzsch, Germany). In total, 5mg starch was weighed in an aluminum pan and mixed with 15 μL water. The samples were stored at 4 °C for 12 h. After incubating at room temperature for 1 h, the samples were then heated from 20 °C to 130 °C at a rate of 10 °C min^{-1}. The samples were analyzed in triplicate.

2.4. Statistical Analyses

The data were analyzed using Microsoft Excel 2013 and SPSS 17.0 (SPSS, Chicago, IL, USA). Following the ANOVA, the means comparison was based on a least significant difference (LSD) test at $p < 0.05$. The data of the grain yield components were analyzed as the average of two years; only the 2016 data were used for the analysis of the grain quality and starch properties.

3. Results

3.1. Grain Yield and Yield Components

The grain yield and yield components of both types of *japonica* rice cultivars varied with the N level (Table 2). Compared with the yield (kg ha^{-1}) in WN, the grain yield of both types of *japonica* rice cultivars increased significantly in NF in 2016 and 2017. In the same N treatment, there were no significant differences in the grain yield between the NAC and LAC cultivars. Of the yield components, the panicles and spikelets per panicle increased significantly, and the grain weight decreased significantly in NF compared with those in WN. In both NF and WN, the yield components were similar between the NAC and LAC cultivars.

3.2. Milling and Appearance Quality

Compared with those in WN, the measures of the milling quality (brown rice, milled rice, and head-milled rice) and appearance quality (chalkiness and chalky grain percentage) in both types of rice cultivars increased slightly in NF (Table 3). Therefore, with N application, the milling quality improved, but the appearance quality deteriorated in both types of rice cultivars. In the same N treatment, the percentages of milled rice and head-milled rice and the measures of the appearance quality were lower in NAC than in LAC cultivars (Table 3), whereas the brown rice rate was higher in NAC than in LAC cultivars. Therefore, the LAC cultivars had a better milled rice quality but inferior appearance quality.

Table 2. Yield and yield components of the *japonica* rice cultivars with different contents of amylose, grown with or without nitrogen fertilization.

Nitrogen Treatment [1]	Cultivar Type [2]	Cultivar	Panicles ($\times 10^4$ ha^{-1})	Spikelets per Panicle	Filled Grain Percentage (%)	Grain Weight (mg)	Yield in 2016 (kg·ha^{-1})	Yield in 2017 (kg·ha^{-1})
NF	NAC	Jingeng 818	379.80	88.60	94.89	27.80	8.46	8.80
		Sugeng 815	443.10	86.50	84.94	26.76	8.30	8.59
		Liangeng 7	427.50	91.70	93.64	25.33	9.44	9.05
		Wuyungeng 27	427.50	97.60	85.13	27.05	8.73	9.28
		Mean	419.48 ± 23.78 a	91.10 ± 4.18 a	89.65 ± 4.64 b	26.73 ± 0.90 b	8.73 ± 0.44 a	8.93 ± 0.26 a
	LAC	Songzaoxiang 1	410.40	82.67	94.95	25.30	7.95	7.93
		Yanggeng 239	347.70	86.70	94.78	30.65	8.42	8.14
		Zaoxianggeng 1	370.50	80.43	95.42	28.30	7.74	7.87
		Ning 4725	401.10	88.43	95.68	25.58	8.10	7.96
		Mean	382.43 ± 24.90 a	84.56 ± 3.17 a	95.21 ± 0.36 ab	27.46 ± 2.18 b	8.06 ± 0.25 a	7.98 ± 0.10 a
WN	NAC	Jingeng 818	288.60	69.60	97.31	31.40	6.12	5.84
		Sugeng 815	342.00	71.50	98.21	29.93	6.45	6.64
		Liangeng 7	316.65	76.84	97.68	26.84	6.15	6.53
		Wuyungeng 27	320.70	71.00	98.67	30.32	6.43	6.60
		Mean	316.99 ± 19.01 b	72.24 ± 2.75 b	97.97 ± 0.52 a	29.62 ± 1.70 a	6.29 ± 0.15 b	6.40 ± 0.33 b
	LAC	Songzaoxiang 1	307.80	68.00	97.12	28.10	5.39	4.81
		Yanggeng 239	293.70	76.30	98.05	32.44	6.23	6.17
		Zaoxianggeng 1	324.90	75.50	94.65	28.89	6.52	6.23
		Ning 4725	404.70	71.60	91.86	26.72	6.83	6.49
		Mean	332.78 ± 42.97 b	72.85 ± 3.32 b	95.42 ± 2.40 ab	29.04 ± 2.11 a	6.24 ± 0.53 b	5.93 ± 0.66 b

[1] NF, nitrogen fertilizer; WN, without nitrogen fertilizer. [2] NAC, *japonica* rice with a normal amylose content; LAC, *japonica* rice with a low amylose content. Different lowercase letters within the same column indicate significantly different means at the 0.05 probability level. The data presented are the mean ± standard deviation, n = 3.

Table 3. Milling and appearance quality of *japonica* rice cultivars with different contents of amylose, grown with or without nitrogen fertilizer.

Nitrogen treatment [1]	Cultivar Type [2]	Cultivar	Milling Quality			Appearance Quality	
			Brown Rice (%)	Milled Rice (%)	Head-Milled Rice (%)	Chalkiness (%)	Chalky Grain Percentage (%)
NF	NAC	Jingeng 818	85.40	76.55	72.05	4.83	18.49
		Sugeng 815	85.65	73.80	68.03	7.57	25.10
		Liangeng 7	85.90	73.91	69.85	11.15	36.87
		Wuyungeng 27	85.45	74.55	69.13	8.56	27.37
		Mean	85.60 ± 0.20 a	74.70 ± 1.10 a	69.76 ± 1.47 a	8.03 ± 2.26 a	26.96 ± 6.59 a
	LAC	Songzaoxiang 1	83.61	76.10	71.15	12.03	51.34
		Yanggeng 239	86.60	77.50	72.07	11.39	51.76
		Zaoxianggeng 1	85.25	76.35	71.37	10.63	36.79
		Ning 4725	83.55	74.50	70.55	8.93	26.85
		Mean	84.75 ± 1.27 a	76.11 ± 1.07 a	71.28 ± 0.54 a	10.75 ± 1.16 a	41.68 ± 10.48 a
WN	NAC	Jingeng 818	84.50	75.75	69.05	4.37	17.39
		Sugeng 815	85.90	72.35	64.52	7.49	22.85
		Liangeng 7	85.55	73.10	66.17	7.94	29.68
		Wuyungeng 27	85.20	74.65	70.53	8.24	24.53
		Mean	85.29 ± 0.52 a	73.96 ± 1.32 a	67.57 ± 2.36 a	7.01 ± 1.55 a	23.61 ± 4.39 a
	LAC	Songzaoxiang 1	81.95	73.25	64.82	7.63	23.96
		Yanggeng 239	84.95	75.95	69.75	13.35	47.80
		Zaoxianggeng 1	84.35	74.70	71.79	8.55	35.28
		Ning 4725	82.05	72.70	69.52	7.38	29.19
		Mean	83.33 ± 1.34 a	74.15 ± 1.27 a	68.97 ± 2.55 a	9.23 ± 2.42 a	34.05 ± 8.89 a

[1] NF, nitrogen fertilizer; WN, without nitrogen fertilizer. [2] NAC, *japonica* rice with a normal amylose content; LAC, *japonica* rice with a low amylose content. Different lowercase letters within the same column indicate significantly different means at the 0.05 probability level. The data presented are the mean ± standard deviation, n = 3.

3.3. Eating Qualities and Their Relationships

Compared with that in WN, the amylose content was significantly lower in both types of rice cultivars in NF (Table 4). By contrast, the protein content was higher in NF than in WN. In both WN and NF, the protein content was higher in NAC than in LAC cultivars. In a comparison of the variability in the amylose content between LAC and NAC cultivars in NF, the coefficient of variation (CV) of the NAC cultivars (CV = 5.09%) was higher than that of the LAC cultivars (CV = 4.49%). Therefore, the response of the amylose content of the NAC cultivars to N fertilization was more variable.

Table 4. Protein content, amylose content, and eating quality of *japonica* rice cultivars with different contents of amylose, grown with or without nitrogen fertilizer.

Nitrogen Treatment [1]	Cultivar Type [2]	Cultivar	Protein Content (%)	Amylose Content (%)	Hardness	Stickiness	Palatability
NF	NAC	Jingeng 818	9.50	16.30	8.80	2.65	42.00
		Sugeng 815	9.80	15.45	8.10	3.45	48.75
		Liangeng 7	9.45	16.55	8.53	2.73	43.75
		Wuyungeng 27	9.05	17.80	8.15	3.65	49.50
		Mean	9.45 ± 0.27 a	16.53 ± 0.84 b	8.39 ± 0.29 a	3.12 ± 0.44 b	46.00 ± 3.20 c
	LAC	Songzaoxiang 1	9.05	13.15	7.30	5.30	60.50
		Yanggeng 239	9.00	12.85	7.13	5.18	61.50
		Zaoxianggeng 1	8.30	14.25	6.58	7.08	66.50
		Ning 4725	8.10	14.15	6.23	6.78	64.24
		Mean	8.61 ± 0.42 b	13.60 ± 0.61 c	6.81 ± 0.43 b	6.08 ± 0.85 a	63.19 ± 5.52 ab
WN	NAC	Jingeng 818	8.60	18.80	8.35	2.68	44.50
		Sugeng 815	8.45	18.15	7.65	4.16	53.63
		Liangeng 7	8.85	17.65	7.95	3.86	51.13
		Wuyungeng 27	8.15	20.05	7.71	4.28	53.88
		Mean	8.51 ± 0.25 b	18.66 ± 0.90 a	7.92 ± 0.27 a	3.74 ± 0.64 b	50.78 ± 3.78 bc
	LAC	Songzaoxiang 1	8.65	15.95	6.28	7.34	73.38
		Yanggeng 239	8.00	16.15	6.30	6.26	70.00
		Zaoxianggeng 1	8.10	16.25	6.21	6.89	72.50
		Ning 4725	7.80	16.55	6.10	7.33	74.50
		Mean	8.14 ± 0.31 b	16.23 ± 0.22 b	6.22 ± 0.08 b	6.95 ± 0.44 a	72.59 ± 1.66 a

[1] NF, nitrogen fertilizer; WN, without nitrogen fertilizer. [2] NAC, *japonica* rice with a normal amylose content; LAC, *japonica* rice with a low amylose content. Different lowercase letters within the same column indicate significantly different means at the 0.05 probability level. The data presented are the mean ± standard deviation, n = 3.

In WN and NF, the LAC cultivars had the highest palatability, indicating the best eating quality (Table 4). Compared with that in WN, the palatability was lower in both types of rice cultivars in NF. Regardless of the N level, the LAC cultivars had significantly higher palatability than the NAC cultivars. Higher palatability was associated with lower hardness and higher stickiness. Compared with the NAC cultivars, the LAC cultivars had lower hardness and higher stickiness.

The amylose content was negatively correlated with the palatability and protein content (Table 5). The protein content was significantly negatively correlated with palatability and stickiness, but significantly positively correlated with hardness. Hence, the low amylose content and low protein content both contributed to the higher palatability of the cooked rice.

3.4. Pasting Properties in Relation to the Eating Quality

The differences in the pasting properties between the two types of *japonica* rice cultivars in the two N treatments are shown in Table 6. In NF, the peak viscosity, trough viscosity, breakdown and final viscosity decreased compared with those in WN. Regardless of the N treatment, compared with the NAC cultivars, the peak viscosity and breakdown increased, whereas the final viscosity, setback and pasting temperature decreased in the LAC cultivars.

The amylose content was significantly negatively correlated with breakdown, and was significantly positively correlated with final viscosity (Table 5). Palatability was significantly positively correlated with the peak viscosity and breakdown, and was significantly negatively correlated with the final viscosity and setback. Therefore, the characteristics of high eating quality were as follows: a low amylose content, a high peak viscosity and breakdown, and a low final viscosity and setback.

Table 5. Correlation matrix of the protein content, amylose content, eating quality and pasting properties [1] of *japonica* rice cultivars.

	AC	PA	HN	SN	PV	TV	BD	FV	SB	PT
PC	−0.164	−0.726 **	0.756 **	−0.704 **	−0.587 *	−0.480	−0.320	−0.019	0.376	0.231
AC		−0.494 *	0.400	−0.439	−0.083	0.368	−0.513 *	0.775 **	0.693 **	0.141
PA			−0.994 **	0.996 **	0.596 *	0.116	0.723 **	−0.543 *	−0.845 **	−0.254
HN				−0.982 **	−0.584 *	−0.123	−0.700 **	0.532 *	0.829 **	0.222
SN					0.615 *	0.118	0.749 **	−0.535 *	−0.851 **	−0.260
PV						0.713 **	0.658 **	0.113	−0.575 *	−0.215
TV							−0.059	0.709 **	0.108	−0.426
BD								−0.601*	−0.934 **	0.152
FV									0.748 **	−0.210
SB										−0.029

[1] PC, protein content; AC, amylose content; PA, palatability; HN, hardness; SN, stickiness; PV, peak viscosity; TV, trough viscosity; BD, breakdown; FV, final viscosity; SB, setback; PT, pasting temperature. * and ** indicate significance at the $p < 0.05$ and $p < 0.01$ levels, respectively (n = 16).

Table 6. Pasting properties of *japonica* rice cultivars with different contents of amylose, grown with or without nitrogen fertilizer.

Nitrogen Treatment [1]	Cultivar Type [2]	Cultivar	Peak Viscosity (cP)	Trough Viscosity (cP)	Breakdown (cP)	Final Viscosity (cP)	Setback (cP)	Pasting Temperature (°C)
NF	NAC	Jingeng 818	2401.0	1307.0	1094.0	2465.0	64.0	70.45
		Sugeng 815	2244.0	970.0	1274.0	1844.5	−399.5	75.58
		Liangeng 7	2200.5	1124.0	1076.5	2262.0	61.5	69.33
		Wuyungeng 27	2427.0	1374.0	1053.0	2546.0	119.0	69.70
		Mean	2318.1 ± 97.5 b	1193.8 ± 158.3 b	1124.4 ± 87.6 a	2279.4 ± 271.6 ab	−38.8 ± 209.5 a	71.26 ± 2.5 a
	LAC	Songzaoxiang 1	3107.0	1516.0	1591.0	2124.0	−983.0	69.75
		Yanggeng 239	2570.5	1321.0	1249.5	1947.5	−623.0	70.10
		Zaoxianggeng 1	2928.0	1440.5	1487.5	2097.5	−830.5	70.50
		Ning 4725	2692.5	1307.5	1385.0	1926.5	−766.0	70.10
		Mean	2824.5 ± 207.6 a	1396.3 ± 86.4 ab	1428.3 ± 126.3 a	2023.9 ± 87.7 b	−800.6 ± 129.3 b	70.11 ± 0.3 a
WN	NAC	Jingeng 818	2952.0	1904.0	1048.0	3078.0	126.0	71.35
		Sugeng 815	2717.0	1219.5	1497.5	2188.5	−528.5	75.58
		Liangeng 7	2723.5	1556.5	1167.0	2753.0	29.5	69.73
		Wuyungeng 27	2742.0	1676.0	1066.0	2890.0	148.0	70.08
		Mean	2783.6 ± 97.6 a	1589.0 ± 247.2 a	1194.6 ± 180.7 a	2727.4 ± 331.8 a	−56.3 ± 276.3 a	71.68 ± 2.3 a
	LAC	Songzaoxiang 1	3242.0	1565.5	1676.5	2221.5	−1020.5	70.08
		Yanggeng 239	2765.5	1473.5	1292.0	2169.5	−596.0	70.13
		Zaoxianggeng 1	2709.0	1367.0	1342.0	2018.0	−691.0	69.65
		Ning 4725	2931.0	1473.0	1458.0	2125.0	−806.0	70.40
		Mean	2911.9 ± 207.3 a	1469.8 ± 70.3 ab	1442.1 ± 148.1 a	2133.5 ± 74.9 b	−778.4 ± 158.3 b	70.06 ± 0.3 a

[1] NF, nitrogen fertilizer; WN, without nitrogen fertilizer. [2] NAC, *japonica* rice with a normal amylose content; LAC, *japonica* rice with a low amylose content. Different lowercase letters within the same column indicate significantly different means at the 0.05 probability level. The data presented are the mean ± standard deviation, n = 3.

3.5. Thermal Properties of Starch

The starch of both types of rice cultivars in both N treatments showed no significant differences in its gelatinization temperatures and enthalpy (ΔHgel) (Table 7). Compared with those in WN, in NF the retrogradation enthalpy (ΔHret) and degree (%R) increased by 51.6% and 48.8% in the NAC cultivars, and by 31.9% and 30.0% in the LAC cultivars, respectively. Therefore, NF increased the starch ΔHret and %R. Compared with those in the LAC cultivars, ΔHret and %R in the NAC cultivars increased by 49.2% and 55.2% in NF, and by 29.8% and 35.6% in WN, respectively. Hence, the ΔHret and %R were higher in NAC than in the LAC cultivars.

The amylose content was significantly negatively correlated with ΔHgel (Table 8). Palatability and stickiness were both significantly negatively correlated with ΔHret and %R. Hardness was significantly positively correlated with ΔHret and %R.

Table 7. Thermal properties of *japonica* rice cultivar starches with different contents of amylose, grown with or without nitrogen fertilization.

Nitrogen Treatment [1]	Cultivar Type [2]	Cultivar	T_o (°C)	T_p (°C)	T_c (°C)	ΔHgel (J/g)	ΔHret (J/g)	%R
NF	NAC	Jingeng 818	60.55	65.40	73.10	11.02	2.27	20.59
		Sugeng 815	66.15	71.75	78.40	12.43	2.33	18.75
		Liangeng 7	61.50	66.90	74.80	11.50	1.69	14.66
		Wuyungeng 27	61.90	66.45	73.50	10.92	1.13	10.31
		Mean	62.53 ± 2.15 a	67.63 ± 2.44 a	74.95 ± 2.09 a	11.47 ± 0.60 a	1.85 ± 0.49 a	16.16 ± 3.96 a
	LAC	Songzaoxiang 1	63.95	68.15	74.10	12.49	1.46	11.69
		Yanggeng 239	63.45	68.20	74.80	11.72	1.17	10.00
		Zaoxianggeng 1	63.10	67.90	75.00	11.69	1.30	11.16
		Ning 4725	63.55	68.35	75.85	11.61	1.01	8.69
		Mean	63.51 ± 0.30 a	68.15 ± 0.16 a	74.94 ± 0.62 a	11.88 ± 0.36 a	1.24 ± 0.17 ab	10.41 ± 1.15 ab
WN	NAC	Jingeng 818	61.60	67.15	74.75	11.40	1.09	9.54
		Sugeng 815	67.25	72.30	75.85	11.87	1.46	12.27
		Liangeng 7	61.65	66.30	73.80	10.77	1.17	10.88
		Wuyungeng 27	61.55	66.50	74.15	10.82	1.16	10.70
		Mean	63.01 ± 2.45 a	68.06 ± 2.47 a	74.64 ± 0.78 a	11.21 ± 0.45 a	1.22 ± 0.14 ab	10.86 ± 0.97 ab
	LAC	Songzaoxiang 1	63.55	67.80	73.80	12.04	0.72	5.96
		Yanggeng 239	63.05	67.85	75.40	11.63	0.70	6.02
		Zaoxianggeng 1	62.75	67.45	74.95	11.15	1.29	11.61
		Ning 4725	63.55	68.30	75.55	11.95	1.03	8.64
		Mean	63.23 ± 0.34 a	67.85 ± 0.30 a	74.93 ± 0.69 a	11.69 ± 0.35 a	0.94 ± 0.25 b	8.01 ± 2.32 b

[1] NF, nitrogen fertilizer; WN, without nitrogen fertilizer. [2] NAC, *japonica* rice with a normal amylose content; LAC, *japonica* rice with a low amylose content. To, Tp, Tc, ΔHgel, ΔHret and %R correspond to onset temperature, peak temperature, conclusion temperature, gelatinization enthalpy, and retrogradation enthalpy and degree, respectively. Different lowercase letters within the same column indicate significantly different means at the 0.05 probability level. The data presented are the mean ± standard deviation, n = 3.

Table 8. Correlations between the eating quality, starch structure and gelatinization properties [1] of *japonica* rice cultivars.

	T_O	T_P	T_C	ΔHgel	ΔHret	%R	PI	PWHM	PP	LD	RC
AC	−0.261	−0.175	−0.186	−0.587 *	−0.084	−0.011	−0.546 *	−0.041	−0.499 *	0.32	−0.740 **
PA	0.267	0.137	0.143	0.318	−0.614 *	−0.656 **	0.709 **	0.494	0.216	−0.121	0.208
HN	−0.292	−0.171	−0.188	−0.316	0.640 **	0.683 **	−0.725 **	−0.486	−0.24	0.131	−0.188
SN	0.256	0.12	0.107	0.32	−0.590 *	−0.630 **	0.681 **	0.496	0.174	−0.11	0.203

[1] AC, amylose content; PA, palatability; HN, hardness; SN, stickiness; T_O, onset temperature; T_p, peak temperature; T_c, conclusion temperature; ΔHgel, gelatinization enthalpy; ΔHret, retrogradation enthalpy; %R, retrogradation degree; PI, peak intensity; PWHM, peak width at half maximum; PP, peak position; LD, lamellar distance; RC, relative crystallinity. * and ** indicate significance at the $p < 0.05$ and $p < 0.01$ levels, respectively (n = 16).

3.6. Small-Angle X-ray Scattering Variables and the Relative Crystallinity of the Starches

The XRD patterns of the starch of eight rice cultivars affected by nitrogen levels were very similar (Figure 2), showing the typical A-type diffraction pattern. The relative crystallinity calculated from the XRD patterns showed no significant differences between the nitrogen levels or the type of rice cultivar (Table 9). Compared with that in WN, the relative crystallinity increased by 5.0% in the NAC cultivars, and by 3.6% in the LAC cultivars in NF. The relative crystallinity was 2.8% and 4.1% lower in NAC cultivars than in LAC cultivars in NF and WN, respectively. The amylose content was significantly negatively correlated with the relative crystallinity (Table 8). Therefore, the starch samples with a low amylose content had a high relative crystallinity.

The lamellar structures of rice starch were investigated using SAXS (Figure 3). The peak intensity, peak width at half maximum, peak position, and lamellar distance calculated from the SAXS patterns are shown in Table 9. The SAXS variables of both cultivar types were not significantly different between NF and WN. Compared with that in the LAC cultivars, the peak intensity in the NAC cultivars decreased by 17.9% in NF and by 21.6% in WN. The peak width at half maximum, peak position and lamellar distance were similar between the NAC and LAC cultivars in the two N treatments. The peak intensity and peak position were significantly negatively correlated with the amylose content, and the peak intensity was significantly positively correlated with the palatability and stickiness (Table 8).

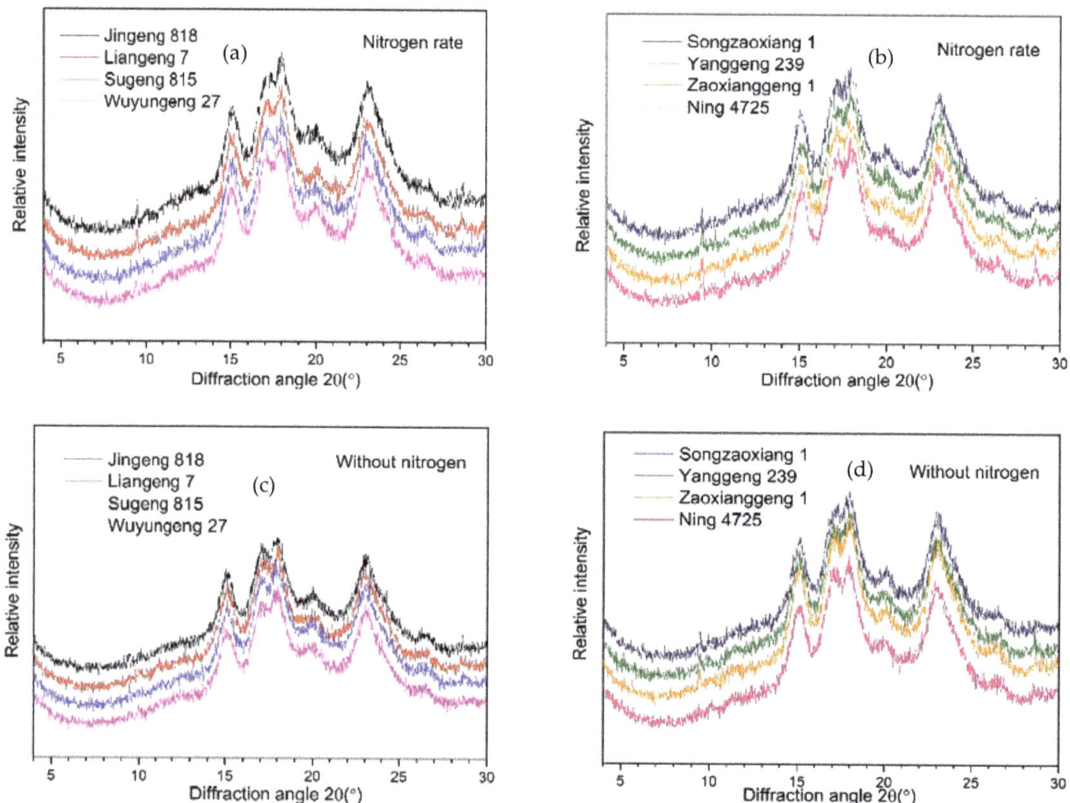

Figure 2. X-ray diffraction patterns of the starch of *japonica* rice cultivars with different contents of amylose, with or without nitrogen fertilization. (**a**) NAC, with fertilization; (**b**) LAC, with fertilization; (**c**) NAC, without fertilization; (**d**) LAC, without fertilization.

Table 9. Small-angle X-ray scattering (SAXS) variables and relative crystallinity of the starch of *japonica* rice cultivars with different contents of amylose, grown with or without nitrogen fertilizer.

Nitrogen Treatment [1]	Cultivar Type [2]	Cultivar	SAXS Variable				Relative Crystallinity (%)
			Peak Intensity (Count)	Peak Width at Half Maximum (Å^{-1})	Peak Position (Å^{-1})	Lamellar Distance (nm)	
NF	NAC	Jingeng 818	108.28	0.019	0.067	9.38	30.4
		Sugeng 815	120.56	0.017	0.069	9.13	32.3
		Liangeng 7	130.55	0.019	0.068	9.24	33.9
		Wuyungeng 27	124.30	0.019	0.068	9.24	30.4
		Mean	120.92 ± 8.12 b	0.018 ± 0.00 a	0.068 ± 0.00 a	9.25 ± 0.09 a	31.75 ± 1.46 a
		Songzaoxiang 1	150.34	0.018	0.068	9.19	33
		Yanggeng 239	157.66	0.020	0.070	9.01	33.3
		Zaoxianggeng 1	128.87	0.019	0.068	9.19	32.9
		Ning 4725	152.40	0.018	0.069	9.38	31.5
		Mean	147.32 ± 10.98 a	0.019 ± 0.00 a	0.069 ± 0.00 a	9.19 ± 0.13 a	32.68 ± 0.69 a

Table 9. *Cont.*

Nitrogen Treatment [1]	Cultivar Type [2]	Cultivar	SAXS Variable					Relative Crystallinity (%)
			Peak Intensity (Count)	Peak Width at Half Maximum (Å^{-1})	Peak Position (Å^{-1})	Lamellar Distance (nm)		
WN	NAC	Jingeng 818	123.24	0.018	0.068	9.23		29.7
		Sugeng 815	129.48	0.018	0.068	9.19		31.6
		Liangeng 7	98.61	0.019	0.068	9.21		30.5
		Wuyungeng 27	118.06	0.019	0.068	9.28		29.2
		Mean	117.35 ± 11.55 b	0.018 ± 0.00 a	0.068 ± 0.00 a	9.23 ± 0.03 a		30.25 ± 0.91 a
	LAC	Songzaoxiang 1	146.48	0.020	0.068	9.21		32.1
		Yanggeng 239	153.79	0.020	0.068	9.21		32.5
		Zaoxianggeng 1	132.39	0.020	0.068	9.23		31.0
		Ning 4725	166.19	0.020	0.068	9.19		30.5
		Mean	149.71 ± 12.23 a	0.020 ± 0.00 a	0.068 ± 0.00 a	9.21 ± 0.01 a		31.53 ± 0.81 a

[1] NF, nitrogen fertilizer; WN, without nitrogen fertilizer. [2] NAC, *japonica* rice with a normal amylose content; LAC, *japonica* rice with a low amylose content. Different lowercase letters within the same column indicate significantly different means at the 0.05 probability level. The data presented are the mean ± standard deviation, n = 3.

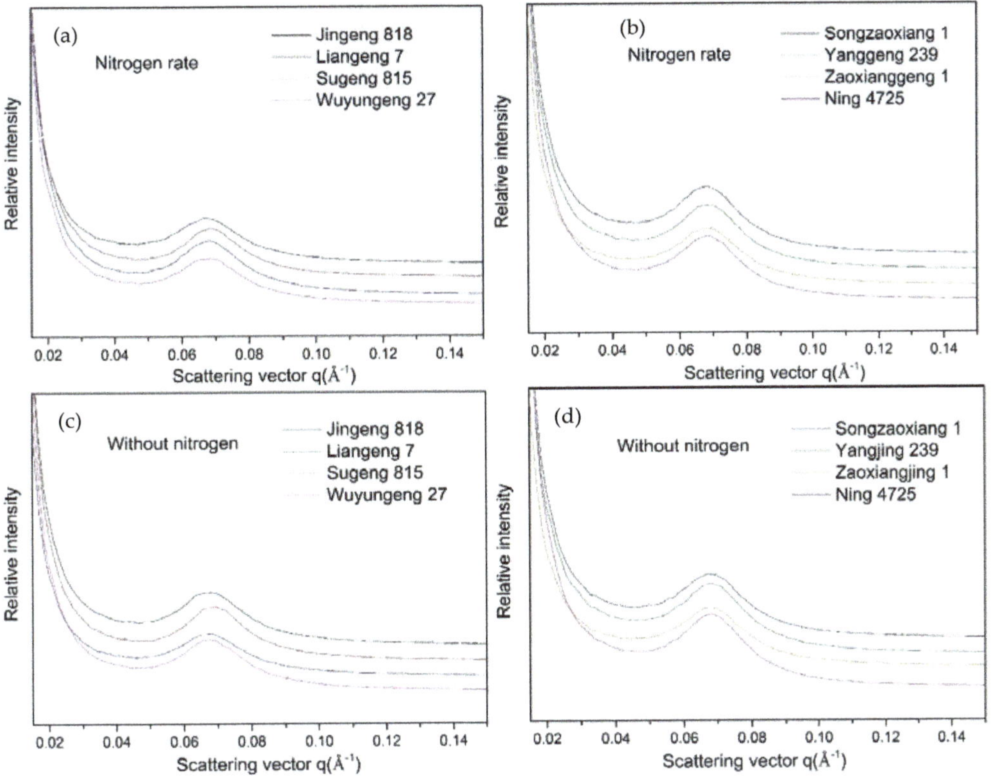

Figure 3. Small-angle X-ray scattering spectra of the starch of *japonica* rice cultivars with different contents of amylose, with or without nitrogen fertilization. (**a**) NAC, with fertilization; (**b**) LAC, with fertilization; (**c**) NAC, without fertilization; (**d**) LAC, without fertilization.

4. Discussion

Rice grain quality is markedly influenced by the cultivar genotype and N application [24]. In previous studies, N fertilizer application improved the milling quality and nutritional quality of rice [11,25]. Our findings are consistent with these results, and in

NF, the milling quality and nutritional quality improved, although the appearance quality and eating quality deteriorated compared to WN (Tables 2 and 3). Top-dressed nitrogen increases the protein content but decreases the amylose content in rice grains [26,27]. Similarly, in our study, the protein content was higher and the amylose content was lower in NF than in WN. Compared with WN, the palatability of the cooked rice decreased significantly in NF because of the higher protein content and hardness and the lower stickiness. Ref. [28] reported similar results in which a high protein content led to a firmer texture of rice, resulting in the insufficient absorption of water for compete gelatinization. Differences in grain quality are found in rice cultivars with different contents of amylose [9]. Zhao et al. [18] reported that rice varieties with lower amylose contents had a lower peak time, trough, final, setback, and consistent viscosity, as well as a lower hardness of the cooked rice and a higher gel consistency, breakdown viscosity, stickiness and comprehensive value of the cooked rice. In this study, compared with those of NAC cultivars, the peak viscosity, breakdown value and stickiness were higher, and the final viscosity, setback value, protein content and hardness were lower in LAC cultivars, leading to the higher palatability of cooked rice.

Yang et al. [29] examined the response of starch physicochemical properties to N application, and according to Zhu et al. [30], an increasing level of N decreased the gelatinization temperature and increased the relative crystallinity and gelatinization enthalpy. Those results are partially consistent with our findings. The highest relative crystallinity (Table 9), gelatinization enthalpy and retrogradation enthalpy (Table 7) were observed in NF, compared with WN. Singh et al. [31] also found that the amylose content decreased, and the pasting temperature, gelatinization temperature and enthalpy increased with N application.

The starch physicochemical properties of different types of rice cultivar have been studied extensively. Cai et al. [15] investigated the structure and functional properties of the starch of 10 rice cultivars with different amylose contents, and found that an increased amylose content increased the gelatinization enthalpy and decreased the relative crystallinity and peak intensity. Compared with NAC cultivars, the relative crystallinity and peak intensity increased (Table 9), and the retrogradation enthalpy and degree decreased (Table 7) in LAC cultivars. However, the gelatinization temperature and enthalpy were similar between LAC and NAC cultivars (Table 7). The intensity of the scattering peaks depends primarily on the order degree in semicrystalline regions, and decreases with an increasing amylose content [32]. Thus, our findings are consistent with those of previous studies [15,32].

The amylose content and protein content in the rice endosperm are the main factors that affect the rice eating quality [28,33,34]. In this study, the amylose content and protein content were significantly negatively correlated with the palatability of cooked rice. According to previous studies, the difference in the amylose content determines the starch crystalline structure and physicochemical properties [35–37]. Kong et al. [35] found that the amylose content was significantly positively correlated with the peak viscosity, hot paste viscosity, cold paste viscosity, setback and hardness, whereas it was negatively correlated with adhesiveness, cohesiveness, gelatinization temperature (To, Tp, Tc) and enthalpy. In our study, the amylose content was significantly positively correlated with the final viscosity and setback, and was significantly negatively correlated with the relative crystallinity, peak intensity, gelatinization enthalpy and breakdown (Table 8); these results are consistent with those of previous studies [15,35].

Pasting properties are also closely associated with the rice eating quality. Our findings showed significant positive correlations between palatability, peak viscosity and breakdown, and significant negative correlations between palatability, final viscosity and setback (Table 5). These results support those of previous studies in which rice with a high eating quality had a high peak viscosity and breakdown, and a low setback (Asante et al., 2013; Zhao et al., 2019). Rice cultivars with low gelatinization temperatures and enthalpy require less water and cooking time, and absorb less thermal energy to reach starch gelatinization, contributing to the high palatability of the cooked rice [38]. In our study, there was no significant correlation between the palatability and gelatinization temperature or enthalpy, but palatability was significantly negatively correlated with the retrogradation enthalpy and degree (Table 8). These results are partially consistent with those of a previous study [33].

Therefore, in this study, the high eating quality of rice was associated with a higher peak viscosity, breakdown, relative crystallinity and peak intensity, as well as a lower protein content, amylose content, final viscosity, setback, and retrogradation enthalpy and degree.

5. Conclusions

Compared with no N application, N fertilizer led to the deterioration of the eating quality of rice cultivars because of the higher protein content, lower pasting viscosity, and higher retrogradation enthalpy and degree. Compared with NAC cultivars, LAC cultivars had a higher pasting viscosity and lower retrogradation enthalpy and degree, leading to a higher eating quality; however, they also tended to have an inferior appearance quality. The amylose content and protein content were significantly negatively correlated with the palatability of cooked rice. The differences in the amylose content determined the starch crystal structure and pasting and gelatinization properties. Significant correlations were detected between the amylose content, eating quality and starch properties. Therefore, in order to increase the eating quality of cooked rice, an LAC rice cultivar should be selected and grown under N deficiency because the amylose content and protein content decrease, and the pasting properties and retrogradation properties improve. These results provide new insight into the grain quality of rice and how to direct breeding and agronomic management to achieve high eating quality.

Author Contributions: Conceptualization, Y.H. and H.Z.; methodology, Y.H.; software, Y.H.; validation, Y.H., S.C. and H.Z.; formal analysis, Y.H.; investigation, Y.H.; resources, H.Z.; data curation, S.C.; writing—original draft preparation, Y.H. and S.C.; writing—review and editing, Y.H. and H.Z.; visualization, Y.H. and H.Z.; supervision, H.Z.; project administration, Y.H. and H.Z.; funding acquisition, Y.H. and H.Z. All authors have read and agreed to the published version of the manuscript.

Funding: This research was funded by National Nature Science Foundation of China (31701350), National Key Research Program of China (2016YFD0300503, 2017YFD0301205), the Earmarked Fund for China Agriculture Research System (CARS-01-27), and the Earmarked Fund for Jiangsu Agriculture Industry Technology System (JATS [2019]444) and the project by the Priority Academic Program Development of Jiangsu Higher Education Institutions.

Institutional Review Board Statement: Not applicable.

Informed Consent Statement: Not applicable.

Data Availability Statement: Not applicable.

References

1. Nguyen, N.V.; Ferrero, A. Meeting the challenges of global rice production. *Paddy Water Environ.* **2006**, *4*, 1–9. [CrossRef]
2. Yu, Y.Q.; Huang, Y.; Zhang, W. Changes in rice yields in China since 1980 associated with cultivar improvement, climate and crop management. *Field Crops Res.* **2012**, *136*, 65–75. [CrossRef]
3. Huang, M.; Zou, Y.B. Integrating mechanization with agronomy and breeding to ensure food security in China. *Field Crops Res.* **2018**, *224*, 22–27. [CrossRef]
4. Yang, J.; Wang, J.; Fan, F.J.; Zhu, J.Y.; Chen, T.; Wang, C.L.; Zheng, T.Q.; Zhang, J.; Zhong, W.G.; Xu, J.L. Development of AS-PCR marker based on a key mutation confirmed by resequencing of Wx-mp in Milky Princess and its application in *japonica* soft rice (*Oryza sativa L.*) breeding. *Plant Breed.* **2013**, *132*, 595–603. [CrossRef]
5. Wang, C.L.; Zhang, Y.D.; Zhu, Z.; Chen, T.; Zhao, Q.Y.; Zhong, W.G.; Yang, J.; Yao, S.; Zhou, L.H.; Zhao, L.; et al. Research progress on the breeding of *japonica* super rice varieties in Jiangsu Province, China. *J. Integr. Agric.* **2017**, *16*, 992–999. [CrossRef]
6. Chen, Y.; Wang, M.; Ouwerkerk, P.B.F. Molecular and envrionmental factors determining grain quality in rice. *Food Energy Secur.* **2012**, *2*, 111–132. [CrossRef]

7.   Liu, Q.H.; Wu, X.; Ma, J.Q.; Xin, C.Y. Effects of cultivars, transplanting patterns, environment, and their interactions on grain quality of japonica rice. *Cereal Chem.* **2015**, *92*, 284–292. [CrossRef]
8.   Chun, A.; Lee, H.J.; Hamaker, B.R.; Janaswamy, S. Effects of ripening temperature on starch structure and gelatinization, pasting, and cooking properties in rice (Oryza sativa). *J. Agric. Food Chem.* **2015**, *63*, 3085–3093. [CrossRef]
9.   Ma, Z.H.; Cheng, H.T.; Nitta, Y.; Aoki, N.; Chen, Y.; Chen, H.X.; Ohsugi, R.; Lyu, W.Y. Differences in viscosity of superior and inferior spikelets of japonica rice with various percentage of apparent amylose content. *J. Agric. Food Chem.* **2017**, *65*, 4237–4246. [CrossRef]
10.   Patindol, J.A.; Siebenmorgen, T.J.; Wang, Y.J. Impact of environmental factors on rice starch structure: A review. *Starch-Starke* **2015**, *67*, 42–54. [CrossRef]
11.   Wei, H.Y.; Chen, Z.F.; Xing, Z.P.; Zhou, L.; Liu, Q.Y.; Zhang, Z.Z.; Jiang, Y.; Hu, Y.J.; Zhu, J.Y.; Cui, P.Y.; et al. Effects of slow or controlled release fertilizer types and fertilization modes on yield and quality of rice. *J. Integr. Agric.* **2018**, *17*, 2222–2234. [CrossRef]
12.   Li, E.P.; Wu, A.C.; Li, J.; Liu, Q.Q.; Gilbert, R.G. Improved understanding of rice amylose biosynthesis from advanced starch structural characterization. *Rice* **2015**, *8*, 1–8. [CrossRef]
13.   Zhang, C.Q.; Zhou, L.H.; Zhu, Z.B.; Lu, H.W.; Zhou, X.Z.; Qan, Y.T.; Li, Q.F.; Lu, Y.; Gu, M.H.; Liu, Q.Q. Characterization of grain quality and starch fine structure of two japonica rice (Oryza Sativa) cultivars with good sensory properties at different temperatures during the filling stage. *J. Agric. Food Chem.* **2016**, *64*, 4048–4057. [CrossRef]
14.   Liu, J.C.; Zhao, Q.; Zhou, L.J.; Cao, L.J.; Cao, Z.Z.; Shi, C.H.; Cheng, F.M. Influence of environmental temperature during grain filling period on granule size distribution of rice starch and its relation to gelatinization properties. *J. Cereal Sci.* **2017**, *76*, 42–55. [CrossRef]
15.   Cai, J.W.; Man, J.M.; Huang, J.; Liu, Q.Q.; Wei, W.X.; Wei, C.X. Relationship between structure and functional properties of normal rice starches with different amylose contents. *Carbohydr. Polym.* **2015**, *125*, 35–44. [CrossRef]
16.   Wang, S.J.; Li, P.Y.; Yu, J.L.; Guo, P.; Wang, S. Multi-scale structures and functional properties of starches from indica hybrid, japoninca and waxy rice. *Int. J. Biol. Macromol.* **2017**, *102*, 136–143. [CrossRef]
17.   Cheng, F.M.; Zhong, L.J.; Zhao, N.C.; Liu, Y.; Zhang, G.P. Temperature induced changes in the starch components and biosynthetic enzymes of two rice varieties. *Plant Growth Regul.* **2005**, *46*, 87–95. [CrossRef]
18.   Zhao, C.F.; Yue, H.L.; Huang, S.J.; Zhou, L.H.; Zhao, L.; Zhang, Y.D.; Chen, T.; Zhu, Z.; Zhao, Q.Y.; Yao, S.; et al. Eating quality and physicochemical properties in Nanjing rice varieties. *Sci. Agric. Sin.* **2019**, *52*, 909–920. (In Chinese)
19.   Gao, H.; Ma, Q.; Li, G.; Yang, X.; Li, X.; Yin, C.; Li, M.; Zhang, Q.; Zhang, H.; Wei, H. Effect of nitrogen application rate on cooking and eating qualities of different growth-development types of japonica rice. *Sci. Agric. Sin.* **2010**, *43*, 4543–4552.
20.   Zhu, D.W.; Zhang, H.C.; Guo, B.W.; Xu, K.; Dai, Q.G.; Wei, H.Y.; Gao, H.; Hu, Y.J.; Cui, P.Y.; Huo, Z.Y. Effects of nitrogen level on yield and quality of japonica soft super rice. *J. Integr. Agric.* **2017**, *16*, 1018–1027. [CrossRef]
21.   Wopereis-Pura, M.M.; Watanabe, H.; Moreira, J.; Wopereis, M.C.S. Effect of late nitrogen application on rice yield, grain quality and profitability in the Senegal river valley. *Eur. J. Agron.* **2002**, *17*, 191–198. [CrossRef]
22.   Wang, Q.; Huang, J.L.; He, F.; Cui, K.H.; Zeng, J.M.; Nie, L.X.; Peng, S.B. Head rice yield of "super" hybrid rice Liangyoupeijiu grown under different nitrogen rates. *Field Crops Res.* **2012**, *134*, 71–79. [CrossRef]
23.   Tan, K.H. *Soil Sampling, Preparation, and Analysis*; CRC Press: Boca Raton, FL, USA, 2005.
24.   Zhou, C.C.; Huang, Y.C.; Jia, B.Y.; Wang, Y.; Wang, Y.; Xu, Q.; Li, R.F.; Wang, S.; Dou, F.G. Effects of cultivar, nitrogen rate, and planting density on rice-grain quality. *Agronomy* **2018**, *8*, 246. [CrossRef]
25.   Leesawatwong, M.; Jamjod, S.; Kuo, J.; Dell, B.; Rerkasem, B. Nitrogen fertilizer increases seed protein and milling quality of rice. *Cereal Chem.* **2005**, *82*, 588–593. [CrossRef]
26.   Champagne, E.T.; Bett-Garber, K.L.; Thomson, J.L.; Fitzgerald, M.A. Unravelling the impact of nitrogen nutrition on cooked rice flavor and texture. *Cereal Chem.* **2009**, *86*, 274–280. [CrossRef]
27.   Gunaratne, A.; Sirisena, N.; Ratnayaka, U.K.; Ratnayaka, J.; Kong, X.L.; Arachchi, L.P.V.; Corke, H. Effect of fertiliser on functional properties of flour from four rice varieties grown in Sri Lanka. *J. Sci. Food Agric.* **2011**, *91*, 1271–1276. [CrossRef] [PubMed]
28.   Martin, M.; Fitzgerald, M.A. Proteins in rice grains influence cooking properties. *J. Cereal Sci.* **2002**, *36*, 285–294. [CrossRef]
29.   Yang, X.Y.; Bi, J.G.; Gilbert, R.G.; Li, G.H.; Liu, Z.H.; Wang, S.H.; Ding, Y.F. Amylopectin chain length distribution in grains of japonica rice as affected by nitrogen fertilizer and genotype. *J. Cereal Sci.* **2016**, *71*, 230–238. [CrossRef]
30.   Zhu, D.W.; Zhang, H.C.; Guo, B.W.; Xu, K.; Dai, Q.G.; Wei, C.X.; Zhou, G.S.; Huo, Z.Y. Effects of nitrogen level on structure and physicochemical properties of rice starch. *Food Hydrocoll.* **2017**, *63*, 525–532. [CrossRef]
31.   Singh, N.; Nisha Pal Mahajan, G.; Singh, S.; Shevkani, K. Rice grain and starch properties: Effects of nitrogen fertilizer application. *Carbohydr. Polym.* **2011**, *86*, 219–225. [CrossRef]
32.   Blazek, J.; Gilbert, E.P. Application of small-angle X-ray and neutron scattering techniques to the characterisation of starch structure: A review. *Carbohydr. Polym.* **2011**, *85*, 281–293. [CrossRef]
33.   Bao, J.S. Toward understanding the genetic and molecular bases of the eating and cooking qualities of rice. *Cereal Foods World* **2012**, *57*, 148–156. [CrossRef]
34.   Balindong, J.L.; Ward, R.M.; Liu, L.; Rose, T.J.; Pallas, L.A.; Ovenden, B.W.; Snell, P.J.; Waters, D.L.E. Rice grain protein composition influences instrumental measures of rice cooking and eating quality. *J. Cereal Sci.* **2018**, *79*, 35–42. [CrossRef]
35.   Kong, X.L.; Zhu, P.; Sui, Z.Q.; Bao, J.S. Physicochemical properties of starches from diverse rice cultivars varying in apparent amylose content and gelatinisation temperature combinations. *Food Chem.* **2015**, *172*, 433–440. [CrossRef] [PubMed]

36. Li, Z.H.; Kong, X.L.; Zhou, X.R.; Zhong, K.; Zhou, S.M.; Liu, X.X. Characterization of multi-scale structure and thermal properties of Indica rice starch with different amylose contents. *RSC Adv.* **2016**, *6*, 107491. [CrossRef]
37. Takahashi, T.; Fujita, N. Thermal and rheological characteristics of mutant rice starches with widespread variation of amylose content and amylopectin structure. *Food Hydrocoll.* **2017**, *62*, 83–93. [CrossRef]
38. Pang, Y.L.; Ali, J.; Wang, X.Q.; Franje, N.J.; Revilleza, J.E.; Xu, J.L.; Li, Z.K. Relationship of rice grain amylose, gelatinization temperature and pasting properties for breeding better eating and cooking quality of rice Varieties. *PLoS ONE* **2016**, *11*, e0168483. [CrossRef]

Article

Relationships Linking the Colour and Elemental Concentrations of Blossom Honeys with Their Antioxidant Activity: A Chemometric Approach

Monika Kędzierska-Matysek [1], Anna Teter [1], Małgorzata Stryjecka [2], Piotr Skałecki [1], Piotr Domaradzki [1], Michał Rudaś [3] and Mariusz Florek [1,*]

[1] Department of Quality Assessment and Processing of Animal Products, University of Life Sciences in Lublin, Akademicka 13, 20-950 Lublin, Poland; monika.matysek@up.lublin.pl (M.K.-M.); anna.wolanciuk@up.lublin.pl (A.T.); piotr.skalecki@up.lublin.pl (P.S.); piotr.domaradzki@up.lublin.pl (P.D.)

[2] Institute of Agricultural Sciences, State School of Higher Education in Chełm, Pocztowa 54, 22-100 Chełm, Poland; mstryjecka@pwsz.chelm.pl

[3] Central Laboratory of Research, University of Life Sciences in Lublin, Głęboka 30 D, 20-612 Lublin, Poland; michal.rudas@up.lublin.pl

* Correspondence: mariusz.florek@up.lublin.pl; Tel.: +48-81-445-6621

Abstract: The antioxidant activity of honey depends on the botanical origin, which also determines their physicochemical properties. In this study, a multivariate analysis was used to confirm potential relationships between the antioxidant properties and colour parameters, as well as the content of seven elements in five types of artisanal honey (rapeseed, buckwheat, linden, black locust, and multifloral). The type of honey was found to significantly influence most of its physicochemical properties, colour parameters, and the content of potassium, manganese and copper. Antioxidant parameters were shown to be significantly positively correlated with redness and concentrations of copper and manganese, but negatively correlated with the hue angle and lightness. The principal component analysis confirmed that the darkest buckwheat honey had the highest antioxidant activity in combination with its specific colour parameters and content of antioxidant minerals (manganese, copper and zinc). The level of these parameters can be potentially used for the identification of buckwheat honey.

Keywords: honey; physicochemical properties; colour; minerals; trace elements; ferric reducing-antioxidant power assay; radical scavenging activity

Citation: Kędzierska-Matysek, M.; Teter, A.; Stryjecka, M.; Skałecki, P.; Domaradzki, P.; Rudaś, M.; Florek, M. Relationships Linking the Colour and Elemental Concentrations of Blossom Honeys with Their Antioxidant Activity: A Chemometric Approach. *Agriculture* **2021**, *11*, 702. https://doi.org/10.3390/agriculture11080702

Academic Editor: Alessandra Durazzo

Received: 26 June 2021
Accepted: 24 July 2021
Published: 26 July 2021

Publisher's Note: MDPI stays neutral with regard to jurisdictional claims in published maps and institutional affiliations.

1. Introduction

Bee products are valuable, natural foodstuffs with a wide range of beneficial properties for human health which are exploited in medicine [1]. Honey is the main product of extensive beekeeping in almost all countries worldwide. Owing to the biodiversity of plants and monocultures, multifloral and unifloral honeys are obtained [2,3].

The basis for the multifaceted use of honey in the human diet and therapy is its complex chemical composition, dominated by carbohydrates (70–80%). In the honeys, enzymes, amino acids, vitamins, carotenoids, organic acids, phenolic acids, polyphenols, and flavonoids are present as well [4]. Honey also contains minerals essential for the functioning of the human body, with the content in fresh honey varying depending on the type, geographic origin, and method of honey harvesting and storage [3,5–8]. Excessively long storage affects the composition of honey and alters the biological activity of its components [3,7]. It should be underlined that honey also demonstrates antioxidant [2,9–11], immunomodulatory, and anti-inflammatory properties [12]. In view of the fact that the ageing process and degenerative diseases have their basis in free radicals, one way to counteract these changes and protect against free-radical diseases is to supply the body with appropriate antioxidants, which reduce the number of superoxide radicals in varying degrees. Substances with a beneficial effect on the human body include natural antioxidants counted among phenolic compounds [13].

Honey intended for human consumption must meet key criteria for content of fructose and glucose (sum of both), sucrose, moisture, water-insoluble content, electrical conductivity, free acid, diastase activity, and hydroxymethylfurfural [14]. One of the most important criteria is the level of 5-hydroxymethylfurfural (HMF $\leq$ 40 mg kg^{-1}), due to the potential risk for bees and humans. Another parameter is the enzymatic activity of α-amylase (diastase activity $\geq$8.0 on the Schade scale) naturally occurring in honey. A high diastase number (DN) and low content of 5-HMF are considered a guarantee of high-quality honey. Negative changes in honey are caused by the effects of high temperature during decrystallization and long-term storage in combination with high temperature and exposure to light [15–17].

The botanical origin of honey determines its diversity in terms of physicochemical, organoleptic and biological properties [5–7]. The type of honey is distinguished by a wide range of colours, a characteristic flavour and aroma, chemical composition, form of crystallization, and specific preventive properties against individual disease conditions. Owing to these varied attributes, each consumer can choose honey according to individual needs and preferences. At the same time, due to the rich content of different sugars, honey can be a healthy substitute for white sugar [10].

Multivariate analysis (e.g., principal component analysis, PCA; hierarchical cluster analysis, HCA; linear discriminate analysis, LDA) were previously very often used to evaluate and/or classify honey in relation to its chemical composition, physicochemical or biological properties. Numerous papers have confirmed the suitability of this method for honey evaluation. [2,3,6–8,18].

The antioxidant activity of honey depends on the botanical origin, which also determines its physicochemical properties. This study had two main objectives: (1) to evaluate and compare the physicochemical properties and antioxidant activity of five Polish blossom and multifloral honeys; (2) to identify/investigate potential relationships between the antioxidant properties of the honeys and their colour parameters and content of macro- and micro-elements using principal component analysis (PCA).

2. Materials and Methods

2.1. Sampling

The study was conducted on 63 honeys harvested in south-eastern Poland (Lublin region) in 2019. The percentage of the predominant pollen grains in the honey samples was determined according to the method recommended by Polish law concerning analytical methods used in the assessment of honey [19]. The following types of honey were distinguished: rapeseed, RS (*Brassica napus* L., n = 10), buckwheat, BW (*Fagopyrum esculentum* Moench, n = 8), linden, LI (*Tilia spp.* L., n = 13), black locust, AC (*Robinia pseudoacacia* L., n = 5), and multifloral, MF (no dominant pollen, n = 27). The honeys were purchased directly from beekeepers and stored in the dark at room temperature 20–25 °C.

2.2. Analyses

2.2.1. Physicochemical Properties

Water content was determined with a refractometer (Abbe Carl Zeiss, Jena, Germany) based on the refractive index of the honey in its liquid state. Water percentage by weight (% m/m) was read from the table as corresponding to the refractive index [20]. The content of reducing sugars and sucrose was determined by the Lane-Eynon method, according to the Polish Committee for Standardization (PN-88/A-77626. Bee honey).

The electrical conductivity (mS cm^{-1}) and pH of the honey were determined according to [20] using a pIONneer 65 Meter (Radiometer Analytical, Villeurbanne, CEDEX-France) with a 4-pole conductivity cell (CDC 30T) and a combined pH electrode (E16M340). The free acidity of the honey was measured by potentiometric titration using a 0.1 M NaOH solution to obtain pH 8.30 and the result was expressed in milliequivalents per kilogram of honey (mval kg^{-1}).

The water activity (aW) of fresh honey was performed using a Rotronic HygroLab C1 analyser (Bassersdorf, Switzerland) equipped with two HC2-AW measurement heads. Duplicate measurements were taken in the AWQ mode with stabilization set to 15 min after previously conditioning the honey samples at room temperature ($20 \pm 1\ ^\circ$C).

The concentration of 5-HMF (5-(hydroxymethyl-)furan-2-carbaldehyde) in mg kg^{-1} was determined according to [21]. The absorbance of a clear solution of honey with water relative to a solution of honey with sodium bisulphate was measured at wavelengths 284 and 336 nm. The measurement was made with a Carry 300 Bio spectrophotometer (Varian Australia Pty, Ltd., Mulgrave, Australia).

Diastase activity (diastase number DN, in Schade units per gram of honey) was determined by photometry using Phadebas tablets (Honey Diastase Test, Magle AB, Lund, Sweden). They contained non-soluble starch conjugated with a blue pigment and hydrolysed by the amylase present in the sample. The resulting water-soluble fragments of the starch chain were dyed blue. The absorbance of the coloured solution was measured with a Varian Carry 300 Bio spectrophotometer (Varian Australia PTY, Ltd.) at 620 nm [20].

The honey's colour was measured according to the procedure of [15] with some modifications. Briefly, the colour of three aliquots of each honey sample (previously equilibrated to room temperature) was measured three times in a round optical glass cell CR-A504 (diameter 34 mm) using a portable CM-600d spectrophotometer (Konica Minolta Sensing, Inc., Osaka, Japan) equipped with cell holder CM-A515. The thickness of the honey layer was 20 mm. Samples of multifloral, rapeseed, buckwheat and linden honey were crystallized naturally (set honey), and the black locust samples were in liquid form (strained honey). The results of the measurements (illuminant D65, observer 10°) were given in the CIE L*a*b* colour space, including the following spectral values: L* (lightness axis), a* (red to green axis), b* (yellow to blue axis), C* (saturation) and h° (hue angle/tint).

For mineral analysis, samples were prepared for mineralization as follows: a 6 mL volume of 65% nitric acid (Suprapur grade; Merck, Germany) was poured over honey samples weighed out to within 0.0001 g and the certified reference material NCS ZC 73014 Tea (to verify the method) in vessels (PFA). All solutions together with a blank sample were mineralized in a MARSXpress 5 microwave digester (CEM Corporation, Matthews, NC, USA). The oven was programmed for mineralization of the samples as follows: power—1600 W/100% max power; temperature increment—20 min/200 °C; holding time—20 min. Then the mineralized samples were transferred to volumetric flasks using ultrapure water produced in an HLP 20UV demineralizer (HYDROLAB, Poland). Schinkel buffer (enth./cont. 10 g L^{-1} CsCl + 100 g L^{-1} La; Merck, Germany) was used to minimize interference during analysis (Mg, K and Na).

K, Na, Mg, Zn, Fe, Mn, and Cu were determined according to the procedure of [3] using a Varian AA240FS spectrometer (Fast Sequential Atomic Absorption Spectrometer, Varian Australia Pty Ltd., Mulgrave, Australia). The elements were atomized in the flame of a burner fed with a mixture of air (oxidizing gas, flow 13 L min^{-1}) and acetylene (combustible gas, flow 2.0 L min^{-1}). The following parameters were used: instrument mode—absorbance; measurement mode—integration; calibration mode—concentration; calibration algorithm—New Rational. Analytical wavelengths (nm): Mg 285.2, Zn 213.9, Fe 248.3, Mn 279.5, Cu 324.7, Na 589.0, and K 766.5. Background correction was used in the determination of Mg, Zn, Fe, Mn and Cu. To plot a standard curve, single-element standard solutions (Merck, Germany) were used for each element (K, Na, Mg, Zn, Cu, Fe and Mn) with a mass concentration of 1000 mg L^{-1}. The following limits of detection (LOD) were used in the analysis: 0.01 mg kg^{-1} for Na, Zn, Mn and Cu; 0.04 mg kg^{-1} for K; 0.09 mg kg^{-1} for Fe; and 0.47 mg kg^{-1} for Mg.

2.2.2. Antioxidant Activity

The capacity of the honeys to scavenge the stable free radical 2,2-diphenyl-1-picrylhydrazyl (DPPH, Sigma Aldrich Co., St. Louis, MO, USA) was tested according to the method of [22]. Samples of the honeys (2 g) were dissolved in 10 mL of distilled water (HLP 20UV, HYDROLAB, Straszyn, Poland), then an aliquot (0.2 mL) of each dilution was mixed with a 1.8 mL solution of 0.1 mM DPPH in methanol (Sigma Aldrich Co., St. Louis, MO, USA). The reaction mixture was left in the dark at room temperature for 60 min. Next, the absorbance of the mixture was measured spectrophotometrically (UV-2600i spectrophotometer, Shimadzu, Japan) at 517 nm against methanol as a blank. All the determinations were performed in triplicates. The calibration curve was plotted in a range from 0.1 to 100 μg mL^{-1} of Trolox (Sigma-Aldrich, Co., St. Louis, MO, USA) solution in ethanol. The results were expressed as mM of Trolox equivalent (TE) per 1 kg of honey (mM TE kg^{-1} honey).

The total ferric reducing antioxidant power assay (FRAP assay) was performed according to [23]. FRAP reagent was prepared by mixing 25 mL of 0.3 M acetate buffer (pH 3.6) with a 2.5 mL solution of 10 mM TPTZ (Sigma Aldrich Co, St. Louis, MO, USA) in 40 mM HCl, and 2.5 mL of 20 mM ferric chloride (Sigma Aldrich Co, USA). An aliquot (0.2 mL) of each honey dilution (1 g in 10 mL distilled water) was mixed with 1.8 mL of FRAP reagent. The resulting mixture was then pre-warmed at 37 °C for 10 min. The absorbance was measured spectrophotometrically (UV-2600i spectrophotometer, Shimadzu, Japan) at 593 nm against a blank that was prepared with distilled water. All the determinations were performed in triplicates. A calibration curve was prepared using the ethanol solution of Trolox (Sigma Aldrich Co., St. Louis, MO, USA) in a range from 25 to 300 nmol mL^{-1}. FRAP values were expressed as mM of Trolox equivalent per kg of honey (mM TE kg^{-1}).

2.3. Statistical Analysis

Statistical analysis of the results was performed in Statistica ver. 13 (TIBCO Software Inc., Palo Alto, CA, USA). One-way analysis of variance (ANOVA) followed by Tukey's (HSD) test was used to compare means of physicochemical traits and colour parameters between honey types (multifloral, rapeseed, buckwheat, linden and black locust). The normality and homogeneity of the variance of minerals were verified by the Kolmogorov–Smirnov test and Levene's F-test, respectively. The influence of the honey type on the elemental concentrations was verified by the Kruskal–Wallis test (comparison of multiple independent groups). Differences between means at confidence levels of 95% and 99% ($p < 0.05$ and $p < 0.01$, respectively) were considered statistically significant. The mean and standard deviation are presented in the tables. The relationship between parameters of antioxidant activity (FRAP and DPPH) and physicochemical traits and mineral contents in the honeys was determined by calculating Pearson's correlation coefficients (r) and Spearman's rank correlation coefficients (rS), respectively. The correlations were further verified by principal component analysis (PCA), separately for two data sets (for antioxidant activity and colour and for antioxidant activity and minerals) to demonstrate the diversity among honey types.

3. Results and Discussion

3.1. Physicochemical Properties

The honey type significantly influenced most of its physicochemical parameters (except the content of water, saccharose and reducing sugars), all colour parameters, and the concentrations of K, Mn and Cu (Table 1).

Although the water content in the honeys did not differ significantly, the value of this parameter indicates the maturity of the honey (readiness for harvest). The mean water content was similar, ranging from 17.11% in the linden honey to 17.93% in the rapeseed honey, and thus did not exceed the maximum acceptable content of 20% [14]. Sugars make up the largest proportion of dry matter in honey, and their qualitative and quantitative composition is an important criterion in the quality assessment of honey. The most important factors for determining the sugar composition of honey include the region of origin, climate conditions, and types of flowers used by the bees [24]. In the present study, the highest content of reducing sugars and the lowest content of saccharose was noted in the black locust honey (77.24% and 1.83%), while the reverse pattern was noted in the linden honey (73.76% and 2.14%). It should be stressed that the saccharose content in the honey did not exceed the acceptable limit of 5 g/100 g of product.

Table 1. Parameters of physicochemical traits and antioxidant activity of the honey types (mean values $\pm$ standard deviation).

Parameters	Multifloral (MF)	Rapeseed (RS)	Buckwheat (BW)	Linden (LI)	Black Locust (AC)
Water (%)	17.53 $\pm$ 1.01 [a]	17.93 $\pm$ 0.89 [a]	17.83 $\pm$ 1.17 [a]	17.11 $\pm$ 1.23 [a]	17.90 $\pm$ 0.58 [a]
Reducing sugars (%)	74.12 $\pm$ 3.18 [a]	74.60 $\pm$ 1.40 [a]	74.93 $\pm$ 2.42 [a]	73.76 $\pm$ 2.39 [a]	77.24 $\pm$ 1.00 [a]
Sucrose (%)	2.09 $\pm$ 0.28 [a]	2.04 $\pm$ 0.13 [a]	2.02 $\pm$ 0.22 [a]	2.14 $\pm$ 0.23 [a]	1.83 $\pm$ 0.09 [a]
pH	3.80 $\pm$ 0.15 [a]	3.83 $\pm$ 0.10 [a]	3.69 $\pm$ 0.17 [a]	3.91 $\pm$ 0.16 [b]	3.68 $\pm$ 0.12 [a]
Free acidity (mval kg^{-1})	32.1 $\pm$ 9.9 [AB]	21.9 $\pm$ 4.2 [A]	44.5 $\pm$ 2.6 [C]	35.1 $\pm$ 9.9 [BC]	29.1 $\pm$ 10.3 [AB]
Electrical conductivity (mS cm^{-1})	0.390 $\pm$ 0.145 [AB]	0.232 $\pm$ 0.037 [A]	0.448 $\pm$ 0.099 [AB]	0.564 $\pm$ 0.163 [B]	0.304 $\pm$ 0.149 [AB]
a_W	0.567 $\pm$ 0.026 [AB]	0.582 $\pm$ 0.017 [B]	0.570 $\pm$ 0.017 [AB]	0.546 $\pm$ 0.023 [A]	0.552 $\pm$ 0.028 [AB]
5-HMF (mg kg^{-1})	5.50 $\pm$ 2.77 [A]	3.67 $\pm$ 3.81 [A]	14.51 $\pm$ 8.54 [B]	3.49 $\pm$ 2.27 [A]	4.20 $\pm$ 1.69 [A]
Diastase number (Schade unit)	27.09 $\pm$ 7.44 [A]	18.58 $\pm$ 5.98 [A]	49.40 $\pm$ 17.66 [B]	29.04 $\pm$ 7.50 [A]	21.45 $\pm$ 4.86 [A]
Colour CIE					
L*	44.12 $\pm$ 7.28 [C]	49.47 $\pm$ 6.60 [D]	30.84 $\pm$ 2.44 [A]	35.64 $\pm$ 3.54 [B]	31.55 $\pm$ 5.42 [AB]
a*	2.36 $\pm$ 1.48 [B]	0.90 $\pm$ 0.36 [A]	3.22 $\pm$ 0.92 [C]	0.88 $\pm$ 0.51 [A]	0.61 $\pm$ 0.28 [A]
b*	16.05 $\pm$ 5.84 [B]	13.29 $\pm$ 2.03 [B]	6.91 $\pm$ 2.90 [A]	8.74 $\pm$ 2.38 [A]	5.09 $\pm$ 2.40 [A]
C*	16.25 $\pm$ 5.96 [B]	13.08 $\pm$ 2.27 [B]	7.64 $\pm$ 2.93 [A]	8.86 $\pm$ 2.36 [A]	5.14 $\pm$ 2.39 [A]
h°	81.97 $\pm$ 3.45 [B]	85.94 $\pm$ 1.39 [C]	62.70 $\pm$ 6.96 [A]	84.40 $\pm$ 3.62 [BC]	82.38 $\pm$ 4.17 [B]
Minerals (mg kg^{-1})					
K	1055.55 $\pm$ 479.47 [ab]	700.94 $\pm$ 399.30 [a]	1155.26 $\pm$ 359.17 [ab]	1258.27 $\pm$ 564.50 [b]	876.26 $\pm$ 435.06 [ab]
Na	29.10 $\pm$ 20.03 [a]	24.21 $\pm$ 12.72 [a]	33.18 $\pm$ 29.98 [a]	25.86 $\pm$ 14.65 [a]	15.09 $\pm$ 1.16 [a]
Mg	28.77 $\pm$ 10.55 [a]	20.15 $\pm$ 4.26 [a]	29.56 $\pm$ 10.36 [a]	27.93 $\pm$ 9.83 [a]	23.19 $\pm$ 9.94 [a]
Zn	1.84 $\pm$ 1.80 [a]	1.26 $\pm$ 0.69 [a]	2.94 $\pm$ 3.03 [a]	1.65 $\pm$ 0.97 [a]	0.88 $\pm$ 0.31 [a]
Fe	1.66 $\pm$ 0.68 [a]	1.26 $\pm$ 0.49 [a]	1.85 $\pm$ 0.92 [a]	1.71 $\pm$ 0.62 [a]	1.39 $\pm$ 0.97 [a]
Mn	1.77 $\pm$ 1.56 [ab]	1.37 $\pm$ 1.50 [ab]	5.58 $\pm$ 4.29 [b]	1.22 $\pm$ 1.36 [ab]	0.72 $\pm$ 0.62 [a]
Cu	0.56 $\pm$ 0.16 [AB]	0.48 $\pm$ 0.17 [A]	0.95 $\pm$ 0.37 [B]	0.52 $\pm$ 0.09 [AB]	0.47 $\pm$ 0.07 [A]
Antioxidant activity					
DPPH (mM TE kg^{-1})	1.12 $\pm$ 0.44 [BC]	0.83 $\pm$ 0.12 [A]	2.33 $\pm$ 0.36 [C]	0.90 $\pm$ 0.16 [AB]	0.63 $\pm$ 0.17 [A]
FRAP (mM TE kg^{-1})	1.28 $\pm$ 0.49 [B]	0.47 $\pm$ 0.10 [A]	2.14 $\pm$ 0.27 [C]	0.53 $\pm$ 0.15 [A]	0.29 $\pm$ 0.01 [A]

a_W, water activity; 5-HMF, 5-(hydroxymethyl-)furan-2-carbaldehyde; L*, lightness; a*, redness; b*, yellowness; C*, saturation; h°, hue angle; K, potassium; Na, sodium; Mg, magnesium; Fe, iron; Zn, zinc; Mn, manganese; Cu, copper; DPPH, scavenging capacity; FRAP, ferric reducing antioxidant power. Means with different letters in rows differ significantly according to Tukey's test: [a],[b]—$p < 0.05$; [A],[B]—$p < 0.01$.

All the honeys had a pH < 4. Although regulations do not specify the pH of honey, it is worth noting that honey pH between 3.2 and 4.5, together with titratable acidity, inhibits the growth of microbes [4]. The higher range of pH for the same honey types in this study (from various regions of Poland), i.e., from 4.07 (buckwheat) to 4.23 (linden) was reported earlier [6]. Lower pH than in the present study was reported for Indian blossom honey (3.5) [7], while higher pH (4.38) was found in Slovakian honey [25].

The honey type significantly influenced its degree of acidity, which ranged from 21.9 mval kg^{-1} (rapeseed) to 44.5 mval kg^{-1} (buckwheat), without exceeding the limit in EU regulations [14], which is 50 mequivalents acid per kg. Acidity results mainly from the presence of organic acids, amino acids and phenolic acids in honey, as well as from processes taking place during its maturation [13]. Free acidity has been shown to remain practically constant in honey, but with an increase observed during storage for 20 months at room temperature [26] or frozen storage for 18 months [15].

Specific conductivity is a parameter that can be used to determine the botanical origin of honey, i.e., to distinguish nectar honey (up to 0.8 mS cm^{-1}) from honeydew honey (over 0.8 mS cm^{-1}). This was also confirmed in the present study, as the specific conductivity of the nectar honeys varied significantly ($p < 0.01$) from 0.232 mS cm^{-1} (rapeseed) to 0.564 mS cm^{-1} (linden). The specific conductivity of Slovakian honey averaged

0.6515 mS cm^{-1}, ranging from 0.1345 to 0.9912 mS cm^{-1} [25]. Multifloral and acacia honeys from India exhibited lower conductivity (0.25–0.26 mS cm^{-1}) [7].

The aW of honey is usually between 0.50 and 0.65, and in the literature, aW values above 0.60 are considered to represent a critical threshold for microbial stability [4] due to the activity of various species of bacteria and osmophilic yeasts resulting in fermentation [12]. In the present study, the honeys did not exceed the critical value for this parameter, but it varied significantly ($p < 0.01$) from 0.546 (linden honey) to 0.582 (rapeseed honey). These honey types also had the lowest (17.11%) and highest (17.93%) water content, which indicates a linear relationship between the two parameters, statistically confirmed in the present study (r = 0.616, p = 0.000). Nonetheless, due to the significant influence of temperature, honey type (flower or honeydew), harvesting year, geographical and botanical origin, a universal linear equation for water activity and moisture content could not be established [27]. The wider range of water activity in honeys ($0.456 \leq$ aW ≤ 0.659) was linked to the varied composition of sugars [25]. Water activity in honey from India ranged from 0.507 to 0.566 [7], and in flower honey from Turkey was between 0.51 and 0.69 [24]. Low water activity (aW) together with low pH, low protein content, and high osmotic pressure has an inhibitory effect on the development of bacteria [13].

An important indicator of the quality and health safety of honey is the content of 5-hydroxymethylfurfural (5-HMF). The values for this parameter in the present study were low, well under the limit of 40 mg kg^{-1} [14]. Significantly ($p < 0.01$) the highest content of 5-HMF was noted in the buckwheat honey (14.51 mg kg^{-1}); it was more than four times as high as in the linden (3.49 mg kg^{-1}) and rapeseed (3.67 mg kg^{-1}) honeys ($p < 0.01$). The 5-HMF content between 0 and 4.12 mg kg^{-1} was reported for Turkish flower honey [24], which was considered fresh honey. Higher 5-HMF content was found in Slovakian honeys 25.76 mg kg^{-1} [25], while the level in Indian honeys ranged from 5.49 mg kg^{-1} in acacia honey to 22.64 mg kg^{-1} in multifloral honey [7]. For Andalusian multifloral honey, the content of 5-HMF was found between 0.19 and 41.16 mg kg^{-1} [28].

The diastase number (DN) indicates the amylolytic activity of honey. Alpha-amylase (diastase) is an enzyme that takes part in the hydrolytic degradation of complex sugars. Like 5-HMF, the diastatic activity of honey can be used as an indicator of adulteration, ageing, overheating (increased temperature), and the degree of preservation [16]. All honey types tested in the present study had a diastase number >8, i.e., above the recommended minimum [14]. DN values obtained for Polish regional honeys [29] confirmed the low enzymatic activity of rapeseed (15.32 DN) and higher activity in multifloral (22.15 DN) and linden (31.99 DN) honey. In turn, the diastase number of polyfloral honeys from Andalusia varied widely from 6.05 to 40.89 [28].

The honey type significantly ($p < 0.01$) influenced all parameters in the instrumental assessment of colour (Table 1). The highest lightness value was noted for the rapeseed honey (L* = 49.47), followed by the multifloral honey (L* = 44.12), while the darkest was black locust honey (L* = 31.55) and buckwheat honey (L* = 30.84). Buckwheat honey also had the highest value for the colour red (a* = 3.22) and the lowest hue angle (h° = 62.70). The hue angle in the other honey types ranged from h° = 81.97 (multifloral) to h° = 85.94 (rapeseed). Hue angle is defined as starting at the +a* axis and is expressed in degrees: 0° is red (+a*), 90° is yellow (+b*), 180° is green (−a*), and 270° is blue (−b*). Values between 0° and 90° are a mixture of red and yellow, resulting in an orange colour.

The amount of red was similar for the black locust, linden and rapeseed honeys ($0.61 \leq$ a* ≤ 0.90), in contrast to the multifloral honey (a* = 2.36) and the aforementioned buckwheat honey (a* = 3.22). In the case of the colour yellow and saturation, the varieties can be divided into two groups with similar values for these parameters, i.e., black locust, buckwheat and linden ($5.09 \leq$ b* ≤ 8.74 and $5.14 \leq$ C* ≤ 8.86) vs. rapeseed and multifloral ($13.29 \leq$ b* ≤ 16.05 and $13.08 \leq$ C* ≤ 16.25).

A wider range of values for colour components for Polish honey was reported earlier [6], respectively: $26 \leq$ L* ≤ 51; $-3.41 \leq$ a* ≤ 7.9; and $5.8 \leq$ b* ≤ 23.7. The differences may have been due to differences in the preparation of samples, which were heated and homogenized prior to analysis. Nectar contains natural plant pigments such as carotenoids, anthocyanins, flavonoids, and chlorophyll, which determine the colour of the honey through various amounts of colours, including yellow, red, brown and green. The colour of the product is also influenced by honey colloids, polyphenols (e.g., tannins), and melanoidins. Phenolic compounds function not only as pigments but also as antioxidants, insecticides and fungicides [6]. The colour of honey is also largely determined by its degree of crystallization and the conditions in which physicochemical changes take place during storage.

In the present study, the dark buckwheat honey (L* = 30.84) had a high free acid value (44.5 mval kg^{-1}), while the light rapeseed honey (L* = 49.47) had significantly ($p < 0.01$) the lowest acidity (21.9 mval kg^{-1}). Dark honeys have been shown to have much higher acidity than light ones [9]. Kaczmarek et al. [6] also reported higher levels of free acids in buckwheat (34.25 meq kg^{-1}), multifloral (34.04 meq kg^{-1}) and linden (31.09 meq kg^{-1}) honey than in acacia (12.8 meq kg^{-1}) and rapeseed (10.5 meq kg^{-1}) honey.

In the present study, the highest content of elements was generally found in the buckwheat honey and the lowest in the black locust and rapeseed honey (Table 1). The type significantly influenced the content of Mn and Cu ($p < 0.01$) and K ($p < 0.05$). The highest content of potassium was noted in the linden honey (1258.3 mg kg^{-1}) and the lowest in the rapeseed honey (700.9 mg kg^{-1}). The potassium content in the honey varieties was similar to that obtained in honey from Poland (892.4 mg kg^{-1}) [3] and Malaysia (904.9 mg kg^{-1}) [30], but lower than in Hungarian honey (397.88 mg kg^{-1}) [31].

The buckwheat honey had significantly the highest content of Mn and Cu (5.58 and 0.95 mg kg^{-1}). The black locust honey had the lowest level of manganese (0.72 mg kg^{-1}), and the black locust and rapeseed honeys had the lowest content of copper (0.47–0.48 mg kg^{-1}). Wieczorek et al. [32] reported lower values for K (233–782 mg kg^{-1}), Mg (11.6–24 mg kg^{-1}), Mn (0.37–1.25 mg kg^{-1}), and Cu (0.04–0.06 mg kg^{-1}) in multifloral, linden and black locust honeys, but higher values for Zn (1.65–6.20 mg kg^{-1}) and Fe (1.9–4.0 mg kg^{-1}) compared to the results of our study, which in turn were in agreement with those obtained by other authors [33] for varietal honeys from Podkarpacie in Poland. They noted the highest content of Zn, Mn and Cu in buckwheat honey, in comparison with linden, rapeseed and multifloral honeys. Our previous research [3] also indicated a higher concentration of these elements in buckwheat honey. For this reason, Deng et al. [34] suggest that the levels of Mn, Zn and Cu in buckwheat honey can potentially be used to distinguish this variety of honey. Many authors also stressed that dark honeys have higher mineral content than light ones [23,30], as well as a stronger flavour [34] and higher content of phenolic compounds, which influence antioxidant activity [7].

3.2. Antioxidant Activity

The buckwheat honey had the highest content of elements considered to be antioxidants (Zn, Mn and Cu). This is also indicated by the results for antioxidant activity measured in the reaction with DPPH and by the FRAP value (Table 1). Significantly ($p < 0.01$) the highest antioxidant activity was noted in the buckwheat honey (DPPH 2.33 mM TE kg^{-1} and FRAP 2.14 mM TE kg^{-1}), and the lowest activity in the black locust (DPPH 0.63 mM TE kg^{-1} and FRAP 0.29 mM TE kg^{-1}), rapeseed (DPPH 0.83 mM TE kg^{-1} and FRAP 0.47 mM TE kg^{-1}) and linden (DPPH 0.90 mM TE kg^{-1} and FRAP 0.53 mM TE kg^{-1}) honeys.

The highest antioxidant activity in buckwheat honey from Poland measured in reactions with DPPH and ABTS and the lowest in rapeseed and acacia honeys were reported previously [11]. In the present study, the mean DPPH values in mM TE kg^{-1} in the rapeseed, buckwheat, linden and black locust honeys were about twice as high as those reported earlier [2]. Škrovánková et al. [35] analysed total antioxidant capacity (TAC) using the DPPH reagent and found that it was highest in honeydew honey, followed by multifloral, forest, and floral honeys, with the lowest values noted in rape and acacia honeys. The DPPH activity in buckwheat honey from China was 0.304 mg Trolox/g, and the FRAP value was 0.355 mg Trolox/g [36]. However, it may be difficult to compare our results for antioxidant activity with values reported by other authors due to the differences in their analysis or presentation. Nevertheless, it should be emphasized that many studies indicate that dark honeys show higher activity than light honeys. Piszcz and Głód [37], based on the assessment of total antioxidant potential (TAP), reported the following order for varietal honeys: buckwheat > honeydew > linden > multifloral > acacia. Dżugan et al. [38] found the highest antioxidant activity (DPPH) in buckwheat honey (82.41%) and the lowest in rapeseed honey (21.81%). Other authors [10] also confirmed high antioxidant activity in buckwheat honey (17.0 mmol Trolox/g) compared to soybean, sweet clover, fireweed, and acacia honeys (8.3, 6.1, 3.1 and 3.0 mmol Trolox/g).

3.3. Correlations

To test the relationships between antioxidant capacity expressed as DPPH and FRAP and physicochemical properties, colour parameters, and concentrations of selected elements, Pearson (r) or Spearman (rS) correlations were calculated. In the case of the first group of properties, DPPH and FRAP were shown to be significantly ($p < 0.001$) and positively correlated with 5-HMF (r = 0.643 and r = 0.676), diastase number (r = 0.612 and r = 0.706), and free acids (r = 0.544 and r = 0.560). The only significant negative correlation was between DPPH and pH (r = -0.320, $p < 0.05$).

Antioxidant activity (DPPH and FRAP) was positively correlated with the colour red (r = 0.452 and r = 0.549, $p < 0.001$) and negatively with lightness (r = -0.309 and r = -0.259, $p < 0.05$) and hue (r = -0.781 and r = -0.706, $p < 0.001$) (Table 2).

Table 2. Correlations between antioxidant activity and colour parameters and minerals.

Parameter	FRAP	DPPH
Colour CIE	Pearson's correlation coefficient (r)	
L*	-0.259 *	-0.309 *
a*	0.549 ***	0.452 ***
b*	0.022	-0.127
C*	0.058	-0.091
h°	-0.706 ***	-0.781 ***
Minerals	Spearman's rank correlation coefficient (rS)	
K	0.155	0.268 *
Na	0.173	0.176
Mg	0.156	0.276 *
Fe	0.247	0.218
Zn	0.212	0.127
Cu	0.386 **	0.522 ***
Mn	0.370 **	0.457 ***

* $p < 0.05$; ** $p < 0.01$; *** $p < 0.001$; L*, lightness; a*, redness; b*, yellowness; C*, saturation; h°, hue angle; K, potassium; Na, sodium; Mg, magnesium; Fe, iron; Zn, zinc; Mn, manganese; Cu, copper; DPPH, scavenging capacity; FRAP, ferric reducing antioxidant power.

Kuś et al. [2], for six Polish single-variety honeys, reported higher correlation coefficients for DPPH and FRAP with colour parameters: for L* r = -0.955 and -0.961, for a* r = 0.943 and 0.964, and for b* r = 0.814 and 0.786. Lower correlation coefficients for these parameters were obtained for honey from Slovenia [23]. The intensity of honey colour may be associated with its antioxidant capacity, as the content of phenols, flavonoids and carotenoids is greater in darker honeys than in lighter ones [5]. Moreover, the acceleration of the Maillard reaction or fructose caramelization can contribute to a darker colour of honey through the production of brown pigments, concomitantly with the formation of HMF as an intermediate product. Similar changes in rape honey, consisting in a reduction in lightness (L*) but an increase in redness (a*) and colour intensity (ABS450, mAU) were observed in our previous study [16], which can be directly linked to the presence of pigments such as terpenes, carotenoids, and some flavonoids [7].

High, positive correlation coefficients between colour evaluation (ΔA) and antioxidant capacity were previously reported for ABTS r = 0.8836 and for DPPH r = 0.8937 [35]. Many authors confirm a strongly positive and significant ($p < 0.01$) relationship between colour intensity, expressed as ABS450, and DPPH and FRAP. For instance, Moniruzzaman et al. [39] reported coefficients of r = 0.938 and r = 0.873, and Beretta et al. [40], reported r = 0.889 and r = 0.933. A lower correlation coefficient (r = 0.68) between ABS450 and DPPH was obtained by Kaczmarek et al. [6].

The DPPH and FRAP parameters characterizing antioxidant activity were most strongly and positively correlated with antioxidant minerals: Cu (rS = 0.522, $p < 0.001$ and rS = 0.386, $p < 0.01$) and Mn (rS = 0.457, $p < 0.001$ and rS = 0.370, $p < 0.01$) (Table 2).

To our knowledge, there are few reports on the correlation between the minerals content and antioxidant activity of honeys. In the present study, low and non-significant ($p > 0.05$) correlation coefficients were obtained for parameters of antioxidant activity and elements Na, Fe and Zn. Perna et al. [41] report significant ($0.01 < p < 0.001$) correlation coefficients for DPPH and FRAP with Fe (r = 0.67 and r = 0.73) and Zn (r = 0.32 and r = 0.48) in honey from Italy.

3.4. Principal Component Analysis

3.4.1. Antioxidant Activity and Colour

For a more in-depth analysis of the results obtained for the antioxidant activity and colour of the honey types, a principal component analysis (PCA) was performed, with 7 variables and 64 cases. Two principal components with eigenvalues exceeding 1 (Kaiser criterion) explained 87.12% of the total variance, with PC 1 accounting for 46.58% and PC 2 for 40.53% (Table S1).

Figure 1a visualizes the projection of variables as a two-factor plane (PC1 × PC2). The first component (PC1) has a positive correlation with L* (0.828) and h° (0.846), but a negative correlation with DPPH (−0.723) and a moderate negative correlation with FRAP (−0628). The second component (PC2) has a negative correlation with most variables, including a* (−0.897), C* (−0.740), b* (−0.708), and FRAP (−0.654) (Table 3).

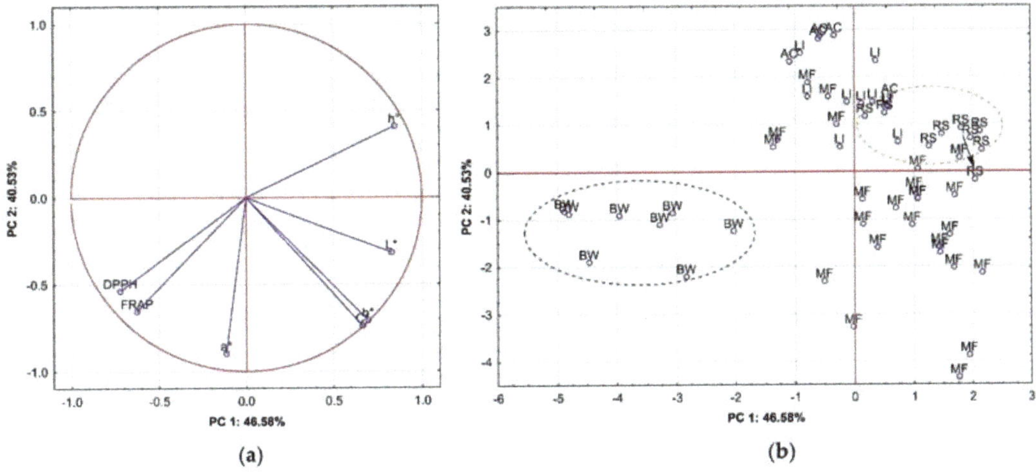

Figure 1. Projection of variables (**a**) and projection of cases (**b**) depending on the botanical origin of the honey in a two-factor plane (PC1 × PC2). (**a**): L*, lightness; a*, redness; b*, yellowness; C*, saturation; h°, hue angle; DPPH, scavenging capacity; FRAP, ferric reducing antioxidant power; (**b**) honey type: RS, rapeseed; BW, buckwheat; LI, linden; AC, black locust; MF, multifloral.

Table 3. Correlations between the principal components and the original variables.

Variable	PC1	PC2
L*	0.828	−0.318
a*	−0.116	−0.897
b*	0.667	−0.708
C*	0.665	−0.740
h°	0.846	0.408
FRAP	−0.628	−0.654
DPPH	−0.723	−0.537

L*, lightness; a*, redness; b*, yellowness; C*, saturation; h°, hue angle; DPPH, scavenging capacity; FRAP, ferric reducing antioxidant power.

As can be seen in Figure 1, three groups of parameters, characterized by their loadings and the length of the directional vectors can be distinguished. The L*, b* and C* variables were distributed in the positive area of PC1 and negative area of PC2, while h° is located in positive areas of both components, and a*, DPPH and FRAP have negative values of PC1 and PC2.

Figure 1b shows the projection of cases depending on the botanical origin of the honey in the coordinate system defined by PC1 × PC2. The buckwheat honey samples are clearly separated and situated in the bottom left area of the plot, i.e., they have negative values of both components. Therefore, the buckwheat honey located in this square of the plot showed the highest values of antioxidant activity and redness (a*), together with the lowest value of h°, which was negatively correlated with FRAP and DPPH (Table 2). Among other honey types, the second group is composed of rapeseed samples (RS) in the upper right square of the plot, which is positively correlated with both components and represents the highest values of L* and h°. In contrast, the samples of multifloral (MF) honey were more scattered, while the black locust (AC) and linden (LI) samples were generally positively correlated with PC2. Summing up, the data presented in Figure 1 confirm the results given in Tables 1 and 2. Buckwheat honey showed the highest antioxidant activity together with the greatest redness and the lowest value of h°.

Kaczmarek et al. [6], based on the cluster analysis dendrogram, have distinguished two well-separated clusters of eight types of Polish honeys. The first cluster included dark honeys (nectar-honeydew, buckwheat, honeydew, and heather), while the second one contained light coloured honeys (acacia, rape, linden, and multiforal).

3.4.2. Antioxidant Activity and Minerals

Another PCA analysis (including 9 variables and 64 cases) was conducted on the results obtained for antioxidant activity and the concentrations of selected minerals and trace elements. Three principal components with eigenvalues exceeding 1 (Kaiser criterion) explained 73.46% of the total variance, with PC1 accounting for 40.65%, PC2 for 18.24%, and PC3 for 14.57% (Table S2). Figure 2a visualizes the projection of variables onto a two-factor plot (PC1 × PC2), explaining 58.89% of the total variance. The variables were distributed in two-quarters of the plot. The first area (bottom right quarter) included Cu, Mn, DPPH, FRAP and Fe, positively correlated with PC1 (0.893, 0.808, 0.772, 0.697 and 0.596) (Table 4). The second area (upper right quarter) included K and Mg, which were positively correlated with PC2 (0.725 and 0.708, respectively). The third component was positively correlated with Na and Zn (0.760 and 0.433).

Figure 2b shows the projection of cases depending on the botanical origin of the honey in the coordinate system defined by the first two principal components (PC1 × PC2). With the exception of buckwheat honey, the honey types were characterized by high dispersion of samples in the plot. Samples of buckwheat honey were found only in the bottom right square, i.e., positively correlated with PC1 and negatively with PC2. Thus, buckwheat (BW) honey showed strong antioxidant activity, together with a high concentration of Mn and Cu. Based on analysis of the content of selected minerals and trace elements in various types of Italian honeys, it was confirmed that their botanical origin significantly influenced their chemical composition, particularly in the case of Ca, Na and Mn [18]. Furthermore, PCA analysis indicated correlations between the concentration of minerals and the type of honey, as the highest element concentrations were found in the darkest honeys (honeydew) and the lowest content in light-coloured samples. Similar relationships between mineral concentration (Fe, Zn and Mn) and pine honeydew (dark) and acacia (light) were reported for honeys from Kashmir [7]. The results of our research confirm these dependencies for buckwheat honey from Poland.

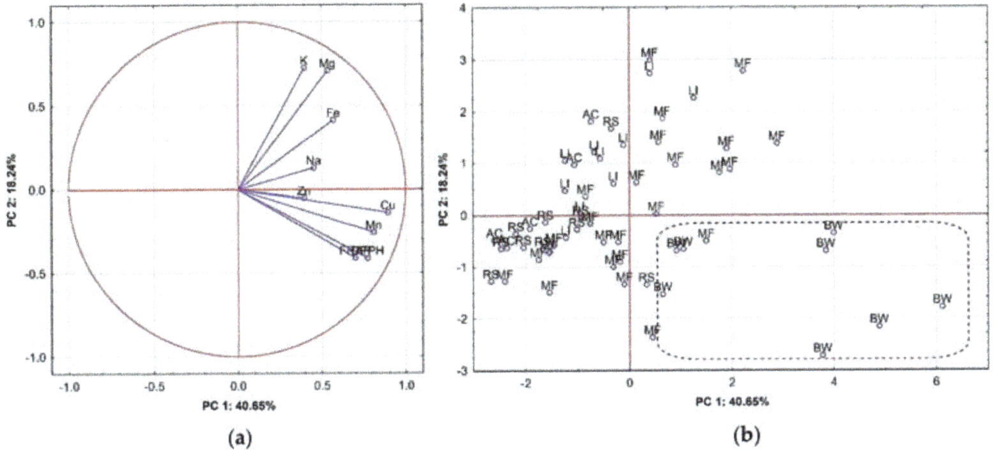

Figure 2. Projection of variables (**a**) and projection of cases (**b**) depending on the botanical origin of the honey in a two-factor plane (PC1 × PC2). (**a**) K, potassium; Na, sodium; Mg, magnesium; Fe, iron; Zn, zinc; Mn, manganese; Cu, copper; DPPH, scavenging capacity; FRAP, ferric reducing antioxidant power; (**b**) honey type: RS, rapeseed; BW, buckwheat; LI, linden; AC, black locust; MF, multifloral.

Table 4. Correlations between the principal components and the original variables.

Variable	PC1	PC2	PC3
K	0.396	0.725	−0.394
Mg	0.535	0.708	−0.264
Na	0.451	0.128	0.760
Fe	0.596	0.413	0.467
Zn	0.395	−0.051	0.433
Cu	0.893	−0.140	−0.093
Mn	0.808	−0.259	−0.170
FRAP	0.697	−0.410	−0.156
DPPH	0.772	−0.413	−0.200

K, potassium; Na, sodium; Mg, magnesium; Fe, iron; Zn, zinc; Mn, manganese; Cu, copper; DPPH, scavenging capacity; FRAP, ferric reducing antioxidant power.

4. Conclusions

The study demonstrated that the physicochemical properties, instrumental colour parameters, content of some elements, and antioxidant activity of five honey types from Poland were significantly influenced by their botanical origin. Strong relationships were shown between antioxidant activity and parameters of instrumental CIE L*a*b* colour analysis, as well as antioxidant minerals. The principal component analysis allowed the types to be classified in terms of their antioxidant activity in combination with their colour characteristics and content of certain elements. The buckwheat honeys, which were the darkest, had the strongest antioxidant activity, which may have been linked to the fact that they had the highest concentrations of copper, manganese and zinc. These parameters can be potentially used for the identification of buckwheat honeys from other blossom types. However, this requires further studies with honeys of different geographical origins.

Supplementary Materials: The following are available online at https://www.mdpi.com/article/10.3390/agriculture11080702/s1, Table S1: Eigenvalues and the proportion of variation (%) explained by 7 principal components, Table S2: Eigenvalues and the proportion of variation (%) explained by 9 principal components.

Author Contributions: Conceptualization, M.K.-M. and M.F.; methodology, M.K.-M., A.T., M.S. and M.R.; investigation, A.T., M.S., P.D. and M.R.; formal analysis, M.K.-M. and M.F.; data curation, M.S., P.S., P.D. and M.R.; writing original draft, M.K.-M., A.T., M.S., P.D. and M.R.; writing—review & editing, P.S. and M.F.; funding acquisition, P.S.; project administration, P.S.; supervision, M.F. All authors have read and agreed to the published version of the manuscript.

Funding: This research was funded under the program of the Minister of Science and Higher Education under the name "Regional Initiative of Excellence" in 2019–2022 project number 029/RID/2018/19 funding amount 11,927,330.00 PLN.

Institutional Review Board Statement: Not applicable.

Data Availability Statement: Data sharing not applicable.

Conflicts of Interest: The authors declare no conflict of interest. The funders had no role in the design of the study; in the collection, analyses, or interpretation of data; in the writing of the manuscript, or in the decision to publish the results.

References

1. Rao, P.V.; Krishnan, K.T.; Salleh, N.; Gan, S.H. Biological and therapeutic effects of honey produced by honey bees and stingless bees: A comparative review. *Rev. Bras. Farmacogn.* **2016**, *26*, 657–664. [CrossRef]
2. Kuś, P.; Congiu, F.; Teper, D.; Sroka, Z.; Jerković, I.; Tuberoso, C.I.G. Antioxidant activity, color characteristics, total phenol content and general HPLC fingerprints of six Polish unifloral honey types. *LWT-Food Sci. Technol.* **2014**, *55*, 124–130. [CrossRef]
3. Kędzierska-Matysek, M.; Florek, M.; Wolanciuk, A.; Barłowska, J.; Litwińczuk, Z. Concentration of Minerals in Nectar Honeys from Direct Sale and Retail in Poland. *Biol. Trace Elem. Res.* **2018**, *186*, 579–588. [CrossRef]
4. da Silva, P.M.; Gauche, C.; Gonzaga, L.V.; Oliveira Costa, A.C.; Fett, R. Honey: Chemical composition, stability and authenticity. *Food Chem.* **2016**, *196*, 309–323. [CrossRef] [PubMed]
5. Álvarez-Suárez, J.M.; Tulipani, S.; Díaz, D.; Estevez, Y.; Romandini, S.; Giampieri, F.; Damiani, E.; Astolfi, P.; Bompadre, S.; Battino, M. Antioxidant and antimicrobial capacity of several monofloral Cuban honeys and their correlation with color, polyphenol content and other chemical compounds. *Food Chem. Toxicol.* **2010**, *48*, 2490–2499. [CrossRef] [PubMed]
6. Kaczmarek, A.; Muzolf-Panek, M.; Tomaszewska-Gras, J.; Konieczny, P. Predicting the Botanical Origin of Honeys with Chemometric Analysis According to Their Antioxidant and Physicochemical Properties. *Pol. J. Food Nutr. Sci.* **2019**, *69*, 191–201. [CrossRef]
7. Nayik, G.A.; Nanda, V. Physico-Chemical, Enzymatic, Mineral and Colour Characterization of Three Different Varieties of Honeys from Kashmir Valley of India with a Multivariate Approach. *Pol. J. Food Nutr. Sci.* **2015**, *65*, 101–108. [CrossRef]
8. Voica, C.; Iordache, A.M.; Ionete, R.E. Multielemental characterization of honey using inductively coupled plasma mass spectrometry fused with chemometrics. *J. Mass Spectrom.* **2020**, *55*, e4512. [CrossRef]
9. Akbulut, M.; Özcan, M.M.; Çoklar, H. Evaluation of antioxidant activity, phenolic, mineral contents and some physicochemical properties of several pine honeys collected from Western Anatolia. *Int. J. Food Sci. Nutr.* **2009**, *60*, 577–589. [CrossRef]
10. Gheldof, N.; Engeseth, N.I. Antioxidant capacity of honeys from various flora sources based on the determination of oxygen radical absorbance capacity and inhibition of in vitro lipoproteid oxidation in human serum samples. *J. Agric. Food Chem.* **2002**, *50*, 3050–3055. [CrossRef]
11. Socha, R.; Juszczak, L.; Pietrzyk, S.; Gałkowska, D.; Fortuna, T.; Witczak, T. Phenolic profile and antioxidant properties of Polish honeys. *Int. J. Food Sci. Technol.* **2011**, *46*, 528–534. [CrossRef]
12. Israili, Z.H. Antimicrobial properties of honey. *Am. J. Therap.* **2014**, *21*, 304–323. [CrossRef]
13. Cianciosi, D.; Forbes-Hernández, T.Y.; Afrin, S.; Gasparrini, M.; Reboredo-Rodriguez, P.; Manna, P.P.; Zhang, J.; Bravo Lamas, L.; Martínez Flórez, S.; Agudo Toyos, P.; et al. Phenolic Compounds in Honey and Their Associated Health Benefits: A Review. *Molecules* **2018**, *23*, 2322. [CrossRef]
14. EU Council Directive 2001/110/EC of 20 December 2001 relating to honey. *Off. J. EU* **2002**, *L10*, 47–52. Available online: http://data.europa.eu/eli/dir/2001/110/2014-06-23 (accessed on 8 November 2020).
15. Kędzierska-Matysek, M.; Florek, M.; Wolanciuk, A.; Skałecki, P. Effect of freezing and room temperatures storage for 18 months on quality of raw rapeseed honey (*Brassica napus*). *J. Food Sci. Technol.* **2016**, *53*, 3349–3355. [CrossRef] [PubMed]

16. Kędzierska-Matysek, M.; Florek, M.; Wolanciuk, A.; Skałecki, P.; Litwińczuk, A. Characterisation of viscosity, colour, 5-hydroxymethylfurfural content and diastase activity in raw rape honey (*Brassica napus*) at different temperatures. *J. Food Sci. Technol.* **2016**, *53*, 2092–2098. [CrossRef] [PubMed]

17. Reshma, K.; Thasniya, M.; Gopika, S.K.; Arunima, S.H. Honey crystallization: Mechanism, evaluation and application. *Pharma Innov.* **2021**, *10*, 222–231. [CrossRef]

18. Pisani, A.; Protano, G.; Riccobono, F. Minor and trace minerals in different honey types produced in Siena County (Italy). *Food Chem.* **2008**, *107*, 1553–1560. [CrossRef]

19. MARD. Regulation of the Minister of Agriculture and Rural Development of 14 January 2009 on the methods of analysis related to the assessment of honey. *J. Laws* **2009**, *17*, 2018–2030.

20. Bogdanov, S.; Martin, P.; Lüllmann, C. Harmonised methods of the European Honey Commission. *Apidologie* **2009**, 1–59. Available online: http://www.ihc-platform.net/ihcmethods2009.pdf (accessed on 28 December 2020).

21. White, J. Spectrophotometric method for hydroxymethylfurfural in honey. *J. Assoc. Off. Anal. Chem.* **1979**, *62*, 509–514. [CrossRef]

22. Blois, M.S. Antioxidant determinations by the use of a stable free radical. *Nature* **1958**, *181*, 1199–1200. [CrossRef]

23. Bertoncelj, J.; Doberšek, U.; Jamnik, M.; Golob, T. Evaluation of the phenolic content, antioxidant activity and colour of Slovenian honey. *Food Chem.* **2007**, *105*, 822–828. [CrossRef]

24. Tornuk, F.; Karaman, S.; Ozturk, I.; Toker, O.S.; Tastemur, B.; Sagdic, O.; Dogan, M.; Kayacier, A. Quality characterization of artisanal and retail Turkish blossom honeys: Determination of physicochemical, microbiological, bioactive properties and aroma profile. *Ind. Crops Prod.* **2013**, *46*, 124–131. [CrossRef]

25. Kacaniova, M.; Knazovicka, W.; Melich, M.; Fikselova, M.; Massanyi, P.; Stawarz, R.; Hascik, P.; Pechociak, T.; Kuczkowska, A.; Putała, A. Environmental concentration of selected elements and relation to physicochemical parameters in honey. *J. Environ. Sci. Health Part A* **2009**, *44*, 414–422. [CrossRef] [PubMed]

26. Cavia, M.M.; Fernández-Muino, M.A.; Alonso-Torre, S.R.; Huidobro, J.F.; Sancho, M.T. Evolution of acidity of honeys from continental climates: Influence of induced granulation. *Food Chem.* **2007**, *100*, 1728–1733. [CrossRef]

27. Chen, C. Relationship between water activity and moisture content in floral honey. *Foods* **2019**, *8*, 30. [CrossRef] [PubMed]

28. Serrano, S.; Espejo, R.; Villarejo, M.; Jodral, M.J. Diastase and invertase activities in Andalusian honeys. *Int. J. Food Sci. Technol.* **2007**, *42*, 76–79. [CrossRef]

29. Wesołowska, M.; Dżugan, M. Activity and thermal stability of diastase present in honey from Podkarpacie Region. *Food Sci. Technol. Qual.* **2017**, *24*, 103–112. Available online: http://wydawnictwo.pttz.org/wp-content/uploads/2018/03/09_Wesolowska.pdf (accessed on 6 December 2020).

30. Kek, S.P.; Chin, N.L.; Tan, S.W.; Yusof, Y.A.; Chua, L.S. Classification of Honey from Its Bee Origin via Chemical Profiles and Mineral Content. *Food Anal. Methods* **2017**, *10*, 19–30. [CrossRef]

31. Ördög, A.; Tari, I.; Bátori, Z.; Poór, P. Mineral content analysis of unifloral honeys from the Hungarian Great Plain. *J. Elem.* **2017**, *22*, 271–281. [CrossRef]

32. Wieczorek, J.; Pietrzak, M.; Pomianowski, J.; Wieczorek, Z. Honey as a source of bioactive compounds. *Pol. J. Nat. Sci.* **2014**, *29*, 275–285. Available online: http://www.uwm.edu.pl/polish-journal/sites/default/files/issues/articles/wieczorek_et_al._2014.pdf (accessed on 6 December 2020).

33. Dżugan, M.; Zaguła, G.; Wesołowska, M.; Sowa, P.; Puchalski, C. Levels of Toxic and Essential Metals in Varietal Honeys from Podkarpacie. *J. Elem.* **2017**, *22*, 1039–1048. [CrossRef]

34. Deng, J.; Liu, R.; Lu, Q.; Hao, P.; Xu, A.; Zhang, J.; Tan, J. Biochemical properties, antibacterial and cellular antioxidant activities of buckwheat honey in comparison to manuka honey. *Food Chem.* **2018**, *252*, 243–249. [CrossRef] [PubMed]

35. Škrovánková, S.; Snopek, L.; Mlček, J.; Volaříková, E. Bioactive Compounds Evaluation in Different Types of Czech and Slovak Honeys. *Potravin. Slovak J. Food Sci.* **2019**, *13*, 94–99. [CrossRef]

36. Cheng, N.; Wu, L.; Zheng, J.; Cao, W. Buckwheat Honey Attenuates Carbon Tetrachloride-Induced Liver and DNA Damage in Mice. *Evid. Based Complementary Altern. Med.* **2015**, *2015*, 987385. [CrossRef]

37. Piszcz, P.; Głód, B.K. Antioxidative Properties of Selected Polish Honeys. *J. Apic. Sci.* **2019**, *63*, 81–91. [CrossRef]

38. Dżugan, M.; Tomczyk, M.; Sowa, P.; Grabek-Lejko, D. Antioxidant Activity as Biomarker of Honey Variety. *Molecules* **2018**, *23*, 2069. [CrossRef] [PubMed]

39. Moniruzzaman, M.; Sulaiman, S.A.; Khalil, M.I.; Gan, S.H. Evaluation of physicochemical and antioxidant properties of sourwood and other Malaysian honeys: A comparison with manuka honey. *Chem. Cent. J.* **2013**, *7*, 138. [CrossRef]

40. Beretta, G.; Granata, P.; Ferrero, M.; Orioli, M.; Maffei Facino, R. Standardization of antioxidant properties of honey by combination of spectrophotometric/fluorimetric assays and chemometrics. *Anal. Chim. Acta* **2005**, *533*, 185–191. [CrossRef]

41. Perna, A.; Simonetti, A.; Intaglietta, I.; Sofo, A.; Gambacorta, E. Metal content of southern Italy honey of different botanical origins and its correlation with polyphenol content and antioxidant activity. *Int. J. Food Sci. Technol.* **2012**, *47*, 1909–1917. [CrossRef]

Article

Milling and Baking Quality of Spring Wheat (*Triticum aestivum* L.) from Organic Farming

Beata Feledyn-Szewczyk [1], Grażyna Cacak-Pietrzak [2], Leszek Lenc [3], Karolina Gromadzka [4] and Dariusz Dziki [5,*]

[1] Department of Systems and Economics of Crop Production, Institute of Soil Science and Plant Cultivation–State Research Institute, Czartoryskich 8 Street, 24-100 Puławy, Poland; bszewczyk@iung.pulawy.pl

[2] Department of Food Technology and Assessment, Warsaw University of Life Sciences, Nowoursynowska, 159C Street, 02-787 Warsaw, Poland; grazyna_cacak_pietrzak@sggw.edu.pl

[3] Department of Phytopathology and Molecular Mycology, University of Technology and Life Sciences, Al. Prof. S. Kaliskiego 7 Street, 85-796 Bydgoszcz, Poland; lenc@utp.edu.pl

[4] Department of Chemistry, Poznan University of Life Sciences, Wojska Polskiego 75 Street, 60-625 Poznań, Poland; karolina.gromadzka@up.poznan.pl

[5] Department of Thermal Technology and Food Process Engineering, Lublin University of Life Sciences, Głęboka 31 Street, 20-612 Lublin, Poland

* Correspondence: dariusz.dziki@up.lublin.pl; Tel.: +48-81-445-6125

Abstract: The quality of grain products from organic agriculture is an important subject of research for food safety and consumer health. The aim of the study was to examine the grain of spring wheat from organic agriculture according to their infestation by *Fusarium* spp., mycotoxin content, and technological value for milling and baking processing. The material was grain of 13 spring wheat varieties cultivated in organic systems in 3 years. The results showed that the intensity of *Fusarium* head blight (FHB) was low and ranged from 0.0% to 5.5% of ears. Grain infestation by *Fusarium* spp. varied between varieties and years from 1.5% to 18.5%. The colonization of grains by *Fusarium* spp. did not reflect the intensity of FHB. The lowest grain infestation by *Fusarium* spp. was noted for the varieties: Waluta, Zadra, and Arabella. Mycotoxin contamination of the grain of tested varieties did not exceed accepted standards. The requirements of the milling and baking industries were generally met by grain and flour of all the tested varieties. On the basis of the 3 year study results related to food safety and processing properties, the varieties most useful for organic production are Arabella, followed by Brawura, Izera, Kandela, Katoda, KWS Torridon, Waluta, and Zadra.

Keywords: spring wheat; organic agriculture; *Fusarium* spp.; mycotoxins; quality of grain; flour yield; technological value

Citation: Feledyn-Szewczyk, B.; Cacak-Pietrzak, G.; Lenc, L.; Gromadzka, K.; Dziki, D. Milling and Baking Quality of Spring Wheat (*Triticum aestivum* L.) from Organic Farming. *Agriculture* **2021**, *11*, 765. https://doi.org/10.3390/agriculture 11080765

Academic Editor: Alessandra Durazzo

Received: 25 June 2021
Accepted: 9 August 2021
Published: 11 August 2021

Publisher's Note: MDPI stays neutral with regard to jurisdictional claims in published maps and institutional affiliations.

1. Introduction

Spring wheat is a popular crop in both conventional and organic farms, as it is an important consumer cereal in Europe [1,2]. However, in organic agriculture, the choice of proper variety is of great importance because it influences the yield [3–7] and quality of grain [8,9]. Spring wheat varieties vary according to their agricultural traits (morphological features, yielding potential, resistance to disease and pests, weed suppression ability) and technological parameters [5,6,10–13]. The information about the suitability of different varieties for organic farming and food processing according to their susceptibility to *Fusarium* sp. disease and mycotoxin contamination, and the technological value of grain and flour, is desired by producers, advisors, and processors. The common wheat pathogens in organic farming are fungi of the genus *Fusarium*. Some of them, such as Fusarium rot pre- and post-emergence, Fusarium foot rot, and Fusarium leaf blight can lead to a significant drop in yields, while Fusarium head blight (FHB) and grain infestation by *Fusarium* spp. (Fusarium disease kernels—FDK) may contribute to a decrease in yields [14]. The infested ears either do not form grains at all or form fewer grains that are smaller and poorly filled that contain less starch and gluten proteins and, therefore, the flour obtained from them has a low baking value [15]. In addition, certain species of fungi of the genus *Fusarium* have the ability to synthesize mycotoxins that accumulate in grains, and these mycotoxins are also present in products derived from them, which can be dangerous to human health [16–18]. The significant contaminants in all cereal grains, including wheat, are deoxynivalenol (DON), zearalenone (ZEA), and fumonisins B_1 and B_2. Zearalenone is one of the strongest non-steroidal estrogenic substances, which can cause functional changes in the reproductive system similar to those of estrogens [19]. Deoxynivalenol and nivalenol are important toxins from the group of trichothecenes. Deoxynivalenol, similar to other trichothecenes, has a significant effect on biochemical processes. The most frequently observed DON toxicosis symptoms in animals include vomiting and body weight loss with successive numerous physiological changes in internal tissues [20]. Fumonisins, especially fumonisin B_1, has caused field outbreaks of leucoencephalomacia in horses, porcine pulmonary oedema in swine, and were found to be hepatotoxic and hepatocancerogenic to rats [21]. Beauvericin and enniatins (EnnA, $EnnA_1$, EnnB, $EnnB_1$) are well-known toxic cyclic hexadepsipeptides with a specific cholesterol acyltransferase inhibitor activity [22]. Beauvericin and enniatins are early emerging mycotoxins, and they are being identified, with an increasing frequency, in cereal grains worldwide. Moniliformin is a potent inhibitor of mitochondrial pyruvate and ketoglutarate oxidation and has caused acute degenerative lesions in the myocardium [23]. EU countries have established and published guidelines for maximum levels of deoxynivalenol (DON), zearalenone (ZEA), and the sum of fumonisins B1 and B2 [24].

The technological value of wheat grains is determined not only by genetic factors but also by habitat conditions and the crop management treatments applied, which are limited in the organic production system [8,16]. Agrotechnical treatments have a significant impact on the quantity as well as the fractional composition of protein, which is commonly considered as one of the basic indicators of the suitability of wheat grains for processing [8,16,25]. The response of individual wheat varieties to the applied cultivation conditions is not the same. It is important to select wheat varieties with the lowest possible variability of grain quality traits for organic farming. In Poland, as in other EU countries, no separate quality requirements have been defined for wheat grain from organic farming, so it has to meet the general quality requirements for wheat grain [26], and the direction of its use should be selected taking into account the requirements of the processing industry [27].

The aim of the study was to examine 13 spring wheat varieties (*Triticum aestivum* L.) cultivated in organic farming due to their susceptibility to *Fusarium* spp. diseases and technological parameters of grain, flour, and bread. Moreover, mycotoxin content in the grain of two selected spring wheat varieties was analyzed.

The hypothesis was that it is possible to obtain wheat grains in organic farming that met processing criteria.

2. Materials and Methods

2.1. Sites Characteristics, Experimental Design and Agronomic Practices

The experiment with the varieties of spring wheat was carried out in the years 2014–2016 in the organic farm of the Institute of Soil Science and Plant Cultivation—State Research Institute (IUNG-PIB), located in Osiny, central-eastern part of Poland (Table 1).

Table 1. The characteristics of habitat conditions of the experiment.

Items	Specification
Localization	Osiny, Lublin province, Poland
Type of organic farm	Experimental Station of the Institute of Soil Science and Plant Cultivation—State Research Institute (IUNG-PIB) started in 1994
Soil type [28]	Luvisol
Texture	loamy sand
pH_{KCl}	5.9
Soil abundance:	
humus (%)	1.4
P_2O_5 (mg $\cdot$ 100 g^{-1} soil)	8.6
K_2O (mg $\cdot$ 100 g^{-1} soil)	10.0
Mg (mg $\cdot$ 100 g^{-1} soil)	9.1
Pre-crop	potato

Spring wheat varieties were cultivated in a randomized complete block design, with four replications. The area of each plot of replication for sowing and harvesting was 30 m^2. Thirteen spring wheat varieties (*Triticum aestivum* L.) included in the Common Catalogue of Varieties of Agricultural Plant Species [29]—Arabella, Brawura, Cytra, Ethos, Izera, KWS Torridon, Kandela, Katoda, Koksa, Korynta, Ostka Smolicka, Waluta, and Zadra—were cultivated in organic system. Sowing treatments were performed in accordance with good agricultural practice at the optimum time for each region. The sowing rates were the same for each variety—450 grains m^{-2}. The row spacing was 12 cm and the planting depth 3.5 cm. According to organic agriculture rules, mineral fertilizers and chemical plant protection products were not used [30]. Harvests were undertaken in the first week of August.

The experimental site belongs to a moderately continental climate zone. The characteristics of the meteorological condition in the location of the experiment in the 3 years of the study are presented in Table 2. In 2014 and 2015, high precipitation occurred in May (more than twice the multi-annual average). The 2016 growing season was warm and dry. The temperatures from March to August exceeded the multi-annual average, and the amount of precipitation was below the average (except in August).

Table 2. Total precipitation and average monthly temperatures in the experiment.

Month	Temperature (°C)				Precipitation (mm)			
	2014	2015	2016	Long-Term Mean	2014	2015	2016	Long-Term Mean
III	6.4	5.2	4.1	1.9	42.7	49.0	25.1	28.1
IV	9.8	8.6	10.2	8.1	72.9	29.0	19.9	42.0
V	13.6	13.0	15.0	13.8	188.9	109.0	38.5	55.0
VI	15.7	17.3	20.7	17.1	118.1	29.0	15.4	71.0
VII	20.7	20.1	19.5	18.6	65.8	52.0	67.9	78.2
VIII	18.3	22.4	18.3	17.8	119.0	4.0	93.5	67.3
Mean	14.1	14.4	14.6	12.9	101.2	45.3	43.4	56.9

2.2. Fusarium sp. Occurrence

An assessment of Fusarium head blight was performed at the milk-dough stage of spring wheat (BBCH 77-83). From each experimental combination, 4 × 50 randomly selected ears were analyzed, and the percentage of their infestation was determined.

After the harvest, a mycological analysis of the grain was performed. From each combination, 4 × 100 grains were taken at random. After rinsing under running water, decontaminating for 2.5 min in 1% NaCl, and rinsing three times in sterile water, the grains were placed into PDA (Potato Dextrose Agar) (pH = 5.5) and incubated at 20 °C for 6 days.

The growing colonies of fungi were grafted onto slants of PDA and identified according to mycological keys [31,32].

2.3. Analysis of Mycotoxins

In the first year of research (2014), five toxins synthesized by *Fusarium* species were analyzed—zearalenone (ZEA), deoxynivalenol (DON), nivalenol (NIV), beauvericin (BEA), and moniliformin (MON)—whereas, in the second and third year (2015–2016), twelve mycotoxins—zearalenone (ZEA), deoxynivalenol (DON), nivalenol (NIV), beauvericin (BEA), moniliformin (MON), enniatins (EnnA, $EnnA_1$, EnnB, $EnnB_1$) and fumonisins (FB_1, FB_2, FB_3)—were measured in the wheat grains of two selected varieties—Kandela and Ostka Smolicka. These varieties were chosen because they showed greater infection of the grain by Fusarium spp. in the first year of research; the same varieties were tested for the next 2 years. Moreover, Ostka Smolicka was the only bristly variety among tested varieties, and we aimed to check its susceptibility to *Fusarium* infection.

In order to achieve effective extraction, the mycotoxins were divided into 3 groups with similar physico-chemical properties: I. ZEA, DON, NIV—extracted with an acetonitrile-water solution (8:2 v/v); II. Enns, FBs—extracted with a methanol-water solution (3:1 v/v); III. MON, BEA—extracted with an acetonitrile–methanol–water solution (16:3:1 $v/v/v$).

Grain material was extracted using 2.5 mL of solvent per 1 g of sample; then, it was homogenized in a Ultraturrax model T25 basic (IKA Werke, Freiburg, Germany) for 4 min at 13,500 rpm. The content of mycotoxins was determined using the chromatographic system with a Waters 2695 high performance liquid chromatograph, a Waters 2475 Multi λ Fluorescence Detector, and/or a Waters 2996 Photodiode Array Detector (Waters Corporation, Milford, PA, USA). The quantification limits (LOQ) were determined by multiplying the detection limits (LOD) by 3.3. The LOD of the methods were calculated by a signal-to-noise ratio of 3:1. Mycotoxin analyses were performed in triplicate. The presented concentration values are the average of the obtained results.

The extract of ZEA was purified on an immunoaffinity column according to the method described by Visconti and Pascale [33]. Zearalenone content was determined using the fluorescence detector. The excitation and emission wavelengths were 274 and 440 nm, respectively. The reserve-phase column was a C-18 Nova Pak column (3.9 × 150 mm), while the mobile phase was acetonitrile–water–methanol (46:46:8, $v/v/v$) at a flow rate of 0.5 mL·min^{-1}. Quantification of ZEA was performed by measuring the peak areas at the ZEA retention time according to the relevant calibration curve (correlation coefficient R = 0.9998). The limit of zearalenone detection was 3 µg·kg^{-1}. The recovery of zearalenone was measured in triplicate by extracting ZEA from blank samples spiked with 5–100 ng·g^{-1} of the compound. The results of the experiments confirmed the literature data on ZEA recovery in the range of 97% to 99%. The relative standard deviation (R.S.D.) was below 1%. In order to confirm the presence of zearalenone, the photodiode array detector was used. To analyze the trichothecenes (DON, NIV), extracts were purified by filtration on a column (Celite 545:charcoal, Darco G-60:neutral alumina, 3:9:5 $w/w/w$) according to the method described by Tomczak et al. [34]. Deoxynivalenol and nivalenol were quantified by HPLC using a C-18 Nova Pak column (3.9 × 300 mm) and a photodiode array detector (λmax = 224 nm for DON and NIV). DON and NIV were eluted from the column with a 25% water solution of methanol (flow rate 0.7 mL·min^{-1}). The detection limit for DON and NIV was 10 µg·kg^{-1}. Positive results (on the basis of retention times) will be confirmed by HPLC analysis and by comparison with the relevant calibration curve (correlation coefficients for NIV and DON are 0.9994 and 0.9997, respectively). The recovery rates for NIV and DON were 75% are 87%, respectively, estimated in triplicate by extracting mycotoxins from blank samples spiked with 10–100 ng·g^{-1} of the compounds. The relative standard deviation (R.S.D.) was below 5% for DON and NIV. Enniatins and beauvericin were identified and quantified as reported by Logrieco et al. [23]. Extracts were prepurified once on a C18 column (500 mg, 3 mL, 40 µm), concentrated to 1 mL, and filtered through an Acrodisk filter (pore size, 0.22 µm) before HPLC analysis. HPLC analyses were performed using a diode array detector and C18 column (250 × 4.6 mm, 5 μm). HPLC conditions included a constant flow at 1.5 mL·min^{-1} and acetonitrile–water (65:35, v/v) as the starting eluent system. The starting ratio was kept constant for 5 min and then linearly modified to 70% acetonitrile over 10 min. After 1 min at 70% acetonitrile, the mobile phase was returned to the starting conditions in 4 min. Beauvericin and enniatins were detected at 205 nm. Mycotoxins were identified by comparing retention times and UV spectra of samples with those of authentic standards. Mycotoxins were quantified by comparing peak areas from samples to a calibration curve of standards. The detection limits were 5.0 μg·kg^{-1} for EnnA, 1.8 μg·kg^{-1} for EnnA$_1$, 1.0 μg·kg^{-1} for EnnB$_1$, 0.5 μg·kg^{-1} for EnnB, and 1.0 μg·kg^{-1} for beauvericin determination. The calculated standard deviation was always less than 5%. The extract of MON was purified on a Florisil column according to the method described by Kostecki et al. [35]. Moniliformin was quantified by HPLC using a C-18 Nova Pak column (3.9 × 300 mm) and a photodiode array detector (λmax = 229 nm). MON was eluted from the column with the acetonitrile–water solvent (15:85, v/v) buffered with 10 mL of 0.1 M K$_2$HPO$_4$ in 40% t-butyl-ammonium hydroxide in 1 L of solvent (flow rate 0.6 mL·min^{-1}). The detection limit for MON was 8 μg·kg^{-1}. Positive results (on the basis of retention times) were confirmed by HPLC analysis and by comparison with the relevant calibration curve (correlation coefficients 0.9990). The recovery rate for MON is 90%, estimated in triplicate by extracting mycotoxin from blank samples spiked with 10–100 ng·g^{-1} of the compound. The relative standard deviation (R.S.D.) was below 7%. The detailed procedure of extraction and purification of FBs was reported by Waśkiewicz et al. [36]. The fumonisin B$_1$, B$_2$, and B$_3$ standards (5 μL) or extracts (20 μL) were derivatized with 20 or 80 μL of the o-phosphoric acid (OPA) reagent. After 3 min, the reaction mixture (10 μL) was injected onto an HPLC column. Methanol sodium dihydrogen phosphate (0.1M in water) solution (77:23, v/v) adjusted to pH 3.35 with o-phosphoric acid, after filtration through a 0.45 μm Waters HV membrane was used as the mobile phase with a flow rate of 0.6 mL·min^{-1}. A HPLC with a C-18 Nova Pak column (3.9 × 150 mm) and a fluorescence detector (λEX = 335 nm and λEM = 440 nm) were used in the metabolite quantitative determination. The detection limit of FB$_1$, FB$_2$, and FB$_3$ determination was 10.0 μg·kg^{-1}. Positive results (on the basis of retention times) were confirmed by HPLC analysis and compared with the relevant calibration curve (the correlation coefficients for FB$_1$, FB$_2$, and FB$_3$ were 0.9967, 0.9983, and 0.9966, respectively). The recovery rates for FB$_1$, FB$_2$, and FB$_3$ were 93, 96, and 87%, respectively. The relative standard deviation (R.S.D.) was below 5%.

2.4. Quality Traits of Grains, Flour, and Bakery Products

The evaluation of the quality traits of the grain of the tested spring wheat varieties, different laboratory tests were carried out according to the methods commonly used to evaluate cereal grain and cereal products [37]. As part of the physico-chemical evaluation of grains, grain selectness, uniformity, glassiness, and hardness were determined using a farinograph adapter (Brabender, Duisburg, Germany).

The laboratory milling of the grain was carried out in a two-passage laboratory mill Quadrumat Senior (Brabender, Duisburg, Germany). Before milling, the grains were cleaned on granules (Brabender, Duisburg, Germany) and conditioned to 14.5% humidity. Based on the milling balance, the total flour yield was calculated. As part of the evaluation of the physical and chemical characteristics of the flour, the following determinations were made: total ash content according to the AACC Method 08-01.01 [33]; total protein content using Kjeldahl's method (N·5,83) in a Kjeltec apparatus 8200 (Foss, Hillerød, Sweden) according to the AACC Method 46-11.02 [38]; wet gluten yield and gluten index (IG) in a Glutomatic 2200 apparatus (Perten Instruments, Hägersten, Stockholm, Sweden) according to the AACC Method 38-12.02 [38]; and the falling number, with the Hagberg–Perten method in a Falling Number 1400 apparatus (Perten Instruments, Hägersten, Stockholm, Sweden) according to the AACC Method 56-81.03 [38].

The suitability of flour for the production of bakery products was determined by conducting a laboratory test baking. Dough (160% yield) was prepared directly from 500 g of flour with a moisture content of 14.0%, 300 cm^3 of water, 15 g of baker's yeast, and 7.5 g of kitchen salt in an SP-800A mixer (SPAR Food Machinery, Taiwan). The time of dough kneading was 4 min. Fermentation was carried out in two stages, with piercing of the dough after 60 and 90 min. The final expansion of the dough was carried out in moulds. Baking took place in a Svena Dahlen DC-32E oven (Sveba Dahlen, Fristad, Sweden) at 230 °C for 30 min.

The evaluation of the baking process was carried out on the basis of calculations of bread yields. The analysis of bread quality was carried out 24 h after baking (the bread was stored in room conditions). It included an evaluation of bread volume, crumb porosity, and an organoleptic evaluation using the point method according to the standard [39]. The assessment team consisted of 20 people. The evaluation included the following quality characteristics of the bread: external appearance of the loaf, colour and thickness of the crust, elasticity, porosity and sliceability of the crumb, taste, and smell.

2.5. Statistical Analysis

For all tested features, analysis of variance (ANOVA) and the post-hoc Tukey's test was done, where varieties and years of the research were the factors of the experiment. The research on *Fusarium* spp. on ears and grain infestation were conducted in 4 replications (n = 3 years × 4 replications = 12), whereas the study of baking quality traits was conducted in 3 replications (n = 3 years × 3 replications = 9). Due to the lack of significant years × cultivar interactions for the features related to the technological value of grain, flour, and bread, these results were presented in the form of means from 3 years of research for each variety. All tests were performed at the significance level of α = 0.05.

3. Results

3.1. Infestation of Spring Wheat Ears and Grain by Fusarium spp.

In the 3 years of the experiment, the intensity of Fusarium head blight (FHB) of spring wheat was low and ranged from 0.0% to 5.5% of ears (Table 3, Figure 1A). The low level of infestation of ears in all years of research was a result of the weather conditions (warm and dry in the phase of ears ripening) (Table 2). The analysis of variance showed significant differences among the wheat varieties only in 2015, when the infestation of Izera variety ears was lower than Katoda variety. No significant interaction year × variety for the occurrence of FHB on the ears was detected.

Table 3. The occurrence of FHB on the ears of spring wheat varieties (% of ears infested by *Fusarium* spp.).

Variety	Years of Research					
	2014		**2015**		**2016**	
Arabella	0.0	a [1]	2.5	ab	0.5	a
Brawura	1.0	a	3.0	ab	0.5	a
Cytra	1.0	a	4.5	ab	0.0	a
Ethos	0.0	a	4.0	ab	0.5	a
Izera	1.0	a	1.5	b	0.5	a
Kandela	0.0	a	4.0	ab	0.5	a
Katoda	0.5	a	5.5	a	1.0	a
Koksa	0.0	a	3.0	ab	1.5	a
Korynta	1.0	a	2.5	ab	0.5	a
KWS Torridon	0.5	a	3.0	ab	2.0	a
Ostka Smolicka	0.5	a	3.5	ab	1.0	a
Waluta	0.5	a	5.0	ab	0.5	a
Zadra	0.0	a	2.0	ab	0.5	a
Mean	0.5 A [2]		3.4 B		0.7 A	

[1] Different letters indicate significant differences between varieties according to Tukey's test at α = 0.05; [2] Different capital letters indicate significant differences between years according to Tukey's test at α = 0.05. Interaction year × variety non-significant (p > 0.05).

Grain infestation by *Fusarium* spp. varied between varieties and years from 1.5% to 18.5% (Table 4, Figure 1B,C). The significant interaction year × variety was found. The highest grain infestation was observed in 2014 (Table 4), when the precipitation was high in May, June, and August (Table 2). Waluta, Zadra, and Arabella varieties in each year of the study belonged to the group of varieties with the lowest percentage of infected grains. In the first year of research, the highest infestation of the grains was noted for Kandela and Ostka Smolicka varieties. On average, the largest amount of *Fusarium* spp. was isolated from grains of KWS Torridon and Cytra (Table 4).

Table 4. The infestation of the grains (%) of spring wheat varieties.

Variety	Years of Research					
	2014		2015		2016	
Arabella	7.0 A	efg [1]	1.5 C	g	4.2 B	def
Brawura	7.0 A	efg	3.8 B	cd	4.8 B	cdef
Cytra	13.0 A	b	11.6 B	a	8.0 C	b
Ethos	11.0 A	bcd	9.8 AB	a	8.3 B	bc
Izera	8.0 A	def	1.8 B	fg	7.3 A	bcd
Kandela	17.0 A	a	3.3 C	def	6.8 B	bcd
Katoda	3.0 B	ij	11.1 A	a	3.8 B	ef
Koksa	5.0 B	ghi	3.5 B	cde	7.7 A	b
Korynta	2.0 B	j	5.8 A	bc	6.7 A	bcd
KWS Torridon	9.0 B	cde	6.6 C	b	18.5 A	a
Ostka Smolicka	12.0 A	bc	5.0 B	bcd	3.3 C	fg
Waluta	6.0 A	fgh	1.8 B	fg	1.8 B	g
Zadra	4.0 B	hi	2.0 C	efg	6.0 A	bcde
Mean	8.0 A [2]		5.2 C		6.7 B	

[1] Different letters indicate significant differences between varieties according to Tukey's test at $\alpha = 0.05$; [2] Different capital letters indicate significant differences between years according to Tukey's test at $\alpha = 0.05$. Significant interaction year × variety ($p = 0.032$).

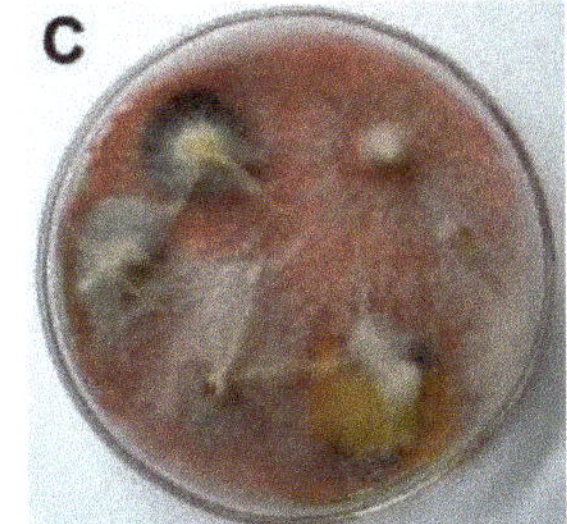

Figure 1. (**A**) Fusarium head blight; (**B**) Fusarium disease on grains; (**C**) infestation of grains by *Fusarium* sp.

The results of the research indicated that the lack of *Fusarium* disease symptoms on the ears does not mean that the grain is not infected by *Fusarium* spp. In all years, *F. poae* was the most frequently isolated species from grains, which does not cause the symptoms on ears (Table 5). The domination of *F. poae* explains the differences between the intensity of *Fusarium* spp. symptoms on ears and the colonization of grain. Other species—*F. avenaceum, F. culmorum, F. equiseti, F. graminearum, F. langsethiae, F. sporotrichioides* and *F. tricinctum*—were isolated occasionally, but to a lesser extent.

Table 5. Species from *Fusarium* genus isolated from grains of spring wheat varieties (% of infested grains).

Variety	*Fusarium avenaceum*	*Fusarium culmorum*	*Fusarium equiseti*	*Fusarium graminearum*	*Fusarium poae*	*Fusarium sporotrichioides*	*Fusarium tricinctum*	Sum
2014								
Arabella	6.0				1.0			7.0
Brawura	1.0			3.0	3.0			7.0
Cytra	4.0			5.0	4.0			13.0
Ethos	1.0			3.0	6.0		1.0	11.0
Izera	4.0			1.0	3.0			8.0
Kandela	5.0	1.0		5.0	6.0			17.0
Katoda	1.0			1.0	1.0			3.0
Koksa	2.0			2.0	1.0			5.0
Korynta				1.0	1.0			2.0
KWS Torridon				1.0	7.0		1.0	9.0
Ostka Smolicka	5.0			2.0	5.0			12.0
Waluta				1.0	3.0	2.0		6.0
Zadra	1.0			1.0	2.0			4.0
Mean	2.3	0.1	0	2.0	3.3	0.1	0.1	8.0
2015								
Arabella					1.5			1.5
Brawura				1.8	2.0			3.8
Cytra	1.5			3.3	6.8			11.6
Ethos	1.5			5.0	3.3			9.8
Izera					1.8			1.8
Kandela	1.8				1.5			3.3
Katoda	1.8			3.3	5.0	1.0		11.1
Koksa				1.5	2.0			3.5
Korynta	2.0				3.8			5.8
KWS Torridon	1.8				3.3		1.5	6.6
Ostka Smolicka	5.0							5.0
Waluta				1.8				1.8
Zadra					2.0			2.0
Mean	1.2	0	0	1.3	2.5	0.1	0.1	5.2
2016								
Arabella					4.2			4.2
Brawura	1.8				1.5		1.5	4.8
Cytra	1.5				6.5			8.0
Ethos		1.5			6.8			8.3
Izera	1.0			1.5	3.8		1.0	7.3
Kandela					6.8			6.8
Katoda					3.8			3.8
Koksa					6.5		1.2	7.7
Korynta	3.2				3.5			6.7
KWS Torridon	3.5			5.0	10.0			18.5
Ostka Smolicka	1.8				1.5			3.3
Waluta	1.8							1.8
Zadra			1.0		5.0			6.0
Mean	1.1	0.1	0.1	0.5	4.6	0.0	0.3	6.7

3.2. Mycotoxin Content in the Spring Wheat Grains

The content of mycotoxins in the tested samples of wheat grain varied depending on the year and variety (Table 6). In 2014, when five toxins were analyzed, the presence of DON was found only in one sample (Ostka Smolicka—181.5 $\mu g \cdot kg^{-1}$). The content of this toxin did not exceed the maximum contamination level of (1250 $\mu g \cdot kg^{-1}$) set by EU regulations [24]. The second of the tested trichothecenes—NIV—was found in the grain of a Kandela variety in the amount of 334.7 $\mu g \cdot kg^{-1}$. Both of these toxins were found in the tested grains only in the first year of the study (Table 6), which was the wettest year (Table 2).

The presence of ZEA was detected in two of the six tested samples; one from 2015 and one from 2016 (Table 6). Contamination of grains with this mycotoxin at levels of 24.1–62.9 $\mu g \cdot kg^{-1}$ did not exceed the accepted standard of 100 $\mu g \cdot kg^{-1}$ [24]. Due to the fact that *F. poae* was the dominant fungus isolated from the grain, accompanied by *F. avenaceum* and *F. tricinctum*, BEA and MON produced by these species were also examined. Mycotoxins BEA and MON were found in larger amounts in the examined grains. The concentration of BEA in the grains of both of the tested cultivars was over 6300 $\mu g \cdot kg^{-1}$, while the concentration of MON in the grains of both tested varieties was over 200 $\mu g \cdot kg^{-1}$ (Table 6).

In the second and third year of the study, 12 metabolites synthesized by *Fusarium* fungi were analyzed (Table 6). Additionally, enniatins (A_1, A, B, and B_1) and fumonisins (FB_1, FB_2, and FB_3) were included in the study. In 2015 and 2016, very high BEA contents were found (respectively, over 13,000 $\mu g \cdot kg^{-1}$ in 2015 and over 89,000 $\mu g \cdot kg^{-1}$ in 2016). No correlation was observed between the number of synthesized toxins and the infestation of variety. No MON was found in 2015, but it was found in both wheat varieties in 2016. Of the analyzed enniatins, only enniatin A_1 was present in all wheat samples from 2015 and 2016. In 2015, this toxin remained relatively stable, and its amount ranged from 23.430 to 25.380 $\mu g \cdot kg^{-1}$. In the last year of the study, its concentration increased significantly (97,426.8 $\mu g \cdot kg^{-1}$ for variety Ostka Smolicka and 81,635.9 $\mu g \cdot kg^{-1}$ for variety Kandela). In 2015 and 2016, the grains were also tested for concentrations of fumonisins. Only one sample from 2015 showed the presence of FB_1 (Kandela 20.3 $\mu g \cdot kg^{-1}$), while none of them had the other fumonisins, FB_2 and FB_3. The correlation analysis showed no relationship between the infestation of grain by *Fusarium* spp. and the concentration of fumonisins.

Table 6. Mycotoxin content in spring wheat grains.

Year	Variety	*Fusarium* spp. [%]	Mycotoxin Content [$\mu g \cdot kg^{-1}$]											
			ZEA	DON	NIV	BEA	MON	Enn A	Enn A_1	Enn B	Enn B_1	FB_1	FB_2	FB_3
Maximum level of mycotoxins in wheat grains *			100	1250	-	-	-	-	-	-	-	-	-	-
2014	Kandela	17.0	nd	nd	334.7	6399.3	482.2	na	na	na	na	na	na	na
	Ostka Smolicka	12.0	nd	181.5	nd	6370.6	335.9	na	na	na	na	na	na	na
2015	Kandela	3.3	nd	nd	nd	7050.0	nd	nd	25,380.0	nd	nd	20.3	nd	nd
	Ostka Smolicka	5.0	62.9	nd	nd	13,180.0	nd	nd	23,430.0	nd	nd	nd	nd	nd
2016	Kandela	6.8	24.1	nd	nd	89,535.8	613.1	nd	81,635.9	nd	nd	nd	nd	nd
	Ostka Smolicka	3.3	nd	nd	nd	56,412.7	228.7	nd	97,426.8	nd	nd	nd	nd	nd

* According to the Commissions Regulations (EC) No 1881/2006 of 19 December 2006, setting maximum levels for certain contaminants in foodstuffs [24]. 'nd'—not detected; 'na'—no analysis; '-' maximum level not set for wheat in the EU regulations.

3.3. Assessment of the Technological Value of Grains of Spring Wheat Varieties and Their Suitability for Processing

The grains of spring wheat cultivars from organic cultivation showed significant differences in terms of all of the assessed physico-chemical characteristics, i.e., selectness and uniformity, glassiness and hardness, and ash content (Table 7). During the three year study period, the average grain selectness ranged from 74.7% to 90.0%. The grain selectness of the majority of the examined wheat varieties (except for Izera and Koksa) coincided with grain uniformity, which proved their high plumpness. The most plump were wheat grains of the varieties Katoda, Waluta, Arabella, KWS Torridon, and Kandela, while the least plump were grains of the varieties Ethos and Koksa.

Grain uniformity ranged from 74.7% to 90.0% on average (Table 7). During the three year research period, the greatest uniformity, as with selectness, was observed in the grains of wheat varieties Katoda, Waluta, Arabella, KWS Torridon, and Kandela.

Table 7. Results of the assessment of grain milling value (averages over 2014–2016).

Variety	Selectness (%)	Uniformity (%)	Glassiness (%)	Hardness (BU) [1]	Ash (% d. m.)	Yield of Flour (%)
Arabella	88.0 a [2]	88.0 a	21 f	605 bcde	1.81 c	77.3 a
Brawura	83.0 bc	83.0 bc	35 cd	623 b	1.79 cd	76.8 a
Cytra	80.7 bc	80.7 bcd	42 b	598 cde	1.90 ab	75.8 a
Ethos	74.7 e	74.7 f	51 a	670 a	1.83 c	77.3 a
Izera	74.0 e	76.0 ef	28 e	585 efg	1.79 c	77.6 a
Kandela	83.7 bc	83.7 bc	15 f	563 g	1.90 ab	77.9 a
Katoda	90.0 a	90.0 a	17 f	608 bcd	1.79 cd	74.7 a
Koksa	74.7 e	77.0 def	38 bc	602 bcde	1.93 a	74.7 a
Korynta	75.3 e	75.3 f	38 bc	613 bc	1.81 c	76.7 a
KWS Torridon	84.0 b	84.0 b	34 cde	665 a	1.89 b	77.2 a
Ostka Smolicka	79.7 cd	79.7 cde	31 de	598 cde	1.89 ab	76.7 a
Waluta	89.0 a	89.0 a	19 f	588 def	1.75 d	77.9 a
Zadra	76.0 de	76.0 ef	33 cde	570 fg	1.88 b	78.0 a
Mean	81.0	81.3	31	607	1.84	76.8

[1] BU—conventional units in Brabender scale; [2] Different letters indicate significant differences between varieties according to Tukey's test at $\alpha = 0.05$.

The glassiness and hardness of grains indicate the structure of the endosperm. In the milling industry, glassy grains with a glassiness above 60% are classified as floury when the number of glassy grains is below 40%. In the 3 year testing period, the glassiness of grains was between 15% and 51% on average (Table 7). The highest percentage of grains with a glassy structure of the endosperm was observed in the following varieties: Ethos, Cytra, Koksa, and Korynta. The most floury grains were the Kandela, Katoda, and Waluta varieties.

During the 3 year testing period, the average grain hardness of the tested wheat varieties ranged from 563 BU to 670 BU (Table 7). Grains of Ethos, KWS Torridon, and Brawura wheat varieties had the hardest white, while grains of the Kandela and Zadra varieties had the softest.

The average content of mineral substances (total ash) in grains of the wheat varieties ranged from 1.75% d.m. to 1.93% d.m. (Table 7). The lowest ash content was found in grains of wheat varieties including Waluta, Brawura, Katoda, Arabella, Izera, and Korynta, while the highest total ash content was found in the grains of wheat varieties Koksa, Kandela, and Cytra.

During the three year research period, the average flour yields (extracts) ranged from 74.7% to 78.0% (Table 7). The obtained flour yields were high and comparable to those obtained in industrial mills. The largest flour extracts (above 77%) were obtained from milling grains of wheat varieties Zadra, Waluta, Kandela, Izera, Arabella, and Ethos.

During the three year testing period, the total protein content of the flours tested was, on average, from 9.5% d.m. to 11.2% d.m. (Table 8). The highest levels of this component were recorded from flours from the grain varieties Koksa, Korynta, Ethos, KWS Torridon, and Cytra.

Table 8. Results of the evaluation of the baking value of flour (averages over 2014–2016).

Variety	Total Protein Content (% d.m.)	Gluten Efficiency (%)	Gluten Index (−)	Falling Number (s)	Yield of Bread (%)	Bread Volume (cm^3 100 g^{-1})	Crumb Porosity (−)	Sensory Evaluation (Points)
Arabella	10.3 cde [1]	23.6 ef	98 a	296 abc	142.0 a	402 ab	68 a	30.6 ab
Brawura	9.9 ef	24.8 de	78 fg	314 a	142.1 a	380 cdef	63 ab	28.3 cd
Cytra	10.6 abc	28. ab	40 i	313 a	141.3 a	380 cdef	60 abc	26.8 d
Ethos	10.7 abc	29.9 a	71 h	284 bcd	142.9 a	369 defg	52 c	28.8 bcd
Izera	10.1 cdef	24.0 e	91 bc	256 f	142.3 a	398 bc	63 ab	30.6 ab
Kandela	9.6 f	21.7 g	95 ab	265 def	141.4 a	390 bcd	66 a	29.9 abc
Katoda	10.0 def	22.2 fg	94 ab	294 abc	141.8 a	420 a	66 a	31.0 a
Koksa	11.2 a	27.1 bc	90 bcd	279 cde	143.0 a	363 fg	57 bc	29.7 abc
Korynta	10.9 ab	27.0 c	86 cde	288 bc	140.9 a	388 bcde	63 ab	30.2 abc
KWS Torridon	10.6 bc	27.5 bc	83 ef	304 ab	141.7 a	364 efg	60 abc	29.8 abc
Ostka Smolicka	9.5 f	21.2 g	92 ab	281 cde	142.3 a	357 g	53 c	27.1 d
Waluta	10. 4 bcde	24.0 e	84 de	288 bc	139.9 a	401 ab	66 a	29.6 abcd
Zadra	10.5 bcd	26.0 cd	75 gh	260 ef	139.1 a	364 efg	53 c	26.9 d
Mean	10.3	25.2	83	286	141.6	383	61	29.2

[1] Different letters indicate significant differences between varieties according to Tukey's test at $\alpha = 0.05$.

During the 3 year study period, the average yields of wet gluten ranged from 21.2% to 29.9% (Table 8). The highest levels of gluten proteins were contained in flours from the grains of the following wheat varieties: Ethos, Cytra, KWS Torridon, Koksa, and Korynta, which were also characterized by the highest total protein content. As required by standards [40], the amount of gluten in low-extraction wheat flours should not be lower than 25%. The gluten yield meeting this requirement was recorded in flours obtained from the grains of varieties Cytra, Ethos, Koksa, Korynta, KWS Torridon, and Zadra.

The average values of the gluten index (GI) ranged from 40 to 98 (Table 8). Flours from most wheat varieties were distinguished by optimal gluten quality (GI values from 60 to 90). Regardless of the year of wheat harvest, exceptionally strong gluten was found in flours from the grains of the Arabella, Kandela, and Waluta varieties, while exceptionally weak gluten was found in the Cytra variety.

During the 3 year study period, the mean values of the falling number, which is the amylolytic enzyme activity index, ranged from 256 s to 314 s (Table 8). The optimal activity of amylolytic enzymes in flour intended for baking should be at an average level (a falling number in the range 220–280 s) [27]. This requirement was met by flours from the grain of Izera, Kandela, Koksa, and Zadra varieties. In wheat grain flours from the remaining varieties, the activity of amylolytic enzymes was lower. Before baking, it can be increased, e.g., by adding malt, which is a source of amylolytic enzymes.

The average bread yield (the amount of bread obtained from 100 parts of flour by weight) was not very diversified, ranging from 139.1% to 143.0% (Table 8). Bread with the highest yield was obtained from flour of the following varieties: Koksa, Ethos, Izera, and Ostka Smolicka.

Breads from the laboratory test baking had the correct taste and smell, typical of wheat bakery products. The shape of the loaves was correct, typical for bread baked in tins. The bread crust had the right thickness, the color ranging from light to dark brown. The growth of the loaves was varied. During the three year research period, the volume of the loaves was, on average, from 357 $cm^3 \cdot 100\ g^{-1}$ to 420 $cm^3 \cdot 100\ g^{-1}$ (Table 8). The largest volume of bread was obtained from flour of the wheat varieties Katoda, Arabella, Waluta, Izera, and Kandela, while the least risen bread loaves were obtained from Ostka Smolicka, Koksa, KWS Torridon, and Zadra.

Bread crumb was characterized by very good or good flexibility. It was varied in its porosity (Table 8, Figure 2). The crumb of bread from Arabella, Kandela, Katoda, and Waluta varieties was the most evenly distributed. The lowest rated in this aspect was the crumb of breads from flour from the varieties Ethos, Ostka Smolicka, and Zadra.

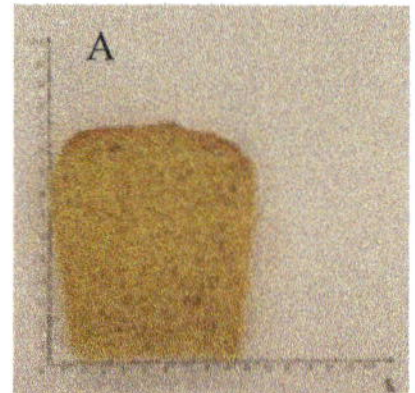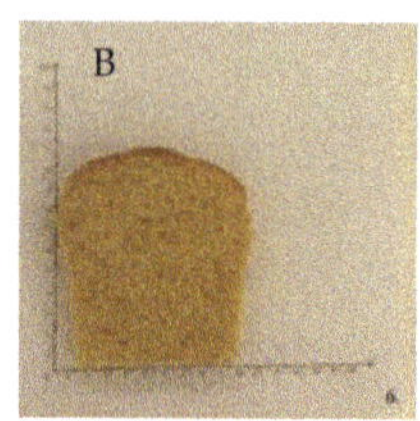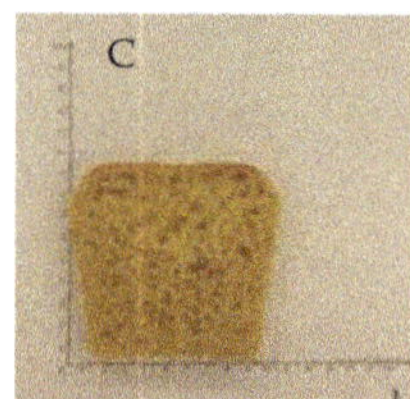

Figure 2. Comparison of crumb porosity in 2016: (**A**) variety Izera, porosity coefficient 70; (**B**) variety Kandela, porosity coefficient 65; (**C**) variety Cytra, porosity coefficient 50; (**D**) variety Zadra, porosity coefficient 45.

During the organoleptic evaluation, the highest value was given to bread made from flour from the grains of the following varieties: Katoda, Arabella, Izera, and Korynta, while the lowest value was from Cytra, Zadra, and Ostka Smolicka (Table 8). Based on the total number of points awarded, breads made from flour from the grains of the Cytra, Zadra, and Ostka Smolicka varieties were graded to the second quality level. Bakery products made of flour from grains of the other wheat varieties were graded to the first quality group.

On the basis of the results of a 3 year study on spring wheat varieties from organic systems that are recommended for the production of bakery products, it was concluded that the requirements of the baking industry were met, to the greatest extent, by flours from the grains of varieties Arabella, Izera, Kandela, Katoda, and KWS Torridon.

4. Discussion

In organic agriculture, chemical products are forbidden; thus, other agrotechnical, mechanical, and biological methods, such as choice of a proper variety, are used to achieve a high yield of good quality [4,16,41]. One of the important features of wheat varieties that should be taken into account when evaluating their usefulness for cultivation in organic agriculture and food processing is their susceptibility to fungal diseases. Fusarium head blight (FHB) and grain infestation by *Fusarium* spp. (FDK—Fusarium disease kernels) may be caused by various species of *Fusarium* that have different climatic requirements and different toxin formations [17,18]. The most dangerous mold fungi for wheat ears include *Fusarium culmorum*, *F. avenaceum*, and *F. graminearum*. Factors that influence the occurrence of FHB and fungal infestation of the grain primarily include the weather (rainfall and temperature) during the stages from BBCH 55 to BBCH 73, as well as crop management treatments applied, crop rotation, nitrogen fertilisation, variety, plant protection products, and the quantity of inoculum of *Fusarium* spp. on a given field [17,42–46]. The severity of the disease symptoms in this experiment in the years 2014–2016 was low, which was probably influenced by the weather conditions, and especially by the low rainfall in the 2015 and 2016. Significant differences among the wheat varieties were observed in each year of the study according to infestation of grain and, in one year, due to the infestation of ears, as other agricultural factors were not differentiated in the experiment.

Mycotoxins most commonly accumulated in grains during wheat growth include deoxynivalenol, zearalenone, T-2 and HT toxins, and nivalenone. The risk is even more significant because certain amounts of mycotoxins have also been found in grains harvested from ears without symptoms of FHB [47]. The content of mycotoxins during the analyzed period was low and did not exceed the standards permitted by the European Union [24]. However, due to their high level of harmfulness, there is a need for constant monitoring of their content.

The profile of fungi inhabiting the territory of Poland has changed in the last years. Therefore, the toxins in cereals have also changed. Many of them were not standardized because they appeared sporadically. Therefore, it is necessary to update these standards allowing for the presence of individual compounds and to determine their toxicity to humans and animals. In Poland, such species as *Fusarium subglutinans*, *Fusarium poae*, and *Fusarium verticilioides* occur, which primarily produce fumonisins, beauvericin, group A and B trichothecenes, and enniatins [48]. The same shift in species was also found in other countries of central Europe [49].

According to Gromadzka et al. [50], the highest amount of BEA content was recorded in 2015, with a maximum concentration of 1,731,550.0 $\mu g \cdot kg^{-1}$. In China, 82.3% of the wheat samples were contaminated by BEA, in the range from 0.04 $\mu g \cdot kg^{-1}$ to 1006.56 $\mu g \cdot kg^{-1}$, followed by EnnA, with the levels ranging from 0.06 $\mu g \cdot kg^{-1}$ to 16.61 $\mu g \cdot kg^{-1}$, and EnnB$_1$, with the levels ranging from 0.07 $\mu g \cdot kg^{-1}$ to 3.33 $\mu g \cdot kg^{-1}$ [51]. According to Chelkowski et al. [52], F. *poae* contributed to an accumulation of a high number of mycotoxins; namely, toxic hexadepsipeptides—both BEA (up to 46,000 $\mu g \cdot kg^{-1}$) and enniatins EnnA (up to 37,000 $\mu g \cdot kg^{-1}$), EnnB (up to 46,000 $\mu g \cdot kg^{-1}$) and EnnB$_1$ (up to 75,000 $\mu g \cdot kg^{-1}$). On the other hand, MON was found in 25,32, and 76% of the Norway samples of barley, oats, and wheat, respectively. The maximum MON concentrations in barley, oats, and wheat were 380, 210, and 950 $\mu g \cdot kg^{-1}$, respectively [53].

In the research of Orlando et al. [54], conducted in France, F. *tricinctum* or F. *poae* affected enniatin content. F. *tricinctum* was the leading enniatin producer in spring barley (23% to 37%). F. *avenaceum* produced large amounts of enniatins in wheat (1% to 18%). F. *poae* made a minor contribution, never accounting for more than 2% of total enniatin content. According to these authors, enniatins were highly prevalent in French small grain cereals and were mostly produced by F. *avenaceum* and F. *tricinctum*.

The presence and concentration of mycotoxins in grains can be influenced by many factors. One of the most important factors is the potential ability of fungi to form mycotoxins, as only a part of the isolates of a given species has the capacity to form secondary metabolites. Moreover, other grain-infesting fungi and weather conditions may affect the number of mycotoxins produced [55]. In plant protection, the breeding of varieties with increased resistance to *Fusarium* fungi has great importance. In our research, the varieties with the smallest colonization of the grain by *Fusarium* spp. were Waluta, Zadra, and Arabella. The grains of the following cultivars were the most infested: Cytra, Kandela, and KWS Torridon.

It could be assumed that crops in organic farming are more vulnerable to diseases as they lack chemical protection. However, there are reports in the literature that, under this cropping system, the intensity of *Fusarium* spp. is not higher and is often even lower than with conventional system [56–60]. According to a review of the literature, *Fusarium* diseases are found less often in cereals grown in the organic system than in the conventional one (this also applies to mycotoxin content) [61,62].

Bernhoft et al. [63,64] report that using crop rotation, as in the organic system, reduced both the occurrence of *Fusarium* spp. in grains, and the mycotoxin content, while the lack of mineral fertilization and herbicides reduced the occurrence of F. *graminearum*. They also found increased levels of *Fusarium* spp. and mycotoxins in the case of plant lodging, which often occurs in conventional cultivation as a result of high mineral fertilization. Other factors determining the lower pressure of *Fusarium* sp. in the organic system, as compared to the conventional one, were: higher overall biodiversity, lower compactness of the canopy for better aeration, and ear structure of individual varieties.

However, it should be taken into account that our research included as many as 12 metabolites of *Fusarium* sp. fungi. Additionally, compounds that were naturally synthesized in significant quantities were analyzed. Therefore, the results obtained constitute an important supplement to the world literature and are the basis for further detailed research.

The basic direction of the use of wheat grains is the production of various types of flour, which are primarily raw materials for the production of bread, but also pastries, pasta, noodles, dumplings, pancakes, etc. [27]. Wheat grains that are commercially marketed, irrespective of the farming system, must be healthy, clean, ripe, and free from foreign odors or pests [26]. The moisture content of the grains must not exceed 15.0%, while the bulk density must not be lower than 72.0 kg/hl. The maximum total content of impurities should not exceed 15% at the grain buying, including harmful and/or toxic seeds 0.5%, and ergot 0.05%. The activity of amylolytic enzymes, determined on the basis of the falling number, should not be lower than 160 s [16,27]. The grains from all the tested organic spring wheat varieties, regardless of the year of harvest, met the general quality requirements, which indicates that they could be marketed.

Depending on the type of intended processing, specific quality requirements for wheat grains are determined. The requirements of the milling industry mainly concern grain size and grain uniformity, the structure of the endosperm (glassiness, hardness), and the maximum ash content [27,65]. The larger (more plump) the grains are, the higher the share of the endosperm and the smaller the share of the peripheral layers. From such raw material, higher flour yields with a low ash content can potentially be obtained [65]. In addition to being more mature, grains should also have a degree of sizing uniformity of not less than 75%. This high level of uniformity facilitates the cleaning (removal of impurities) and conditioning of grains [27]. It could be assumed that grains from organic farming, where the use of mineral fertilizers is forbidden, will be smaller than grain from conventional farming. Data from the literature indicates, however, that organic farming produces wheat grain of comparable size to that of conventional farming [16,25,66–70]. The possibility of obtaining grains of organic wheat of high maturity and uniformity is also indicated by the results obtained in this work, in which the average values of grain selectness and uniformity ranged from 74.0% to 90.0%. As in previous studies [16,25,66], both grain selectness and uniformity were significantly dependent on varietal traits. In terms of these parameters, in this study, the highest ratings were given to the grains of the varieties Katoda, Waluta, and Arabella.

Hardness and glassiness (flouriness) are considered to be the basic indicators for assessing the structural and mechanical properties of grains, especially their basic anatomical part—endosperm. The greater the hardness and glassiness of the grains, the greater the resistance to grinding, which translates into higher energy consumption [69]. The glassiness and flouriness are related to the chemical composition of grains, especially the protein–starch ratio. Glassy grains have a higher protein content than floury grains and a more compact structure, which makes them more resistant to mechanical stress. From glassy grains, large quantities of flour from grinding passages are produced. Flours from these passages exhibit low ashiness and light color [65]. The grains of the tested organic wheat cultivars were characterized by a floury structure of the endosperm; the percentage of glassy grains ranged from 15% to 51% and depended significantly on varietal traits. According to Marzec et al. [25], grains of wheat from organic farming exhibit lower glassiness and hardness of the endosperm than grains from intensive farming due to a lower content of protein substances that form the matrix surrounding the starch grains.

Ash content is one of the basic parameters characterizing wheat grain in terms of milling. The ash content of the grains determines, to a large extent, the ashiness of the flour obtained from it. A good raw material for the milling industry is wheat grain with a low ash content (maximum 1.75–1.80% dry matter), especially with a low ash content of the endosperm [65]. The ash content in the grain of the examined wheat varieties was relatively high, ranging from 1.75% to 1.93% d.m. An ash content not exceeding 1.80% d.m. was found in the grains of four wheat varieties: Brawura, Izera, Katoda, and Waluta. According to the literature data [8,16,25,66], ash content in the grains of wheat from organic farming may be slightly higher than in the grains from intensive farming. Spring wheat grains generally contain more ash than winter wheat [65].

The yields of flours obtained from milling the grains of the tested wheat varieties from organic farming were high (over 74%). According to many authors [16,66,70], the yields of flour obtained from grains of organic wheat are comparable to those of grains from intensive cultivation, which indicates that it is a suitable raw material for milling into low-extraction flours.

Flours obtained from the milling of grains should have the appropriate utility traits that are desirable for further processing. In the case of wheat flour intended for the production of bread, the quantity and quality of protein substances are particularly important [27,71,72]. The protein content of wheat grain, which is the raw material for flour production, is favorably influenced by mineral nitrogen fertilizers, which, however, cannot be used in an organic production system [6]. As a result, the protein content in the flour of organic wheat grains is lower than that of intensive cultivation, as indicated by many authors [16,70,71,73,74]. The flours tested in this study had a relatively low protein content, ranging from 9.5% to 11.2% d.m. Significant differences in the content of this component indicate that some varieties (Koksa, Korynta, Ethos, Cytra, KWS Torridon) better used the nitrogen available in the soil for protein synthesis than other varieties (Ostka Smolicka, Kandela, Brawura).

Among the proteins present in wheat flour, gluten proteins are of particular importance. They have a structuring function in bread due to their ability to form a branched structure, which, when the dough is mixed, envelops the swollen starch grains and enables the retention of gases produced during dough fermentation [72]. The results of many studies [16,68,70,71,73,74] indicate that organic wheat grain flours have a lower gluten protein content, as well as total protein content, than intensive grain flours. In this study, depending on the variety, the amount of wet gluten isolated from the tested flours ranged from 21.2% to even 29.9%. The gluten washed out of most flours, except for grain flour of Cytra variety (GI 40), was of good quality (GI 71 to 98).

An important parameter in assessing the baking value of wheat is also the falling number, which is an indicator of α-amylase activity. Flours for baking purposes should have the average activity of this enzyme (falling number of 220–280 s) [27]. The organic flours tested in this study showed medium or low amylolytic activity, none of them having an increased value of this parameter. According to some authors [70,72], the activity of amylolytic enzymes in organic wheat grain flour may be higher than in intensive wheat grain flour, and, according to others [16,68], it is at a similar level.

The best method of evaluating the value of baked flour is to carry out laboratory baking combined with an evaluation of the quality of the baked goods obtained, including a sensory evaluation. Breads obtained from the tested organic flours were highly rated in terms of sensory quality. They exhibited an intensive smell and taste typical of wheat bread; there were no major objections to the external appearance of the loaf and crumb. The loaves were generally well-risen; the volume per 100 g of bread was in the range of 363–420 cm^3, which met the requirement set out in Polish standard (no less than 260 cm$^3\cdot$100 g^{-1}) [75]. Data from the literature [16,66,67,71] indicate that flours from organic wheat grains can be used to make bread of a similar volume to that made of flour from intensively farmed wheat. However, there are also reports [73,74] that indicate that bread made from organic grains has a smaller volume. In addition to the volume of the loaf, the porosity of the crumb is an important quality feature. Wheat bread should have an even, finely porous crumb with thin walls, which adheres well to the crust. The porosity of the crumb of bread obtained from the tested organic flours varied. The crumb with the best porosity was noted for breads made of flour from grains of the following wheat varieties: Arabella, Kandela, Katoda, and Waluta.

An evaluation of the technological value of the grains showed that the grains of the tested spring wheat varieties generally met the quality requirements for the milling and baking industries. In the presented ranking of spring wheat varieties, the Arabella variety, a quality bread variety, was ranked the highest. Arabella has been noted since 2009 in the Polish Varieties Register carried out by Research Centre for Cultivar Testing (COBORU) [76]. This variety is also useful for late autumn sowing. It tolerates slight soil acidification, which makes it suitable for cultivation on poorer, sandy soils. This variety was also ranked high according to the agricultural traits and yield in our previous research [10] so it can be especially recommended for organic production.

5. Conclusions

The presented study is in line with the current trends in the greening of agriculture and the organic agriculture development according to the recent European Union strategies, such as the European Green Deal, the Farm to Fork strategy, and the European Biodiversity Strategy for 2030, which aim to increase the share of organic farming to 25% area of agricultural land in Europe by 2030. Our results give information about the traits of spring wheat varieties that are suitable for an organic system and food processing. The most resistant varieties to the threat of infection by *Fusarium* spp. were Waluta, Zadra, and Arabella. Mycotoxin contamination of the grain of two examined wheat varieties did not exceed maximum accepted levels. However, we observed that the colonization of grains by *Fusarium* spp. did not reflect the intensity of FHB; thus, the lack of symptoms of disease on ears does not mean the lack of grain infestation by mycotoxins. Therefore, there is a need for constant monitoring of phytosanitary condition of the canopy and wheat grain in organic farming. On the basis of the results of a 3 year study, the varieties most useful for organic production and processing were found to be Arabella, followed by Brawura, Izera, Kandela, Katoda, KWS Torridon, Waluta, and Zadra. The research results could be applied by farmers, advisors, and food processors that carry out crop production and processing in organic agriculture sector. Future research will focus on the suitability of other wheat varieties, in particular winter varieties, for organic farming and food processing.

Author Contributions: Conceptualization, B.F.-S., G.C.-P., L.L., K.G.; methodology, B.F.-S., G.C.-P., L.L., K.G., D.D.; validation, B.F.-S., G.C.-P., D.D.; investigation, B.F.-S., G.C.-P., L.L., K.G.; data curation, B.F.-S., G.C.-P., L.L.; writing—original draft preparation, B.F.-S.; G.C.-P., L.L., K.G.; writing—review and editing, B.F.-S., G.C.-P., L.L., D.D.; visualization, B.F.-S., G.C.-P., L.L.; supervision, B.F.-S., G.C.-P., D.D.; project administration, B.F.-S.; funding acquisition, B.F.-S., D.D. All authors have read and agreed to the published version of the manuscript.

Funding: The study was supported by the Polish Ministry of Agriculture and Rural Development under the grant for the research on organic crop production No HORre-msz-078-23/16(243).

Institutional Review Board Statement: Not applicable.

Informed Consent Statement: Not applicable.

Data Availability Statement: The data presented in this study are available upon request from the first author.

Conflicts of Interest: The authors declare no conflict of interest. The funders had no role in the design of the study; in the collection, analyses, or interpretation of data; in the writing of the manuscript, or in the decision to publish the results.

References

1. Łączyński, A. *Land Use and Sown Area in 2018*; Central Statistical Office: Warsaw, Poland, 2019; p. 74. Available online: http://www.stat.gov.pl (accessed on 20 May 2020).
2. Food and Agriculture Organization of the United Nations. Crops. Available online: http://www.fao.org/faostat/en/#data/QC (accessed on 20 May 2020).
3. Hoad, S.; Topp, C.; Davies, K. Selection of cereals for weed suppression in organic agriculture: A method based on cultivar sensitivity to weed growth. *Euphytica* **2008**, *163*, 355–366. [CrossRef]
4. Lammerts van Bueren, E.T.; Jones, S.S.; Tamm, L.; Murphy, K.M.; Myers, J.R.; Leifert, C.; Messmer, M.M. The need to breed crop varieties suitable for organic farming, using wheat, tomato and broccoli as examples: A review. *NJAS Wagening. J. Life Sci.* **2011**, *58*, 193–205. [CrossRef]
5. Liatukas, Ž.; Leistrumite, A. Selection of winter wheat for organic growing. *Agron. Res.* **2009**, *7*, 381–386.
6. Przystalski, M.; Osman, A.; Thiemt, E.M.; Rolland, B.; Ericsson, L.; Østergård, H.; Levy, L.; Wolfe, M.; Büschse, A.; Piepho, H.P.; et al. Comparing the performance of cereal varieties in organic and non-organic cropping systems in different European countries. *Euphytica* **2008**, *163*, 417–433. [CrossRef]
7. Wolfe, M.S.; Baresel, J.P.; Desclaux, D.; Goldringer, I.; Hoad, S.; Kovacs, G.; Löschenberger, F.; Miedaner, T.; Østergård, H.; Lamberts van Bueren, E.T. Developments in breeding cereals for organic agriculture. *Euphytica* **2008**, *163*, 323–346. [CrossRef]
8. Dziki, D.; Cacak-Pietrzak, G.; Gawlik-Dziki, U.; Świeca, M.; Miś, A.; Różyło, R.; Jończyk, K. Physicochemical Properties and Milling Characteristics of Spring Wheat from Different Farming Systems. *J. Agr. Sci. Tech.* **2017**, *19*, 1253–1266. Available online: http://jast.modares.ac.ir/article-23-7491-en.html (accessed on 22 April 2021).
9. Cacak-Pietrzak, C.; Ceglińska, A.; Jończyk, K. Value of baking flour wheat variety with grain grown the organic system of production. *Zesz. Probl. Postępów Nauk. Rol.* **2014**, *576*, 23–32. (In Polish)
10. Feledyn-Szewczyk, B.; Cacak-Pietrzak, G.; Lenc, L.; Stalenga, J. Rating of Spring Wheat Varieties (*Triticum aestivum* L.). According to Their Suitability for Organic Agriculture. *Agronomy* **2020**, *10*, 1900. [CrossRef]
11. Feledyn-Szewczyk, B.; Nakielska, M.; Jończyk, K.; Berbeć, A.K.; Kopiński, J. Assessment of the Suitability of 10 Winter Triticale Cultivars (x *Triticosecale* Wittm. ex A. Camus) for Organic Agriculture: Polish Case Study. *Agronomy* **2020**, *10*, 1144. [CrossRef]
12. Tamm, I.; Tamm, Ü.; Ingver, A. Spring cereals performance in organic and conventional cultivation. *Agronomy Res.* **2009**, *7*, 522–527.
13. Kronberga, A. Selection criteria in triticale breeding for organic farming. *Agron. Vestis* **2008**, *11*, 89–94.
14. Lenc, L. Fusarium head blight and *Fusarium* spp. occuring on grain of spring wheat in an organic farming system. *Pytopathologia* **2011**, *62*, 31–39.
15. Siuda, R.; Grabowski, A.; Lenc, L.; Ralcewicz, M.; Spychaj-Fabisiak, E. Influence of the degree of fusariosis on technological traits of wheat grain. *Int. J. Food Sci. Tech.* **2010**, *45*, 2596–2604. [CrossRef]
16. Cacak-Pietrzak, G. Studies on the effect of organic and conventional system of plant production on the technological value of selected varieties of winter wheat. In *Treatises and Monographs*; Publications of Warsaw University of Life Sciences: Warsaw, Poland, 2011; pp. 1–83.
17. Parry, D.W.; Jenkinson, P.; Mc Leod, L. Fusarium ear blight (scab) in small grain cereals—A review. *Plant Pathol.* **1995**, *44*, 207–238. [CrossRef]
18. Wiśniewska, H.; Stępień, Ł.; Waśkiewicz, I.; Beszterda, M.; Góral, T.; Belter, J. Toxicenic *Fusarium* species infecting wheat heads in Poland. *Cent. Eur. J. Biol.* **2014**, *9*, 163–172.
19. Gromadzka, K.; Waśkiewicz, A.; Chełkowski, J.; Goliński, P. Zearalenone and its metabolites—Occurrence, detection, toxicity and guidelines. *World Mycotoxin J.* **2008**, *1*, 209–220. [CrossRef]

20. Wiśniewska, H.; Perkowski, J.; Kaczmarek, Z. Scab response and deoxynivalenol accumulation in spring wheat kernels of different geographical origins following inoculation with *Fusarium culmorum*. *J. Phytopathol.* **2004**, *152*, 613–621. [CrossRef]
21. Desjardins, A.E. *Fusarium Mycotoxins, Chemistry, Genetics and Biology*; The American Phytopathological Society: St. Paul, MN, USA, 2006; p. 220.
22. Tomoda, H.X.; Huang, H.; Nishida, H.; Nagao, R.; Okuda, S.; Tanaka, H.; Omura, S.; Arai, H.; Inoe, K. Inhibition of acyl-coA: Cholesterol acyltransferase activity by cyclodepsipeptide antibiotics. *J. Antibiot.* **1992**, *45*, 1626–1632. [CrossRef]
23. Logrieco, A.; Rizzo, A.; Ferracane, R.; Ritieni, A. Occurrence of Beauvericin and Enniatins in Wheat Affected by *Fusarium avenaceum* Head Blight. *Appl. Environ. Microbiol.* **2002**, *68*, 82–85. [CrossRef] [PubMed]
24. Commissions Regulations (EC) No 1881/2006 of 19 December 2006 Setting Maximum Levels for Certain Contaminants in Foodstuffs. Available online: https://eur-lex.europa.eu/legal-content/EN/TXT/PDF/?uri=CELEX:02006R1881-20140901&rid=9 (accessed on 28 May 2020).
25. Marzec, A.; Cacak-Pietrzak, G.; Gondek, E. Mechanical and acoustic properties of spring wheat versus its technological quality factors. *J. Texture Stud.* **2011**, *42*, 319–329. [CrossRef]
26. The Polish Committee for Standardization. *PN-R-74103: 1996. Cereal Grains. Common Wheat*; The Polish Committee for Standardization: Warsaw, Poland, 1996.
27. Cacak-Pietrzak, G. The use of wheat in various branches of the food industry—Technological requirements. *Przegląd Zboż. Młyn.* **2008**, *52*, 11–13. (In Polish)
28. IUSS Working Group WRB. International soil classification system for naming soils and creating legends for soil maps. In *World Reference Base for Soil Resources*; World Soil Resources Reports, No. 106; Food and Agriculture Organization of the United Nations: Rome, Italy, 2014; pp. 1–191.
29. EU Common Catalogue of Varieties of Agricultural Plant Species. Consolidated Version, 20.09.2019. 2019. Available online: https://ec.europa.eu/food/sites/food/files/plant/docs/plant_variety_catalogues_agricultural-plant-species.pdf (accessed on 28 May 2020).
30. Council Regulation (EC) No 834/2007 of 28 June 2007 on Organic Production and Labelling of Organic Products and Repealing Regulation (EEC) No 2092/91. Available online: http://eur-lex.europa.eu/legal-content/EN/TXT/PDF/?uri=CELEX:32007R0834&from=EN\T1\textgreater{} (accessed on 28 May 2020).
31. Booth, C. *The Genus Fusarium*; Commonwealth Mycological Institute: Surrey, UK, 1971; p. 237.
32. Kwaśna, H.; Chełkowski, J.; Zajkowski, P. *Fungi, XII*; Polish Academy of Sciences: Warsaw-Cracow, Poland, 1991; p. 136.
33. Visconti, A.; Pascale, M. Determination of zearalenone in corn by means of immunoaffinity clean-up and high-performance liquid chromatography with fluorescence detector. *J. Chromatogr. A* **1998**, *815*, 133–140. [CrossRef]
34. Tomczak, M.; Wiśniewska, H.; Stepień, Ł.; Kostecki, M.; Chełkowski, J.; Goliński, P. Deoxynivalenol, nivalenol and moniliformin in wheat samples with head blight (scab) symptoms in Poland (1998-2000). *Eur. J. Plant Pathol.* **2002**, *108*, 625–630. [CrossRef]
35. Kostecki, M.; Wiśniewska, H.; Perrone, G.; Ritieni, A.; Goliński, P.; Chełkowski, J.; Logrieco, A. The effects of cereal substrate and temperature on production of beauvericin, moniliformin and fusaproliferin by *Fusarium subglutinans* ITEM-1434. *Food Addit. Contam.* **1999**, *16*, 361–365. [CrossRef]
36. Waśkiewicz, A.; Goliński, P.; Karolewski, Z.; Irzykowska, L.; Bocianowski, J.; Kostecki, M.; Weber, Z. Formation of fumonisins and other secondary metabolites by *Fusarium oxysporum* and *F. proliferatum* a comparative study. *Food Addit. Contam.* **2010**, *27*, 608–615. [CrossRef] [PubMed]
37. Cacak-Pietrzak, G. Cereal grains. In *Selected Issues from Plant-Based Food Technology*; Mitek, M., Leszczyński, K., Eds.; SGGW Publication: Warsaw, Poland, 2014; pp. 261–273. (In Polish)
38. AACC. AACC Method 08-01.01: Total ash. Method 38-12.02: Wet gluten, Dry Gluten, Water—Binding Capacity, and Gluten Index. Method 46-11.02: Crude Protein—Improved Kjeldahl Method, Copper Catalyst Modyfication. Method 5681.03: Determination of Falling Number, 11th ed. In *Official Methods of the American Association of Cereal Chemists*; AACC: St. Paul, MN, USA, 2010.
39. The Polish Committee for Standardization. *PN-A-74108: 1996. Bread. Methods of Research*; The Polish Committee for Standardization: Warsaw, Poland, 1996.
40. The Polish Committee for Standardization. *PN-91/A-74022:1992. Cereal Preparations. Wheat Flour*; The Polish Committee for Standardization: Warsaw, Poland, 1992.
41. Finney, D.M.; Creamer, N.G. *Weed Management on Organic Farms*; The Organic Production Publication Series; Center for Environmental Farming Systems (CEFS), North Carolina Cooperative Extension Service: Raleigh, NC, USA, 2008; pp. 1–34.
42. Jennings, P.; Turner, J.A. Towards the Prediction of Fusarium Ear Blight Epidemics in the UK—The Role of Humidity in Disease Development. In *Brighton Crop Protection Conference: Pest & Diseases 1996*; 3D-12; British Crop Protection Council: Farnham, UK, 1996; pp. 233–238.
43. Lemmens, M.; Buerstmayr, H.; Krska, R.; Schuhmacher, R. The effect of inoculation treatment and long-term application of moisture on Fusarium head blight symptoms and deoxynivalenol contamination in wheat grains. *Eur. J. Plant Pathol.* **2004**, *110*, 299–308. [CrossRef]
44. De Wolf, E.D.; Madden, L.V.; Lipps, P.E. Risk assessment models for wheat Fusarium head blight epidemics based on within-season weather data. *Phytopathology* **2003**, *93*, 428–435. [CrossRef]

45. Xu, X.M.; Nicholson, P.; Thomsett, M.A.; Simpson, D.; Cooke, B.M.; Doohan, F.M.; Brennan, J.; Monaghan, S.; Moretti, A.; Mule, G.; et al. Relationship between the fungal complex causing Fusarium head blight of wheat and environmental conditions. *Phytopathology* **2008**, *98*, 69–78. [CrossRef]
46. Cowger, C.; Patton-Özkurts, J.; Brown-Guedira, G.; Perugini, L. Postanthesis moisture increased Fusarium head blight and deoxynivalenol levels in North Carolina winter wheat. *Phytopathology* **2009**, *99*, 320–327. [CrossRef] [PubMed]
47. Lacey, J.; Bateman, G.L.; Mirocha, C.J. Effects of infection time and moisture on development of ear blight and deoxynivalenol production by Fusarium spp. in wheat. *Ann. Appl. Biol.* **1999**, *134*, 277–283. [CrossRef]
48. Gromadzka, K.; Górna, K.; Chełkowski, J.; Waśkiewicz, A. Mycotoxins and related Fusarium species in preharvest maize ear rot in Poland. *Plant Soil Environ.* **2016**, *62*, 348–354. [CrossRef]
49. Adler, A.; Lew, H.; Brunner, S.; Oberforster, M.; Hinterholzer, J.; Kulling-Gradinger, C.M.; Mach, R.L.; Kubicek, C.P. Fusaria in Austrian cereals—Change in species and toxins spectrum. *J. Appl. Genet.* **2002**, *43A*, 11–16.
50. Gromadzka, K.; Błaszczyk, L.; Chełkowski, J.; Waśkiewicz, A. Occurrence of Mycotoxigenic Fusarium Species and Competitive Fungi on Preharvest Maize Ear Rot in Poland. *Toxins* **2019**, *11*, 224. [CrossRef]
51. Han, X.; Xu, W.; Zhang, J.; Xu, J.; Li, F. Natural occurrence of beauvericin and enniatins in corn- and wheat-based samples harvested in 2017 collected from Shandong Province, China. *Toxins* **2019**, *11*, 9. [CrossRef] [PubMed]
52. Chełkowski, J.; Ritieni, A.; Wiśniewska, H.; Mule, G.; Logrieco, A. Occurrence of toxic hexadepsipeptides in preharvest maize ear rot infected by Fusariun poae in Poland. *J. Phytopathol.* **2007**, *155*, 8–12. [CrossRef]
53. Uhlig, S.; Torp, M.; Jarp, J.; Parich, A.; Gutleb, A.C.; Krska, R. Moniliformin in Norwegian grain. *Food Addit. Contam.* **2004**, *21*, 598–606. [CrossRef]
54. Orlando, B.; Grignon, G.; Vitry, C.; Kashefifard, K.; Valade, R. *Fusarium* species and enniatin mycotoxins in wheat, durum wheat, triticale and barley harvested in France. *Mycotoxin Res.* **2019**, *35*, 369–380. [CrossRef] [PubMed]
55. Stępień, Ł.; Chełkowski, J. Fusarium head blight of wheat: Pathogenic species and their mycotoxins. *World Mycotoxin J.* **2010**, *3*, 107–119. [CrossRef]
56. Dexter, J.E.; Nowicki, T.W. Safety assurance and quality assurance issues associated with Fusarium head blight in wheat. In *Fusarium Head Blight of Wheat and Barley*; Leonard, K.J., Bushnell, W.R., Eds.; APS Press: St. Paul, MN, USA, 2003; pp. 420–460.
57. Champeil, A.; Fourbet, J.F.; Dore, T.; Rossignol, L. Influence od cropping system of Fusarium head blight and mycotoxin levels in winter wheat. *Crop Prot.* **2004**, *23*, 6, 531–537. [CrossRef]
58. Lenc, L. Fusarium head blight (FHB) and Fusarium populations in grain of winter wheat grown in different cultivation systems. *J. Plant Prot. Res.* **2015**, *55*, 94–109. [CrossRef]
59. Stein, J.M.; Osborne, L.E.; Bondalapati, K.D.; Glover, K.D.; Nelson, C.A. Fusarium head blight and deoxynivalenol concentration in wheat response to Gibberella zeae inoculum concentration. *Phytopathology* **2009**, *99*, 759–764. [CrossRef] [PubMed]
60. Schneweis, I.; Meyer, K.; Ritzmann, M.; Hoffmann, P.; Dempfle, L.; Bauer, J. Influence of organically or conventionally produced wheat on health, performance and mycotoxin residues in tissues and bile of growing pigs. *Arch. Anim. Nutr.* **2005**, *59*, 155–163. [CrossRef]
61. Sadowski, C.; Lenc, L.; Kuś, J. Fusarium head blight and *Fusarium* spp. on grain of winter wheat, a mixture of cultivars and spelt grown in organic system. *J. Res. Appl. Agric. Eng.* **2010**, *55*, 79–84.
62. Benbrook, C.M. Breaking the mold—Impacts of organic and conventional farming systems on mycotoxins in food and livestock feed. In *An Organic Center State of Center Review*; The Organic Center: Washington, DC, USA, 2005; pp. 1–58.
63. Bernhoft, A.; Clasen, P.E.; Kristoffersen, A.B.; Torp, M. Less Fusarium infestation and mycotoxin contamination in organic than in conventional cereals. *Food Addit. Contam. Part A Chem. Anal. Control Expo. Risk Assess.* **2010**, *27*, 842–852. [CrossRef] [PubMed]
64. Bernhoft, A.; Torp, M.; Clasen, P.E.; Løes, A.K.; Kristoffersen, A.B. Influence of agronomic and climatic factors on Fusarium infestation and mycotoxin contamination of cereals in Norway. *Food Addit. Contam. Part A Chem. Anal. Control Expo. Risk Assess.* **2012**, *29*, 1129–1140. [CrossRef] [PubMed]
65. Cacak-Pietrzak, G.; Ceglińska, A.; Torba, J. The milling value of some wheat cultivars from Breeding "Nasiona Kobierzyc". *Pam. Puławski.* **2005**, *139*, 27–38. (In Polish)
66. Cacak-Pietrzak, G.; Ceglińska, A.; Jończyk, K. Technological value of selected winter wheat varieties cultivated in different crop production systems. *Pam. Puławski.* **2003**, *133*, 17–25. (In Polish)
67. Mäder, P.; Hahn, D.; Dubois, D.; Gunst, L.; Alföldi, T.; Bergmann, H.; Oehme, M.; Amadò, R.; Schneider, H.; Graf, U.; et al. Wheat quality in organic and conventional farming: Results of 21 year field experiment. *J. Sci. Food Agric.* **2007**, *87*, 1826–1835. [CrossRef]
68. Lacko-Bartošovà, M.; Korczyk-Szabó, J. Winter wheat productivity and quality in sustainable farming systems. In *Scientific Papers Agriculture*; Agroprint: Timisoara, Romania, 2008; Volume 40, pp. 9–14.
69. Dziki, D.; Cacak-Pietrzak, G.; Miś, A.; Jończyk, K.; Gawlik-Dziki, U. Influence of wheat kernel physical properties on the pulverizing process. *J. Food Sci. Technol.* **2014**, *51*, 2648–2655. [CrossRef] [PubMed]
70. Mazurkiewicz, J. Comparison of the technological quality of wheat and rye cultivated in conventional and ecological farm conditions. *Acta Agroph.* **2005**, *6*, 729–741. (In Polish)
71. Cacak-Pietrzak, G.; Ceglińska, A.; Jończyk, K.; Kuś, J. Utilization on chosen winter wheat cultivars from ecological cropping system to the bread production. *Fragm. Agron.* **2008**, *1*, 67–75. (In Polish)
72. Wieser, H. Chemistry of gluten proteins. *Food Microbiol.* **2007**, *24*, 115–119. [CrossRef]

73. Kihlberg, I.; Öström, A.; Johansson, L.; Rosvik, E. Sensory qualities of plain white pan bread: Influence of farming system, milling and baking technique. *J. Cer. Sci.* **2006**, *43*, 15–30. [CrossRef]
74. Krejčiřová, L.; Capouchová, I.; Petr, J.; Bicanová, E.; Famera, O. The effect of organic and conventional growing systems on quality and storage protein composition of winter wheat. *Plant. Soil Environ.* **2007**, *53*, 499–505. [CrossRef]
75. The Polish Committee for Standardization. *PN-92/A-74105: 1993. Plain and Choice Wheat Bread*; The Polish Committee for Standardization: Warsaw, Poland, 1993.
76. Research Centre for Cultivar Testing (COBORU). National List of Varieties. Available online: https://coboru.gov.pl/English/Rejestr_eng/gat_w_rej_eng.aspx (accessed on 20 March 2021).

Article

Microbiome Analysis of the Rhizosphere from Wilt Infected Pomegranate Reveals Complex Adaptations in Fusarium—A Preliminary Study

Anupam J. Das [1,2], Renuka Ravinath [1], Talambedu Usha [2], Biligi Sampgod Rohith [3], Hemavathy Ekambaram [4], Mothukapalli Krishnareddy Prasannakumar [5], Nijalingappa Ramesh [1] and Sushil Kumar Middha [4,*]

[1] School of Applied Sciences, Reva University, Rukmini Knowledge Park, Bangalore 562149, Karnataka, India; anupam.das@molsys.in (A.J.D.); renukamadhu.rism@gmail.com (R.R.); dean.tpp@reva.edu.in (N.R.)
[2] Department of Biochemistry, Maharani Lakshmi Ammanni College for Women, Bangalore 560012, Karnataka, India; ushatalambedu@mlacw.edu.in
[3] Molsys Pvt. Ltd., Yelahanka, Bangalore 561249, Karnataka, India; rohith.molsys@gmail.com
[4] DBT-BIF Facility, Department of Biotechnology, Maharani Lakshmi Ammanni College for Women, Bangalore 560012, Karnataka, India; hemavathykmbrm@yahoo.co.in
[5] Department of Plant Pathology, University of Agricultural Sciences, Bangalore 560065, Karnataka, India; babu_prasanna@rediffmail.com
* Correspondence: drsushilmiddha@mlacw.edu.in; Tel.: +91-988-6098267

Citation: Das, A.J.; Ravinath, R.; Usha, T.; Rohith, B.S.; Ekambaram, H.; Prasannakumar, M.K.; Ramesh, N.; Middha, S.K. Microbiome Analysis of the Rhizosphere from Wilt Infected Pomegranate Reveals Complex Adaptations in Fusarium—A Preliminary Study. *Agriculture* 2021, 11, 831. https://doi.org/10.3390/agriculture11090831

Academic Editors: Alessandra Durazzo and Anna Andolfi

Received: 19 June 2021
Accepted: 18 August 2021
Published: 30 August 2021

Publisher's Note: MDPI stays neutral with regard to jurisdictional claims in published maps and institutional affiliations.

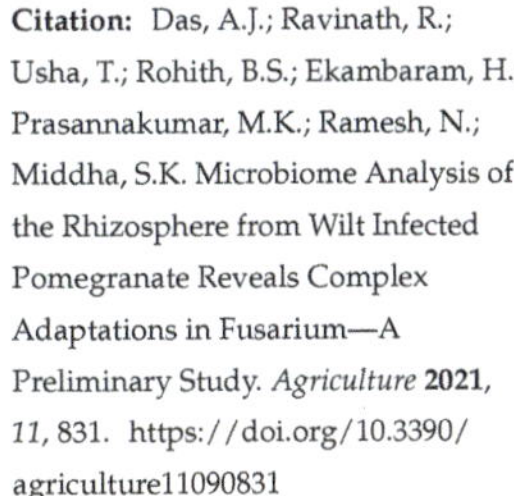

Abstract: Wilt disease affecting pomegranate crops results in rapid soil-nutrient depletion, reduced or complete loss in yield, and crop destruction. There are limited studies on the phytopathogen *Fusarium oxysporum* prevalence and associated genomic information with respect to *Fusarium* wilt in pomegranate. In this study, soil samples from the rhizosphere of different pomegranate plants showing early stage symptoms of wilt infection to an advanced stage were collected from an orchard situated in Karnataka, India. A whole metagenome sequencing approach was employed to gain insights into the adaptations of the causative pathogen *F. oxysporum*. Physicochemical results showed a drop in the pH levels, N, Fe, and Mn, and increase in electrical conductivity, B, Zn, Cl, Cu was observed in the early and intermediate stage samples. Comparative abundance analysis of the experimental samples ESI and ISI revealed an abundance of Proteobacteria phyla *Achromobacter* sp. 2789STDY5608625, *Achromobacter* sp. K91, and *Achromobacter aegrifaciens* and Eukaryota namely *Aspergillus arachidicola*, *Aspergillus candidus*, and *Aspergillus campestris*. Functional pathway predictions implied carbohydrate binding to be significant ($p < 0.05$) among the three experimental samples. Microbiological examination and whole microbiome analysis confirmed the prevalence of *F. oxysporum* in the soil samples. Variant analysis of *F. oxysporum* revealed multiple mutations in the 3IPD gene with high impact effects. 3-Isopropylmalate dehydratase and carbohydrate-active enzymes could be good targets for the development of antifungals that could aid in biocontrol of *F. oxysporum*. The present study demonstrates the capabilities of the whole metagenome sequencing approach for rapid identification of potential key players of wilt disease pathogenesis wherein the symptomatology is complex.

Keywords: microbiomics; soil metagenomics; DNA sequencing; wilt; rot; *Punica granatum*

1. Introduction

Pomegranate is a widely cultivated fruit crop with its origins traced to Turkey and Iran. The crop is extensively cultivated in various parts of India and India has emerged as the leading producer of pomegranate globally [1–3]. There are extensive reports on the medicinal properties of the pomegranate viz it's antimicrobial [4], antihyperglycemic [5], anticancerous [6,7], its nutraceutical [8], pharmaceutical [9], and cosmeceutical [10] applications due to the presence of a wide range of nutrients, secondary metabolites such as alkaloids and flavonoids [2]. There are also reports of anti-inflammatory properties and the potential of the pomegranate juice and peel against various disorders [2,3,11–13] and protection from UV photodamage [14]. However, cultivating the crop has been challenging due to its susceptibility to diseases and pest infestations which results in a drastic reduction in the yield and quality of the fruit. Major diseases are caused as a result of bacterial and fungal infections. Some of the fungal pathogens reported are *Colletotrichum acutatum* [15] *Trichoderma* spp., *Botrytis cinerea*, *Aspergillus niger*, *Penicillium* spp., *Alternaria* spp., *Colletotrichum gloeosporioides*, *Pestalotia brevista,* and *Pilidiella granati* [16–19] anthracnose disease of the flower. The infection results in the abortion of the flower leading to a reduction in the yield [15]. The manifestations of infection caused by *Pilidiella granati* are crown rot, twig blight, and dieback with common symptoms of necrosis in fruits and twigs respectively during the early stage of infection [15,19]. Among the bacterial diseases, blight disease is one of the serious challenges faced by farmers in India. The causative organism has been identified as *Xanthomonas campestris.* pv. *punicae.* Yield loss of up to 80% has been reported in Bangalore, Karnataka as a result of an epidemic outbreak. The pathogen infects the entire plant. Epidemic outbreaks are also reported in Andhra Pradesh, Maharashtra, and Delhi [20]. The detection of the fungal and bacterial pathogens is generally done through the isolation of the organism followed by culturing them. The identification is done based on the morphological characters and physiological and biochemical tests. These methods are highly labor-intensive, time-consuming, need expertise [21] and only the cultivable organisms can be identified. These limitations were later overcome by PCR based diagnostic method. The identification and detection of *P. granati* were carried out through the nested PCR method. Species-specific primers were designed and the method could effectively detect the pathogen in the fruits of pomegranate [22]. In another study, the phytopathogen *Xanthomonas campestris* pv. *punicae*, causing blight disease in pomegranate was detected using ERIC-PCR-Generated genomic fingerprints. A relationship was established between the fingerprints and virulence pattern of the blight-causing pathogen [20]. These methods have limitations with respect to specificity as they are not based on DNA sequencing.

16S rRNA gene sequencing is an excellent approach to reveal the identity of the pathogen as they are signature specific sequences in bacterial species with higher accuracy. Bacterial wilt disease in *Cucurbita maxima* in China caused by *Ralstonia solanacearum* was identified by 16S rRNA gene sequencing of the isolates obtained from the plants infected with wilt. Pathogenicity analysis revealed that all the isolates belonged to *Ralstonia solanacearum* [23]. Investigations on determining the microbiota associated with symptomatic and non-symptomatic bacterial wilt-diseased banana plants were also done using 16S rRNA metagenome sequencing. Illumina MiSeq platform was used for sequencing. The results revealed the predominance of *Ralstonia* in the pseudostem of the symptomatic diseased plant compared to non-symptomatic [24]. The findings could also throw light on the role of endophytic microbes revealed through sequencing studies in conferring tolerance to the disease. Many successful studies have been carried out in fruit crops and vegetables, where 16S rRNA gene sequencing has emerged as an excellent tool for the detection of the associated plant pathogens. Another newly reported disease in pomegranate is the Bacterial root-bark necrosis disease and wilt in pomegranate, which was found to affect the plant entirely. A recent study by Ajaysree and Borkar, 2018, shed light on the symptomatology of the disease that includes symptoms of wilt disease on the leaves and stem such as yellowing of leaves, followed by leaf fall and wilting of branches. The study reports complete death of the plant with no recovery in a period of 2–3 months. On the other hand, reports suggest that the roots of plants show symptoms of root-bark necrosis. 16S rRNA sequencing facilitated the identification of the pathogenic bacterium *Klebsiella pneumoniae* [25].

To explore the correlation between monocropping followed in banana and the Fusarium wilt incidence, the soil samples from such fields were subjected to sequencing of 16S rRNA genes for bacteria and internal transcribed spacer using the MiSeq platform for fungal identification. The findings led to the conclusion that monocropping significantly increased the incidence of Fusarium wilt [26], with the help of 16S metagenomics the role of the cropping system in disease management.

16S rRNA sequencing using the 454 platforms could accurately reveal the bacteria associated with the nematodes infesting pine trees. This association is responsible for the wilt disease of pine. 25 Operational Taxonomic Units could be analyzed based on 97% of similarity in the sequences of the library. The microbial diversity revealed Alphaproteobacteria, Betaproteobacteria, Gammaproteobacteria, and Bacteroidetes [27]. These findings are vital for adopting the proper control measures for wilt disease as it is influenced by the nematodes as well as the associated microbiota establishing a unique ecosystem.

However, despite all the benefits of employing the 16S metagenomics approach, there are certain limitations to consider prior to planning a soil metagenomics study. Firstly, soil microbial diversity is vast, and exploring the soil communities with a targeted approach that considers quantifying relative abundances of taxa may remain incomplete in terms of its functional potential. Secondly, the resolution is dependent on the databases employed. There are large scale efforts put towards developing database and tools to improve classification of bacterial communities and their diversity [28–30] The emergence of long read platforms have offered potential solutions to help sequence the entire 16S rRNA using the Nanopore or PacBio platforms. Some studies have reported higher microbial identification and taxonomic resolution as compared to the short amplicon sequencing despite the higher depth from platforms such as Illumina [31]. To specifically identify fungi, the gene cluster within the 18S ribosomal RNA is considered and the repetitive internal transcribed spacer (ITS) sequences are used. Furthermore, 18S rRNA sequencing comes with its limitations particularly with extraction methods showing biased results [32], primer-biases in PCR resulting in amplification of certain taxa preferentially [33], the copy numbers of the small subunit (SSU) rRNA genes [34], sequencing errors [35,36] and remnant DNA amplification [37].

The whole metagenome sequencing approach has helped address some of these limitations pertaining to targeted metagenome approaches. Microbe-pest-host associations are complex and their adaptations remain elusive. The whole metagenome approach is a powerful method to not only screen microbes but also facilitate understanding of plant-microbe-soil interactions and the disease pathogenesis in plants. The advanced Illumina Novaseq 6000 (Illumina) offers a unique possibility to perform soil microbial characterization [38,39]. Despite the large data outputs from these platforms, bioinformatics analysis of the data employing server and cloud-based analytics services have enhanced the speed and efficiency of analysis [40]. There are limited studies on the phytopathogen *Fusarium oxysporum* prevalence and associated genomic information with respect to Fusarium wilt in pomegranate. In the present study, we demonstrate the implementation of shotgun sequencing using the whole metagenome approach to study the pathogenomics of wilt disease in *Punica granatum* caused by Fusarium, wherein the symptomatology is complex.

2. Materials and Methods

The present study involved screening the physiochemical parameters of the soil samples from the rhizosphere of the infected plants. The total microbial counts were estimated. Employing conventional microbiological methods the soil samples were plated on specialized media to confirm the presence of the pathogen *Fusarium oxysporum* and *Aspergillus niger*. Following which genomic DNA was isolated from soil samples and the quality control of the samples was performed. Whole metagenome shotgun sequencing was carried out. Thereafter, the data was subject to bioinformatics analysis to estimate the relative abundances of the microbes, and the functional predictions were performed. Finally, variant analysis was carried out to screening for possible targets that could provide key leads to understand the adaptations of the pathogen. An overview of the methodology adapted in the study is provided in Figure 1.

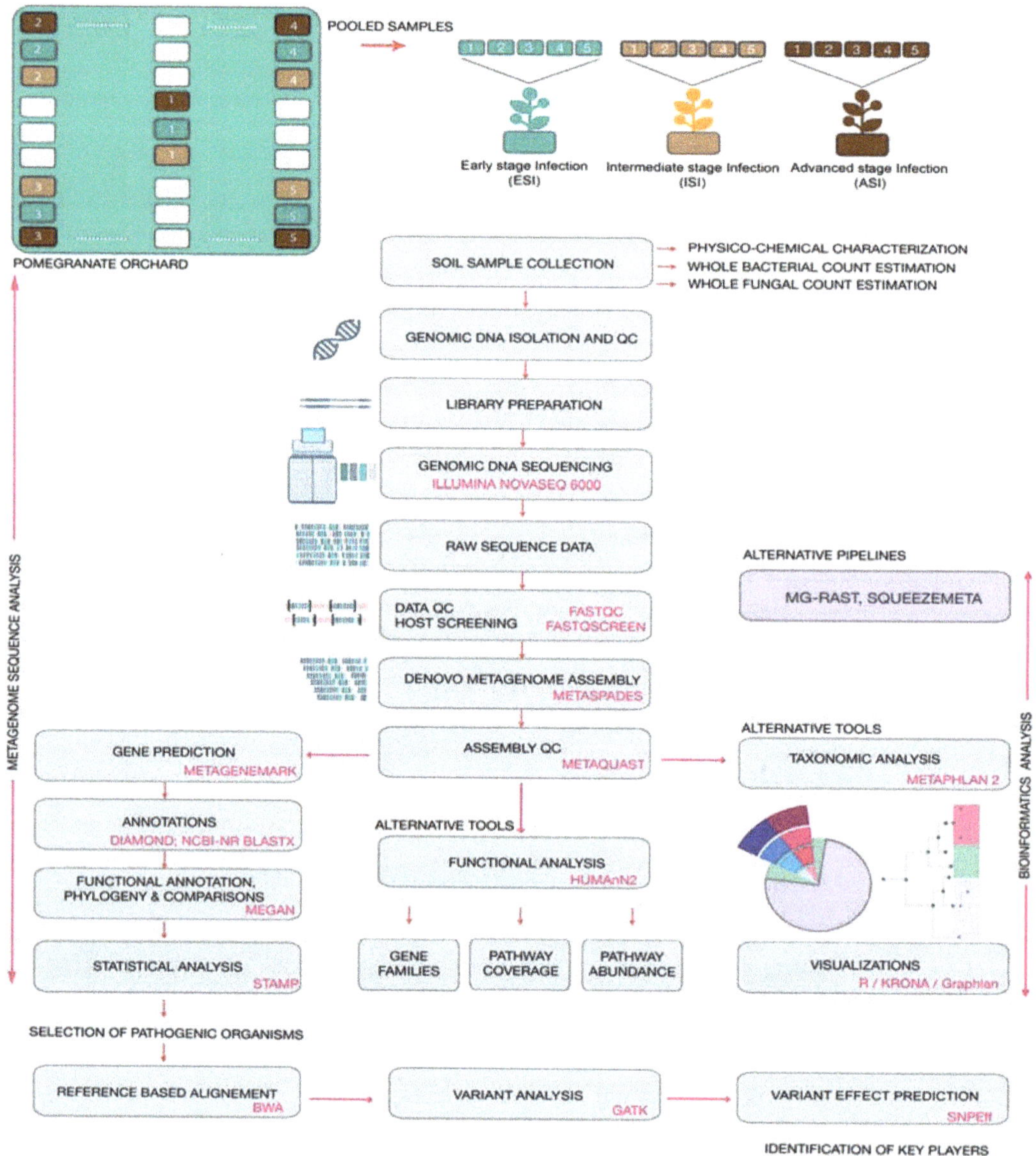

Figure 1. Overview of the experimental workflow from sample collection to identifying key players in microbial adaptation. The figure depicts the wet laboratory and dry laboratory methods employed in the present study.

2.1. Site Description and Sampling

Soil samples were collected from a pomegranate orchard close to Chikkaballapur (13.3907° N, 77.6880° E) from Karnataka, India. The orchard has been used for cultivating the crop over a span of 5 years and the farmer suffered huge losses due to the reducing fruit yield attributed to severe pest infestations. In the past year alone, the farmer suffered a loss in fruit yield by over 36%. The land was surveyed and post-harvest, without the application of any pesticides or antibacterial or antifungal agents, samples were collected in December 2019 from the rhizosphere of 5 plants from each category showing similar symptoms. The symptoms were categorized as early signs of infection, moderate signs of infection, on and severe infections, in triplicates and pooled (Figure 1). The plants were identified as early-stage infection (ESI) with early symptoms of wilt, intermediate stage of infection (ISI), and advanced stage of infection (ASI) on the basis of physical examination of the leaves, stem, fruits, and roots (Figure 2).

Figure 2. Physical symptoms of the plants, figures depict the physical symptoms of the infected plants—ESI; Early-stage infection, ISI; Intermediate stage infection, and ASI; Advanced stage infection (**a–d**). Other wilt-associated symptoms such as yellowing of leaves and root knots observed (**e**), rotting of fruit (**g**), complete defoliation (**f**), brown decay, and sporulation (**h**) observed are depicted.

2.2. Physicochemical Characterization and Total Microbial Count Estimation

All the physical and chemical characterizations were carried out based on the procedures provided by [41]. pH values, electrical conductivity, were estimated by the electrometric method [41]. The total microbial counts were estimated using the protocols provided in IS 5402 and IS 5403 for the total bacterial and total fungal count respectively [42,43]. Each reading was collected in duplicates.

2.3. Isolation of Fusarium oxysporum, Aspergillus niger from Soil Samples

39 g of potato dextrose agar powder (catalog no. M096, HiMedia) was added in 1 L sterile water and it was thoroughly mixed. The media was autoclaved at 15 psi pressure at 121 °C for 15 min. Test Samples (1 mL) were 10-fold diluted in 9 mL of water (10^{-1}). From that sample was serially diluted up to (10^{-2}) and ($10^{-3)}$. All three dilutions were plated on Selective media by spread plate technique. The plates were incubated in both aerobic chambers at 37 °C for 24–48 h for bacteria and 27 °C for 48–72 h. After 24–48 h incubation colonies were observed and recorded [42] (Tables S1 and S2).

2.4. DNA Extraction and Quality Control

DNA extraction was carried out based on the protocol by Amorim et al. [44]. Nanodrop was used initially to test the purity of DNA (OD260/OD280) (NanoDrop, Wilmington, DE, USA). Agarose Gel Electrophoresis was performed to assess DNA degradation and potential contaminations (Figure S1) and finally, Qubit 2.0 was used to quantify the DNA concentration precisely.

2.5. Library Construction and Quality Control

Qualified DNA was cut into fragments by the restriction enzyme. The construction of the DNA libraries is through the processes of end repairing, adding A to tails, purification, PCR amplification, and Libraries were sequenced by Illumina high-throughput sequencer with paired-end sequencing strategy. The libraries, that passed the QC, were then fed into sequencers after pooling according to their effective concentration and expected data volume.

2.6. Whole Meta-Genome Sequencing

The qualified libraries are fed into sequencer Illumina Novaseq 6000 (sequencing facility of Novogene Co. Ltd., Beijing, China) after pooling according to its effective concentration and expected data volume. The detailed protocol is provided in the supplementary data, Table S3.

2.7. Data Analysis

Raw Data QC of individual samples was conducted using FastQC (parameters: default) (https://www.bioinformatics.babraham.ac.uk/projects/fastqc/; accessed on 10 January 2020, Table S4). FastQ Screen v0.14.0 (https://www.bioinformatics.babraham.ac.uk/projects/fastq_screen/; accessed on 11 January 2020) was employed to screen the host genome sequences (References: GCF_007655135.1, GCA_002837095.1, GCA_002864125.1, GCA_002201585.1) from the raw data (parameters: –tag –filter '00000'; configured with bowtie2-2.3.5.1 and default Adapters). Host screened data was further validated using fastq-pair v1.0 and then assembled using metaSPAdes v3.13.0 (parameters: default) [45], metagenomic classification and visualizations using Kraken2 (parameters: –use-names –paired –gzip-compressed; database: built on 12 November 2020) [46] and pavian (https://github.com/fbreitwieser/pavian/; accessed on 10 January 2020) respectively. SPAdes assembled genome was subjected to gene prediction using MetaGeneMark (parameters: -f 3 -a -d -k -v -m MetaGeneMark_v1.mod) [47]. The predicted nucleotides were searched against NCBI NR database (Downloaded in March 2020) using Diamond v0.9.30 (parameters: -k 1) BlastX. Annotations were further meganized (parameters: default) using MEGAN v6 (relevant database were Downloaded in March 2020) [48].

Genome resolved metagenomics of Individual samples was performed using the SqueezeMeta pipeline v1.3.0 in sequential mode with MegaHIT assembler (default parameters). Short contigs (<200 bps) were removed and contig statistics were estimated using prinseq. RNAs were predicted using Barrnap. 16S rRNA sequences were taxonomically classified using the RDP classifier. tRNA/tmRNA sequences were predicted using Aragorn [49]. ORFs were predicted using Prodigal. Similarity searches for GenBank, eggNOG, KEGG, were done using Diamond HMM homology searches were done by HMMER3 [50] for the Pfam database. Read mapping against contigs was performed using Bowtie2. Binning was done using MaxBin2 [51] and Metabat2 [52]. A combination of binning results was performed using the DAS Tool [53]. Bin statistics were computed using CheckM. Bins with at least 50% completeness and <20% contamination were selected and subjected to annotation using enrichM's annotation module (https://github.com/geronimp/enrichM, accessed on 20 May 2021) against KO, PFAM, EC, and CAZY databases. The generated annotation matrices were subjected to enrichment using enrichM's enrich module. Alternatively, functional analysis of the metagenomic samples was also done using MG-RAST online server (https://www.mg-rast.org/; accessed on 22 January 2020) with default parameters.

2.8. Prediction of Protein Functions

Processed FASTQ reads of individual samples were used to search most popular databases that include protein databases, protein databases with functional hierarchy information, and ribosomal RNA databases namely RefSeq, IMG, TrEMBL, Subsystems, KEGG, GenBank, SwissProt, PATRIC, eggNOC, KO, GO, COG, RDP, LSU, SSU and NOG as a part of the MG-RAST analysis.

2.9. Variant Analysis

Based on the analysis of the processed reads and the resultant assemblies obtained from metaspades, the organisms *Fusarium oxysporum* and *Aspergillus niger* were used from the samples ESI and ISI respectively. BWA [53] was utilized for indexing (parameters: *index*) and mapping (parameters: mem) the pre-processed reads to *A. niger CBS 513.88* (NCBI Accession: GCF_000002855.3), *and F. oxysporum* f. sp. *lycopersici* 4287 (NCBI Accession: GCF_000149955.1) reference genomes. Then the aligned reads were converted to bam and sorted using Samtools v.1.6 [54] (parameters: *sort -l 9*) and duplicate reads were marked using GATK MarkDuplicates v.4.1.9.0 [55], followed by adding and replacing the read groups using GATK AddOrReplaceReadGroups. The reference dictionary was created for the genome using Samtools (parameters: *dict)* and the reference genome was also indexed using Samtools (parameters: *faidx*) to aid the variant calling process. GATK HaplotypeCaller (parameters: *-ERC GVCF -GQB 50*) and GATK GenotypeVCFs to generate known variants for base quality score recalibration using GATK BaseRecalibrator, followed by GATK ApplyBQSR. Furthermore, the recalibrated alignments were again run through GATK HaplotypeCaller (parameters: *-ERC GVCF -GQB 50*) and GATK GenotypeVCFs to identify the variants, followed by selecting identified SNPs and INDELs separately using GATK SelectVariants (parameters: *–select-type-to-include SNP)* (Parameters: *–select-type-to-include INDEL*) and extracted SNPs were further masked using GATK VariantFiltration (parameters: *-mask -mask-extension 5 –mask-name "INDEL"*) to tag the SNPs identified in and around INDELs. Finally, the masked SNPs and filtered INDELs were merged using GATK MergeVcfs [56]. In order to predict the variant effects, reference genomes were prepared using snpEff v.5.0d [57] (parameters: *build-genbank*) and variant summary and effect predictions were obtained.

2.10. Statistical Analysis

Output from MEGAN in. SPF format was used for statistical analysis and was performed using STAMP 2.1.3 (http://kiwi.cs.dal.ca/Software/STAMP; accessed on 10 January 2020, [58]). The one-sided G-test (w/Yates' + Fisher's) with asymptomatic confidence intervals (0.95) using the Benjamini–Hochberg FDR method was implemented [59].

3. Results

3.1. Physical Examination

Physical examination of the pomegranate plants, with respect to their, roots, leaves, stems, and fruits showed symptoms of wilt were considered for the study. Examination of the fruits showed the presence of rot disease was deemed for study.

Considering the ESI plant sample, few leaves showed mild yellowing (Figure 2b,e) and on examination of the roots, root knots were observed (Figure 2e). A few of the fruits showed black spots with mild discoloration. In the ISI, the dark brown coloration of the stem was observed, the fruits were darkly colored irregular spots with cracking (Figure 2f) and leaves showed yellowing, presence of moist, dark-colored irregular spots (Figure 2b). The ASI sample showed complete defoliation (Figure 2c) fruits that had completely turned dry with dark-brown pigmentation (Figure 2h), the root systems were dry and reduced with elongated galls, and dark brown coloration of the stem which has turned completely dry was observed (Figure 2f). ASI sample had complete yield reduction with no recovery (Figure 2).

3.2. Physiochemical Properties

The pH of the ESI sample was found to be 7.73 and electrical conductivity was estimated to be 135 µs/cm. Macronutrient and micronutrient analysis of ESI revealed 0.20% of total nitrogen (N), 0.0084% Phosphorous (P), 0.011% Potassium (K), 0.92% organic Carbon (C), 15 ppm Chloride(Cl), 0.98% Iron (Fe), 9.5 ppm Manganese (Mn), 26.9 ppm Copper (Cu), 24.8 ppm Zinc (Zn), and 3.4 ppm Boron (B). The pH and electrical conductivity of the ISI sample were estimated to be 6.35 and 139 µs/cm, respectively. In ISI samples the total N was calculated to be 0.19%, followed by P (0.010%), K (0.011%), C (0.93%), Cl (18 ppm), Fe (0.93%), Mn (9.1 ppm), Cu (29.4 ppm), Zn (30.9 ppm), B (4.1 ppm). The pH of the ASI sample was found to be 6.63 and electrical conductivity was estimated to be 180 µs/cm, followed by total N (0.20%), P (0.011%), K (0.014%), C (0.97%), Cl (21 ppm), Fe (0.98%), Mn (9.6 ppm), Cu (31.4 ppm), Zn (33.2 ppm), B (4.3 ppm). The pH was found to be altered in the ISI sample (6.35) and ASI sample (6.63) as compared to that of the ESI sample (7.73). Furthermore, reduced levels of N, Fe, and Mn micronutrients were reported in the ISI sample as compared to the ESI sample. Whereas, the micronutrients B, Zn, Cl, Cu, were found to be higher in the ISI sample when compared to the ESI sample (Tables 1 and S5).

Table 1. Physicochemical Characteristics and total microbial count of the soil samples.

Sample	pH	EC	N	P	K	OC	Cl	Fe	Cu	Mn	Zn	B	Microbial Count/g	
		(µs/cm)	%					ppm					Bacterial (cfu)	Fungal (cfu)
ESI	7.73	135	0.20	0.0084	0.011	0.92	15	0.98	26.9	9.5	24.8	3.4	1968	154
ISI	6.35	139	0.191	0.010	0.011	0.93	18	0.93	29.4	9.1	30.9	4.1	2240	170
ASI	6.63	180	0.20	0.011	0.014	0.97	21	0.98	31.4	9.6	33.2	4.3	2126	154

EC—Electrical conductivity.

3.3. Total Microbial Counts

The total bacterial counts and total fungal were estimated to be 1968 g/cfu, 2240 g/cfu and 2126 g/cfu for the ESI, ISI, and ASI, respectively. The total bacterial counts in the ISI sample showed a significant increase as compared to ESI and ASI. Similarly, the total fungal counts showed a significant increase in the ISI sample, which was estimated to be 170 g/cfu, as compared to ESI (154 g/cfu) and ASI (154 g/cfu) (Tables 1 and S5).

3.4. Sequence Information

13 GB of high-quality raw data per sample for ESI, ISI, and ASI was generated using the Illumina Novaseq 6000 sequencer. The complete protocol information and sequencing results are tabulated in Tables 2 and S4. The BioProject Id is PRJNA701747. The BioSample Ids for the samples ESI, ISI, and ASI are SAMN17910186, SAMN17910187, and SAMN17910188, respectively. The rarefaction curve of the experimental data sets ESI, ISI and ASI are provided in Figure S2.

Table 2. Sequence Information.

Sample	Raw Reads	Raw Data (Gb)	Sequence Count	BioProject	BioSample	SRA
ESI	44869321	13.5	5,924,482		SAMN17910186	SRR13705840
ISI	44773336	13.4	5,165,924	PRJNA701747	SAMN17910187	SRR13705839
ASI	44063752	13.2	5,379,446		SAMN17910188	SRR13705838

3.5. Microbial Abundance Analysis

Table 3 shows an overview of the taxonomic hits distribution in all three samples of ESI, ISI, and ASI. The ESI sample showed early symptoms of wilt disease. Employing the MG-Rast server, authors obtained maximum reads that mapped to Bacteria—2,076,360 (96.54%), followed by Eukaryota—63,248 (2.94%), Archaea—8979 (0.42%), unclassified sequences 1311 (0.06%), and Viruses 935 (0.04%) at the kingdom level. Actinobacteria 1,026,641 (57.52%), Proteobacteria 450,739 (25.25%), Planctomycetes 55,347 (3.10%), Ascomycota

52,339 (2.93%), Chloroflexi 37,935 (2.13%), Bacteroidetes 33,727 (1.89%), Firmicutes 29,671 (1.66%), Verrucomicrobia 21,659 (1.21%), Acidobacteria 20,072 (1.12%), Cyanobacteria 11,919 (0.67%), unclassified (derived from Bacteria) 7962 (0.45%), Gemmatimonadetes 7030 (0.39%), Deinococcus-Thermus 4694 (0.26%) and Euryarchaeota 4231 (0.24%) were mapped at the phylum level.

Table 3. Taxonomic hits distribution.

	Taxonomic Hits Distribution **Domain Level Microbial Abundance**		
	ESI **Percent of Reads**	**ISI** **Percent of Reads**	**ASI** **Percent of Reads**
Bacteria	2,076,360 (96.54%)	2,192,380 (96.08%)	2,348,985 (97.61%)
Eukaryota	63,248 (2.94%)	79,978 (3.50%)	43,462 (1.81%)
Archaea	8979 (0.42%)	7089 (0.31%)	10,928 (0.45%)
Unclassified sequences	1311 (0.06%)	1496 (0.07%)	2003 (0.08%)
Viruses	935 (0.04%)	939 (0.04%)	1001 (0.04%)
	Phylum level Microbial Abundance		
Actinobacteria	1,026,641 (57.52%)	995,904 (52.92%)	955,903 (49.62%)
Proteobacteria	450,739 (25.25%)	517,416 (27.49%)	562,154 (29.18%)
Plantomycetes	55,347 (3.10%)	72,331 (3.84%)	102,778 (5.34%)
Ascomycota	52,339 (2.93%)	69,904 (3.71%)	34,814 (1.81%)
Chloroflexi	37,935 (2.13%)	39,681 (2.11%)	54,360 (2.82%)
Bacteroidetes	33,727 (1.89%)	38,024 (2.02%)	36,944 (1.92%)
Firmicutes	29,671 (1.66%),	33,232 (1.77%)	37,538 (1.95%)
Verrucomicrobia	21,659 (1.21%)	28,599 (1.52%)	33,042 (1.72%)
Acidobacteria	20,072 (1.12%)	25,715 (1.37%)	33,599 (1.74%)
Cyanobacteria	11,919 (0.67%)	13,685 (0.73%)	18,085 (0.94%)
Unclassified (from Bacteria)	7962 (0.45%)	9248 (0.49%)	11,289 (0.59%)
Gemmatimonadetes	7030 (0.39%),	9358 (0.50%)	8945 (0.46%)
Deinococcus-Thermus	4694 (0.26%)	4978 (0.26%)	6399 (0.33%)
Euryarchaeota	4231 (0.24%).	4290 (0.23%)	5509 (0.29%)

The ISI sample showed further symptoms of wilt disease and maximum reads from this sample also mapped to Bacteria—2,192,380 (96.08%), followed by Eukaryota—79,978 (3.50%), Archaea—7089 (0.31%), unclassified sequences—1496 (0.07%) and Viruses—939 (0.04%) at the kingdom level. At the phylum level the mapped reads revealed Actinobacteria—995,904 (52.92%), Proteobacteria—517,416 (27.49%), Planctomycetes—72,331 (3.84%), Ascomycota—69,904 (3.71%), Chloroflexi—39,681 (2.11%), Bacteroidetes—38,024 (2.02%), Firmicutes—33,232 (1.77%), Verrucomicrobia—28,599 (1.52%), Acidobacteria—25,715 (1.37%), Cyanobacteria—13,685 (0.73%), Gemmatimonadetes—9358 (0.50%), unclassified (derived from Bacteria)—9248 (0.49%), Deinococcus-Thermus—4978 (0.26%), and Euryarchaeota—4290 (0.23%). The all major symptoms of wilt disease were seen in ASI sample. Our analysis revealed that the most reads were mapped to Bacteria 2,348,985 (97.61%), followed by Eukaryota 43,462 (1.81%), Archaea 10,928 (0.45%), unclassified sequences 2003 (0.08%) and Viruses 1001 (0.04%) at the kingdom level. Furthermore, at the phylum level, Actinobacteria 955,903 (49.62%), Proteobacteria 562,154 (29.18%), Planctomycetes 102,778 (5.34%), Chloroflexi 54,360 (2.82%), Firmicutes 37,538 (1.95%), Bacteroidetes 36,944 (1.92%), Ascomycota 34,814 (1.81%), Acidobacteria 33,599 (1.74%), Verrucomicrobia 33,042 (1.72%), Cyanobacteria 18,085 (0.94%), unclassified (derived from Bacteria) 11,289 (0.59%), Gemmatimonadetes 8945 (0.46%), Deinococcus-Thermus 6399 (0.33%), and Euryarchaeota 5509 (0.29%) were identified. The alpha diversity for ESI was 351, ISI was 367 and ASI was 401 species.

Comparative statistical analysis of the samples ESI and ISI revealed 35,554 features in all after filtering out unclassified reads, of which 29,747 mapped to Bacteria, 4582 to Eukaryota, 1214 Archaea, 1 to none. From the total number of features, 79 were found to be significant with corrected *q*-value =< 0.05 (Figure 3b). Considering which of the microbes were more abundant in the ESI sample, the top differentially abundant microbes belonged to Proteobacteria phyla *Achromobacter* sp. 2789STDY5608625 with a count of 299 in ESI and 8 in ISI samples. *Achromobacter* sp. K91 from the phyla Proteobacteria followed by the *Achromobacter* sp. 2789STDY5608625 with 236 (ESI) and 4 (ISI). *Achromobacter aegrifaciens* followed the two Proteobacteria phyla members with 23. The parent sequence count 2975 (ESI) and 588 (ISI). *Microbacterium* sp. SUBG005 from phyla Actinobacteria and *Agrobacterium larrymoorei* from the Proteobacteria phyla were the subsequently most abundant bacteria (Table 4). Furthermore, comparing the ISI and ASI samples, *Streptomyces* sp. FxanaC1, *Streptomyces* sp. F12 and *Rhizobium* sp. NFACC06-2 was estimated as being the top three differentially abundant species. Amongst the Eukaryota, the *Aspergillus arachidicola* was found to be differentially abundant, followed by *Aspergillus candidus* and *Aspergillus campestris* all from the phyla Ascomycota. With the ISI and ASI comparison, *Aspergillus nomius* and *Aspergillus ochraceoroseus* were found to be species that were prevalent in the ISI sample and significantly lower in the ASI sample (Tables 4 and S6).

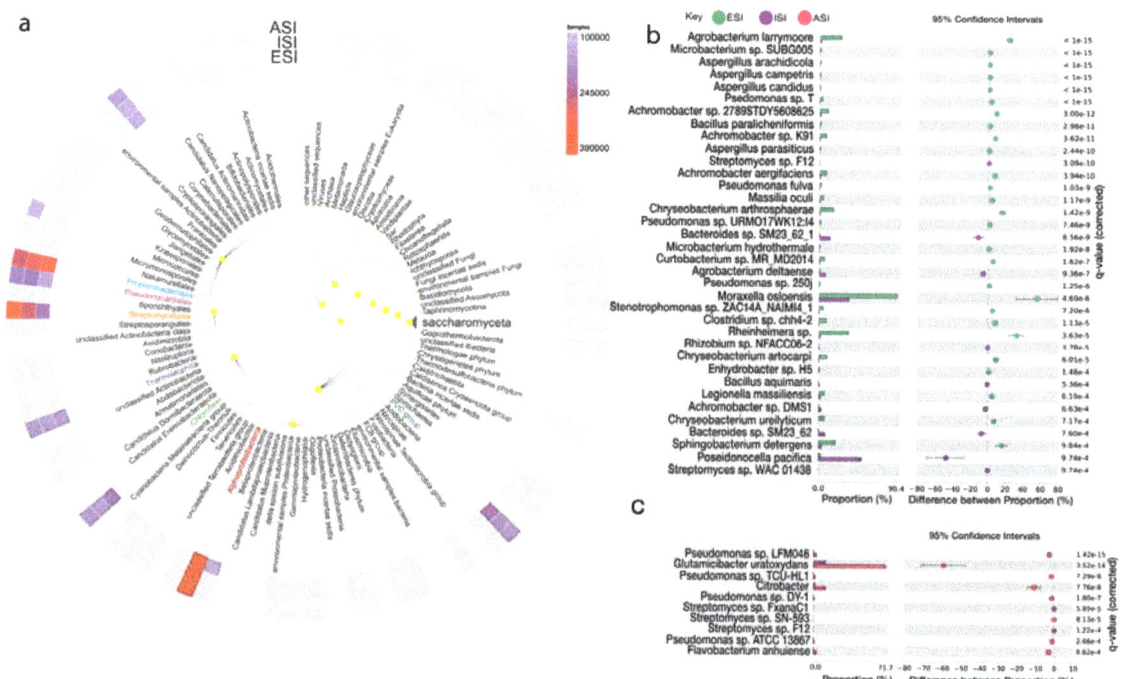

Figure 3. Microbial diversity and abundance. (**a**) The top-ranked microbes at the phylum level are depicted along with a comparison between ESI (blue) and ISI (purple); (**b**) and, ISI and ASI (pink); (**c**) depict the top-ranked microbes and the proportions and difference between the proportions with 95% confidence intervals along with the *q*-value (corrected) (*q* value < 0.001).

Table 4. Relative Abundance of Microbial species.

Species			Parent Sequence Count		Relative Frequency %		*p*-Values		Effect Size
	ESI	ISI	ESI	ISI	ESI	ISI	PVal	corrected	
Achromobacter sp. 2789STDY5608625	299	8	2975	588	10.05	1.36	5.90×10^{-16}	3.00×10^{-12}	8.69
Achromobacter sp. K91	236	4	2975	588	7.93	0.68	9.17×10^{-15}	3.62×10^{-11}	7.25
Achromobacter aegrifaciens	233	5	2975	588	7.83	0.85	1.33×10^{-13}	3.94×10^{-10}	6.98
Microbacterium sp. SUBG005	196	6	19,463	13,719	1.01	0.04	1.22×10^{-37}	2.17×10^{-33}	0.96
Agrobacterium larrymoorei	174	14	691	741	25.18	1.89	6.24×10^{-44}	2.22×10^{-39}	23.29
Curtobacterium sp. MR_MD2014	153	6	2534	695	6.04	0.86	8.66×10^{-11}	1.62×10^{-7}	5.17
Pseudomonas sp. T	140	7	4318	3374	3.24	0.21	2.75×10^{-27}	1.63×10^{-23}	3.03
Moraxella osloensis	132	9	146	26	90.41	34.62	2.90×10^{-9}	4.69×10^{-6}	55.80

The Figure 3 depicts an overview of the kingdoms and corresponding microbes. The heatmap in the outer circle represents the prevalence among the three experimental samples as per the scale provided (a). The top-ranked microbes at the phylum level are depicted along with a comparison between ESI (blue) and ISI (purple) (b) and, ISI and ASI (pink) (c) depict the top-ranked microbes and the proportions and difference between the proportions with 95% confidence intervals along with the *q*-value (corrected) (*q* value < 0.001).

3.6. Pathway Predictions

The output files from MEGAN were used to search the InterPro2GO, a resource of protein information [60], under the Molecular functions category. The top hits included catalytic activity—oxidoreductase activity, transferase activity, and hydrolase activity. The other major functions were transporter activity, ion binding, and nucleotide binding. Under biological process, the top hits were mapped to Metabolic processes, transport, DNA, and RNA metabolic processes. Under cellular component, the intrinsic component of membrane and membrane functions were highlighted (Figure 4). eggNOC, a database for functional annotations, orthology, and gene evolution [61] revealed the major pathways related to metabolism, information storage, and processing, cellular processing, and signaling. The major pathways in metabolism included amino acid metabolism and transport, energy production and conversion, and carbohydrate metabolism and transport, inorganic ion and transport, lipid transport and metabolism, secondary metabolites biosynthesis, transport and catabolism, coenzyme transport and metabolism, and nucleotide transport and metabolism. In information storage and processing, the major pathways were transcription, replication, recombination and repair, and translation, ribosomal structure, and biogenesis. Under the cellular processes and signaling category, the pathways were signal transduction and mechanisms, cell wall/membrane/envelope biogenesis, Post-translational modifications, protein turnover, chaperones, and defense mechanisms. SEED functional annotation using SEED [62] highlighted Metabolism and stress response, defense, and virulence. Under metabolism, the predictions showed fatty acids, lipids, and isoprenoids mainly indicating acyl carrier protein (Figure S3). Statistical analysis of the predicted pathways revealed significant hits obtained from the InterPro2GO database. Carbohydrate binding was found to be significant (*p* < 0.05) between ESI and ISI. Kyoto Encyclopedia of Genes and Genomes (KEGG) [63] hits included K03088; RNA polymerase sigma-70 factor ECF sub-family, K12132; Eukaryotic-like serine/threonine protein kinase and K01990; ABC-2 type transport system ATP-binding protein (Tables 5, S7 and S8).

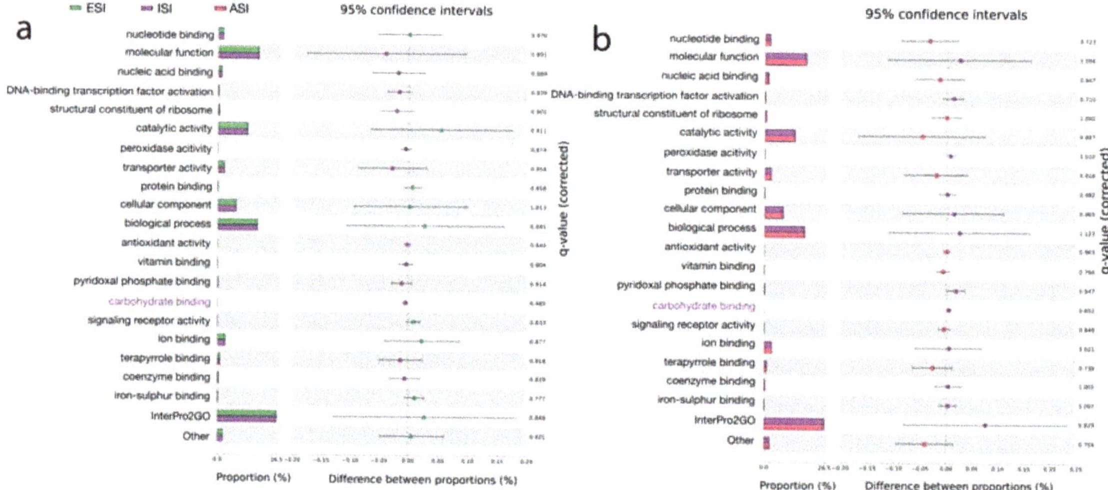

Figure 4. Pathway hits from InterPro2GO, databases. The top-ranked pathways are represented as per the color scheme—three samples ESI (blue), ISI (purple), and ASI (pink) (**a**). The top-ranked pathways are represented (**a**). A comparison between ESI (blue) and ISI (purple) (**b**) and, Significant hits represented show the top-ranked pathways along with the proportions and difference between the proportions with 95% confidence intervals considering a *q*-value (corrected) (*p* < 0.05).

Table 5. Pathway predictions. The top hits from the most popular databases have been furnished in the table below.

| | | | Pathway Predictions | | |
InterPro2GO [60]	KEGG [63]	SEED [62]	COG [64]	Pfam [65]	eggNOC [61]
GO: 0030246; Carbohydrate binding. GO: 0046906; Tetrapyrrole binding. GO: 0030170; Pyridoxal phosphate binding.	K03088; RNA polymerase sigma-70 factor ECF sub-family. K12132; Eukaryotic-like serine/threonine Protein kinase. K01990; ABC-2 type transport system ATP-binding protein.	Acyl carrier protein. Stress response, defense virulence.	ENOG410XNMH; Histidine kinase. COG1012; NAD-dependent aldehyde dehydrogenases. COG1960; Acyl-CoA Dehydrogenases. COG0515; Serine/threonine Protein kinase.	PF00005; ABC transporter. PF07690; Major Facilitator superfamily. PF00528; Binding protein-dependent Transport system Inner membrane component.	ISP *: Transcription. Replication, recombination, and repair. CSP +: Cell wall/Membrane/Envelope biogenesis, signal transduction mechanisms. Metabolism: Amino acid transport and metabolism, carbohydrate transport and metabolism, energy production and conversion.

* Information storage and processing; +Cellular processes and signaling.

3.7. Genome Resolved Metagenomics

Following confirmation with microbiological methods, normalized read counts obtained for *F. oxysporum* in the ESI were found to be 209, in ISI it was 120 and 94 in ASI. Using the SqueezMeta pipeline, a total of 13, 8, and 5 genome bins were recovered from samples ISI, ASI, and ESI respectively. Based on CheckM analysis, 4, 4, 2 bins from samples ISI, ASI, and ESI were found to be at least 50% complete and with less than 15%. Significantly enriched terms (Mann–Whitney U test; *p* value < 0.05) of these bins against KO, PFAM, and other databases are shown in Table 5 and complete enrichment results are provided as Supplementary Figures S4–S6; Tables S7 and S8.

3.8. Variant Analysis

F. oxysporum genome in the sample ESI was analyzed to screening for variants using the GCF_000149955.1 as reference (Tables S9 and S10). The variant analysis showed 81 SNPs out of which multiple mutations were observed in the 3-Isopropylmalate dehydratase (IPMD) predicted as high impact effects, which is required for fungal pathogenicity [66]. IPMD is encoded by LEU2 and involved in the leucine biosynthetic pathway.

Variant Analysis of *A. niger* genome from the ISI sample with reference (GCF_00000285 5.3) showed three genes mutated with high impact names histidine kinase J7, MFS transporter, and NADH-ubiquinone oxidoreductase subunit. While two of the genes, histidine kinase and the MFS transporter gene, showed multiple mutations with high impact variations namely frameshift mutations, NADH-ubiquinone oxidoreductase subunit gene showed one variation leading to a frameshift mutation.

ANI_1_1000064 | histidine kinase J7 showed an insertion at position 1764099 of A/ACG AGT. The other high-impact variant was a deletion at 1764101 GTCCTT/G. Predictions revealed three high-impact variants in the ANI_1_2008144 | MFS transporter gene. The first one was an insertion at 1799089 position A/ACGCGCTTC, the second one was again an insertion at position 1799092 G/GTGCGT and the third one was an insertion at position 1799094 C/CG. In the ANI_1_742164 | NADH-ubiquinone oxidoreductase subunit a large insertion was found at the position 1286777 T/TCGAGAACTCGAAGTTCGGACCCTCGACG ATGGCATCGACC.

4. Discussion

Pomegranate has been used for a wide range of health benefits making it a commercially important crop. India is the leading producer of pomegranate with Maharashtra, Karnataka, Odisha, Tamil Nadu, Gujarat, Rajasthan, Chattisgarh, Telangana, and Nagaland states contributing to India's major producer of the fruit crop. Karnataka, which is the second-largest contributor of fruit produce to India, faces a number of challenges in crop management due to wilt, anthracnose, bacterial blight, and heart rot. In the present study, we explored the soil samples from an Orchard in the Chikkaballapur district of Karnataka. Wilt infection in the orchard resulted in a 36% yield loss to the farmer. A shotgun metagenomics approach was employed and the microbial communities in soil samples were screened.

Comparative analysis of the samples ESI and ISI revealed 35,554 features in all after filtering out unclassified reads, of which the majority of the features mapped to Bacteria (29,747), followed by Eukaryota (4582) and Archaea (1214). In our analysis, we reported 79 features to be significant (corrected q-value =< 0.05). The top differentially abundant microbes prevalent in the ESI sample to Proteobacteria phyla *Achromobacter* sp. 2789STDY5608625, ESI (299), and ISI (8) samples. *Achromobacter* sp. 2789STDY5608625 was followed member of the sample phyla, *Achromobacter* sp. K91, ESI (236), and ISI (4). *Achromobacter aegrifaciens* followed the two Proteobacteria phyla members with 23. *Microbacterium* sp. SUBG005 from phyla Actinobacteria and *Agrobacterium larrymoorei* from the Proteobacteria phyla were the subsequently most abundant bacteria. There have been reports of members of the genus *Achromobacter* employed as biocontrol agents against *Fusarium oxysporum* causing wilt in other plants [67,68]. The role of the microbes from this genus could be explored for their biocontrol potential against *F. oxysporum*. Furthermore, comparing the ISI and ASI samples, *Streptomyces* sp. FxanaC1, *Streptomyces* sp. F12 and *Rhizobium* sp. NFACC06-2 was estimated as being the top three differentially abundant species. Amongst the Eukaryota, the *Aspergillus arachidicola* was found to be differentially abundant, followed by *Aspergillus candidus* and *Aspergillus campestris* all from the phyla Ascomycota. With the ISI and ASI comparison, *Aspergillus nomius* and *Aspergillus ochraceoroseus* were found to be species that were prevalent in the ISI sample and significantly lower in the ASI sample.

We particularly screened *F. oxysporum* as a causative pathogen for Wilt disease in pomegranate after assessing the physical symptoms of the plant. The presence of *F. oxysporum* was confirmed both with microbial isolation and metagenomics validation. The presence of early symptoms of Wilt was reported in the ESI sample in our study. Furthermore, the fruits showed early symptoms of rot disease. The sample was screened for *A. niger* the causative pathogen for rot disease. The plant with more noticeable symptoms of the disease in our study was the ISI sample. In the rhizospheric samples of the ISI sample, we reported an abundance of *A. niger* with a decrease in *F. oxysporum*. A decline in Fusarium species with an increase in Aspergillus species was observed in the plants from ESI and ISI samples respectively. Variant analysis of *F. oxysporum* showed multiple high-impact mutations on the IPMD gene. IPMD has been reported in other studies for its role in fungal pathogenesis. IPMD is encoded by LEU2 and involved in the catalysis of leucine biosynthesis particularly in the conversion of 3-isopropylmalate (3-IPPM) to 2-ketoisocaproate (2-KIC). Intriguingly, another important finding from this study is with respect to the carbohydrate binding pathway which is one of the significant hits. Phytopathogens are known to synthesize carbohydrate-active enzymes (CAZymes) also known as plant cell wall degrading enzymes (PCWDE) [69], which can also function as Carbohydrate binding modules (CBM). CAZymes are required for pathogenesis as well as growth [70,71]. It may be reasonable to assume that targeting IPMD and CAZymes could be a good strategy for the development of antifungals which could aid in biocontrol of *F. oxysporum*.

The present study took advantage of the current state-of-the-art sequencing platform, the Illumina Novaseq 6000 platform that provides higher resolution in screening and identification microbial communities. The approach aided in the identification of certain key targets that are linked to the pathogenicity of Fusarium. However, further research is being carried out to particularly validate the key findings of this study. In this study, we demonstrate the capabilities of the whole metagenome sequencing approach in identifying potential key players of wilt disease affecting the pomegranate plant, wherein the symptomatology is complex.

Supplementary Materials: The following are available online at https://www.mdpi.com/article/10.3390/agriculture11090831/s1, Figure S1: DNA Extraction, Figure S2: Rarefaction curves from experimental data sets ESI, ISI and ASI, Figure S3: Pathway predictions, Figure S4: KEGG Annotations, Figure S5: COG Annotations, Figure S6: PFAM Annotations. Table S1: Isolation of *Fusarium oxysporum*, Table S2: Isolation of *Aspergillus niger*, Table S3: DNA Sequencing Information, Table S4: Data Quality Summary (R2-ESI, NP-ISI and NC-ASI), Table S5: Physicochemical and Microbial Analysis of the soil samples, Table S6: Relative Abundance of Microbial species, Table S7: Pathway predictions, Table S8: PFAM Abundance Overview, Table S9: Variant Analysis, Table S10: Number variants by type.

Author Contributions: The idea and Funding, S.K.M.; Writing, original draft preparation, experimental work, A.J.D.; review, R.R.; analytics support, B.S.R.; supervision, N.R. and S.K.M.; funding acquisition, M.K.P. review, T.U.; wet laboratory methods support, H.E.; This work is a part of the Ph.D. work of A.J.D. All authors have read and agreed to the published version of the manuscript.

Funding: This research received no external funding.

Institutional Review Board Statement: Not applicable.

Informed Consent Statement: Not applicable.

Data Availability Statement: The sequence information is available publically for this project at https://www.ncbi.nlm.nih.gov/bioproject/?term=PRJNA701747/, accessed on 10 January 2020 (Bioproject number PRJNA701747).

Acknowledgments: We thank the DST-FIST program level O under Govt of India and BISEP program under Govt of Karnataka at Maharani Lakshmi Ammanni College for Women for extending their generous support for this study. This manuscript was funded by University of Agricultural Sciences, Bangalore.

Conflicts of Interest: The authors declare no conflict of interest.

References

1. Teksur, P.K. Alternative technologies to control postharvest diseases of pomegranate. *Stewart Postharvest Rev.* **2015**, *11*, 1–7. [CrossRef]
2. Middha, S.K.; Usha, T.; Pande, V. A Review on Antihyperglycemic and Antihepatoprotective Activity of eco-friendly *Punica granatum* peel waste. *Evid. Based Complement. Alt Med.* **2013**, *2013*, 1–10. [CrossRef] [PubMed]
3. da Silva, J.A.T.; Rana, T.S.; Narzary, D.; Verma, N.; Meshram, D.T.; Ranade, S.A. Scientia Horticulturae Pomegranate biology and biotechnology: A review. *Sci. Hortic.* **2013**, *160*, 85–107. [CrossRef]

4. Chen, J.; Liao, C.; Ouyang, X.; Kahramanoğlu, I.; Gan, Y.; Li, M. Antimicrobial Activity of Pomegranate Peel and Its Applications on Food Preservation. *J. Food Qual.* **2020**, *2020*, 1–8. [CrossRef]

5. Alexandre, E.; Silva, S.; Santos, S.; Silvestre, A.J.; Duarte, M.F.; Saraiva, J.A.; Pintado, M. Antimicrobial activity of pomegranate peel extracts performed by high pressure and enzymatic assisted extraction. *Food Res. Int.* **2018**, *115*, 167–176. [CrossRef]

6. Usha, T.; Middha, S.K.; Sidhalinghamurthy, K.R. Pomegranate peel and its anticancer activity: A mechanism-based review. In *Plant-derived Bioactives: Chemistry and Mode of Action*; Springer: Berlin/Heidelberg, Germany, 2020; ISBN 9789811523618.

7. Sharma, P.; McClees, S.F.; Afaq, F. Pomegranate for Prevention and Treatment of Cancer: An Update. *Molecules* **2017**, *22*, 177. [CrossRef] [PubMed]

8. Caruso, A.; Barbarossa, A.; Tassone, A.; Ceramella, J.; Carocci, A.; Catalano, A.; Basile, G.; Fazio, A.; Iacopetta, D.; Franchini, C.; et al. Pomegranate: Nutraceutical with Promising Benefits on Human Health. *Appl. Sci.* **2020**, *10*, 6915. [CrossRef]

9. Elshafie, H.; Caputo, L.; de Martino, L.; Sakr, S.; de Feo, V.; Camele, I. Study of Bio-Pharmaceutical and Antimicrobial Properties of Pomegranate (*Punica granatum* L.) Leathery Exocarp Extract. *Plants* **2021**, *10*, 153. [CrossRef]

10. Aslam, M.N.; Lansky, E.P.; Varani, J. Pomegranate as a cosmeceutical source: Pomegranate fractions promote proliferation and procollagen synthesis and inhibit matrix metalloproteinase-1 production in human skin cells. *J. Ethnopharmacol.* **2006**, *103*, 311–318. [CrossRef]

11. Al-Zoreky, N. Antimicrobial activity of pomegranate (*Punica granatum* L.) fruit peels. *Int. J. Food Microbiol.* **2009**, *134*, 244–248. [CrossRef]

12. Usha, T.; Goyal, A.K.; Lubna, S.; Prashanth, H.P.; Mohan, T.M.; Pande, V.; Middha, S.K. Identification of anti-cancer targets of eco-friendly waste *Punica granatum* peel by dual reverse virtual screening and binding analysis. *Asian Pac. J. Cancer Prev.* **2015**, *15*, 10345–10350. [CrossRef] [PubMed]

13. Salgado, J.; Ferreira, T.R.B.; Biazotto, F.D.O.; Dias, C. Increased Antioxidant Content in Juice Enriched with Dried Extract of Pomegranate (Punica granatum) Peel. *Plant Foods Hum. Nutr.* **2012**, *67*, 39–43. [CrossRef] [PubMed]

14. Henning, S.M.; Yang, J.; Lee, R.-P.; Huang, J.; Hsu, M.; Thames, G.; Gilbuena, I.; Long, J.; Xu, Y.; Park, E.H.; et al. Pomegranate Juice and Extract Consumption Increases the Resistance to UVB-induced Erythema and Changes the Skin Microbiome in Healthy Women: A Randomized Controlled Trial. *Sci. Rep.* **2019**, *9*, 14528. [CrossRef] [PubMed]

15. Bellé, C.; Moccellin, R.; Souza-Júnior, I.T.; Maich, S.L.P.; Neves, C.G.; Nascimento, M.B.; Barros, D.R. First Report of Colletotrichum acutatum Causing Flower Anthracnose on Pomegranate (*Punica granatum*) in Southern Brazil. *Plant Dis.* **2018**, *102*, 2373. [CrossRef]

16. Kanetis, L.; Testempasis, S.; Goulas, V.; Samuel, S.; Myresiotis, C.; Karaoglanidis, G.S. Identification and mycotoxigenic capacity of fungi associated with pre- and postharvest fruit rots of pomegranates in Greece and Cyprus. *Int. J. Food Microbiol.* **2015**, *208*, 84–92. [CrossRef] [PubMed]

17. Cintora-Martínez, E.A.; Leyva-Mir, S.G.; Ayala-Escobar, V.; Quezada, G.D.A.; Camacho-Tapia, M.; Tovar-Pedraza, J.M. Pomegranate fruit rot caused by Pilidiella granati in Mexico. *Australas. Plant Dis. Notes* **2017**, *12*, 4. [CrossRef]

18. Palou, L.; Crisosto, C.H.; Garner, D. Combination of postharvest antifungal chemical treatments and controlled atmosphere storage to control gray mold and improve storability of 'Wonderful' pomegranates. *Postharvest Biol. Technol.* **2007**, *43*, 133–142. [CrossRef]

19. Thomidis, T. Pathogenicity and characterization of Pilidiella granati causing pomegranate diseases in Greece. *Eur. J. Plant Pathol.* **2014**, *141*, 45–50. [CrossRef]

20. Mondal, K.K.; Mani, C. ERIC-PCR-Generated Genomic Fingerprints and Their Relationship with Pathogenic Variability of Xanthomonas campestris pv. punicae, the Incitant of Bacterial Blight of Pomegranate. *Curr. Microbiol.* **2009**, *59*, 616–620. [CrossRef]

21. Grote, D.; Olmos, A.; Kofoet, A.; Tuset, J.; Bertolini, E.; Cambra, M. Specific and Sensitive Detection of Phytophthora nicotianae By Simple and Nested-PCR. *Eur. J. Plant Pathol.* **2002**, *108*, 197–207. [CrossRef]

22. Yang, X.; Hameed, U.; Zhang, A.-F.; Zang, H.-Y.; Gu, C.-Y.; Chen, Y.; Xu, Y.-L. Development of a nested-PCR assay for the rapid detection of Pilidiella granati in pomegranate fruit. *Sci. Rep.* **2017**, *7*, 40954. [CrossRef]

23. She, X.; Yu, L.; Lan, G.; Tang, Y.; He, Z. Identification and Genetic Characterization of Ralstonia solanacearum Species Complex Isolates from Cucurbita maxima in China. *Front. Plant Sci.* **2017**, *8*, 1794. [CrossRef]

24. Suhaimi, N.S.M.; Goh, S.-Y.; Ajam, N.; Othman, R.Y.; Chan, K.-G.; Thong, K.L. Diversity of microbiota associated with symptomatic and non-symptomatic bacterial wilt-diseased banana plants determined using 16S rRNA metagenome sequencing. *World J. Microbiol. Biotechnol.* **2017**, *33*, 168. [CrossRef] [PubMed]

25. Ts, A.; Sg, B. Bacterial root bark necrosis and wilt of pomegranate, hereto a new disease. *J. Appl. Biotechnol. Bioeng.* **2018**, *5*, 329–332. [CrossRef]

26. Shen, Z.; Penton, C.R.; Lv, N.; Xue, C.; Yuan, X.; Ruan, Y.; Li, R.; Shen, Q. Banana Fusarium Wilt Disease Incidence Is Influenced by Shifts of Soil Microbial Communities Under Different Monoculture Spans. *Microb. Ecol.* **2017**, *75*, 739–750. [CrossRef]

27. Tian, X.; Zhang, Q.; Chen, G.; Mao, Z.; Yang, J.; Xie, B. Diversity of bacteria associated with pine wood nematode revealed by metagenome. *Acta Microbiol. Sin.* **2010**, *50*, 909–916.

28. Reddy, T.; Thomas, A.D.; Stamatis, D.; Bertsch, J.; Isbandi, M.; Jansson, J.; Mallajosyula, J.; Pagani, I.; Lobos, E.A.; Kyrpides, N.C. The Genomes OnLine Database (GOLD) v.5: A metadata management system based on a four level (meta)genome project classification. *Nucleic Acids Res.* **2014**, *43*, D1099–D1106. [CrossRef] [PubMed]

29. Pruitt, K.D.; Tatusova, T.; Brown, G.R.; Maglott, D.R. NCBI Reference Sequences (RefSeq): Current status, new features and genome annotation policy. *Nucleic Acids Res.* **2011**, *40*, D130–D135. [CrossRef] [PubMed]

30. Yarza, P.; Yilmaz, P.; Pruesse, E.; Glöckner, F.O.; Ludwig, W.; Schleifer, K.-H.; Whitman, W.; Euzéby, J.; Amann, R.; Rossello-Mora, R. Uniting the classification of cultured and uncultured bacteria and archaea using 16S rRNA gene sequences. *Nat. Rev. Genet.* **2014**, *12*, 635–645. [CrossRef]

31. Singer, E.; Bushnell, B.; Coleman-Derr, D.; Bowman, B.; Bowers, R.M.; Levy, A.; Gies, E.A.; Cheng, J.-F.; Copeland, A.; Klenk, H.-P.; et al. High-resolution phylogenetic microbial community profiling. *ISME J.* **2016**, *10*, 2020–2032. [CrossRef]

32. Santos, S.S.; Nielsen, T.; Hansen, L.H.; Winding, A. Comparison of three DNA extraction methods for recovery of soil protist DNA. *J. Microbiol. Methods* **2015**, *115*, 13–19. [CrossRef] [PubMed]

33. Medinger, R.; Nolte, V.; Pandey, R.V.; Jost, S.; Ottenwälder, B.; Schlötterer, C.; Boenigk, J. Diversity in a hidden world: Potential and limitation of next-generation sequencing for surveys of molecular diversity of eukaryotic microorganisms. *Mol. Ecol.* **2010**, *19*, 32–40. [CrossRef] [PubMed]

34. Gong, J.; Dong, J.; Liu, X.; Massana, R. Extremely High Copy Numbers and Polymorphisms of the rDNA Operon Estimated from Single Cell Analysis of Oligotrich and Peritrich Ciliates. *Protist* **2013**, *164*, 369–379. [CrossRef] [PubMed]

35. Kunin, V.; Engelbrektson, A.; Ochman, H.; Hugenholtz, P. Wrinkles in the rare biosphere: Pyrosequencing errors can lead to artificial inflation of diversity estimates. *Environ. Microbiol.* **2010**, *12*, 118–123. [CrossRef]

36. Lee, C.; Herbold, C.; Polson, S.; Wommack, K.E.; Williamson, S.J.; McDonald, I.; Cary, S. Groundtruthing Next-Gen Sequencing for Microbial Ecology–Biases and Errors in Community Structure Estimates from PCR Amplicon Pyrosequencing. *PLoS ONE* **2012**, *7*, e44224. [CrossRef]

37. Pawlowski, J.W.; Christen, R.; Lecroq, B.; Bachar, D.; Shahbazkia, H.R.; Amaral-Zettler, L.; Guillou, L. Eukaryotic Richness in the Abyss: Insights from Pyrotag Sequencing. *PLoS ONE* **2011**, *6*, e18169. [CrossRef]

38. Abraham, B.S.; Caglayan, D.; Carrillo, N.V.; Chapman, M.C.; Hagan, C.T.; Hansen, S.T.; Jeanty, R.O.; Klimczak, A.A.; Klingler, M.J.; Kutcher, T.P.; et al. Shotgun metagenomic analysis of microbial communities from the Loxahatchee nature preserve in the Florida Everglades. *Environ. Microbiome* **2020**, *15*, 1–10. [CrossRef] [PubMed]

39. Babalola, O.O.; Fadiji, A.E.; Ayangbenro, A.S. Shotgun metagenomic data of root endophytic microbiome of maize (*Zea mays* L.). *Data Brief* **2020**, *31*, 105893. [CrossRef]

40. Das, A.J.; Ravinath, R.; Shilpa, B.R.; Rohith, B.S.; Goyal, A.K.; Ramesh, N.; Ekambaram, H.; Prasannakumar, M.K.; Usha, T.; Middha, S.K. Microbiomics and cloud-based analytics advance sustainable soil management. *Front. Biosci.* **2021**, *26*, 478–495. [CrossRef] [PubMed]

41. Weston, W.H. Soil and Plant Analysis. A Laboratory Manual of Methods for the Examination of Soils and the Determination of the Inorganic Constituents of Plants. C. S. Piper. *Q. Rev. Biol.* **1945**, *20*, 272–273. [CrossRef]

42. Bureau of Indian Standards Microbiology. *General Guidance for the Enumeration of Micro-organisms Colony Count Technique at 30 °C*; Bureau of Indian Standards Microbiology: New Delhi, Indian, 2002.

43. Bureau of Indian Standards Microbiology. *Method for Yeast and Mould Count of Foodstuffs and Animal Feeds*; Bureau of Indian Standards Microbiology: New Delhi, Indian, 1999.

44. Amorim, J.H.; Macena, T.; Lacerda-Junior, G.; Rezende, R.; Dias, J.; Brendel, M.; Cascardo, J. An improved extraction protocol for metagenomic DNA from a soil of the Brazilian Atlantic Rainforest. *Genet. Mol. Res.* **2008**, *7*, 1226–1232. [CrossRef]

45. Nurk, S.; Meleshko, D.; Korobeynikov, A.; Pevzner, P.A. MetaSPAdes: A new versatile metagenomic assembler. *Genome Res.* **2017**, *27*, 824–834. [CrossRef]

46. Wood, D.E.; Lu, J.; Langmead, B. Improved metagenomic analysis with Kraken 2. *Genome Biol.* **2019**, *20*, 1–13. [CrossRef]

47. Zhu, W.; Lomsadze, A.; Borodovsky, M. Ab initio gene identification in metagenomic sequences. *Nucleic Acids Res.* **2010**, *38*, e132. [CrossRef]

48. Huson, D.H.; Beier, S.; Flade, I.; Górska, A.; El-Hadidi, M.; Mitra, S.; Ruscheweyh, H.-J.; Tappu, R. MEGAN Community Edition—Interactive Exploration and Analysis of Large-Scale Microbiome Sequencing Data. *PLoS Comput. Biol.* **2016**, *12*, e1004957. [CrossRef] [PubMed]

49. Laslett, D. ARAGORN, a program to detect tRNA genes and tmRNA genes in nucleotide sequences. *Nucleic Acids Res.* **2004**, *32*, 11–16. [CrossRef] [PubMed]

50. Eddy, S.R.; Wheeler, T. HMMER-biosequence analysis using profile hidden Markov models. *Howard Hughes Medical Institute* **2015**.

51. Wu, Y.-W.; Simmons, B.; Singer, S.W. MaxBin 2.0: An automated binning algorithm to recover genomes from multiple metagenomic datasets. *Bioinformatics* **2015**, *32*, 605–607. [CrossRef] [PubMed]

52. Kang, D.D.; Li, F.; Kirton, E.; Thomas, A.; Egan, R.; An, H.; Wang, Z. MetaBAT 2: An adaptive binning algorithm for robust and efficient genome reconstruction from metagenome assemblies. *PeerJ* **2019**, *7*, e7359. [CrossRef]

53. Sieber, C.; Probst, A.; Sharrar, A.; Thomas, B.C.; Hess, M.; Tringe, S.G.; Banfield, J.F. Recovery of genomes from metagenomes via a dereplication, aggregation and scoring strategy. *Nat. Microbiol.* **2018**, *3*, 836–843. [CrossRef] [PubMed]

54. Li, H.; Durbin, R. Fast and accurate short read alignment with Burrows-Wheeler transform. *Bioinformatics* **2009**, *25*, 1754–1760. [CrossRef]

55. Li, H.; Handsaker, R.; Wysoker, A.; Fennell, T.; Ruan, J.; Homer, N.; Marth, G.; Abecasis, G.; Durbin, R. The Sequence Alignment/Map format and SAMtools. *Bioinformatics* **2009**, *25*, 2078–2079. [CrossRef]

56. McKenna, A.; Hanna, M.; Banks, E.; Sivachenko, A.; Cibulskis, K.; Kernytsky, A.; Garimella, K.; Altshuler, D.; Gabriel, S.B.; Daly, M.J.; et al. The Genome Analysis Toolkit: A MapReduce framework for analyzing next-generation DNA sequencing data. *Genome Res.* **2010**, *20*, 1297–1303. [CrossRef]

57. Cingolani, P.; Platts, A.; Wang, L.L.; Coon, M.; Nguyen, T.; Wang, L.; Land, S.J.; Lu, X.; Ruden, D.M. A program for annotating and predicting the effects of single nucleotide polymorphisms, SnpEff. *Fly* **2012**, *6*, 80–92. [CrossRef]

58. Parks, D.H.; Tyson, G.; Hugenholtz, P.; Beiko, R.G. STAMP: Statistical analysis of taxonomic and functional profiles. *Bioinformatics* **2014**, *30*, 3123–3124. [CrossRef]

59. Parks, D.H.; Beiko, R.G. Identifying biologically relevant differences between metagenomic communities. *Bioinformatics* **2010**, *26*, 715–721. [CrossRef] [PubMed]

60. Camon, E.B.; Barrell, D.G.; Dimmer, E.C.; Lee, V.; Magrane, M.; Maslen, J.; Binns, D.; Apweiler, R. An evaluation of GO annotation retrieval for BioCreAtIvE and GOA. *BMC Bioinform.* **2005**, *6*, 1–11. [CrossRef]

61. Huerta-Cepas, J.; Szklarczyk, D.; Heller, D.; Hernández-Plaza, A.; Forslund, S.K.; Cook, H.V.; Mende, D.R.; Letunic, I.; Rattei, T.; Jensen, L.J.; et al. eggNOG 5.0: A hierarchical, functionally and phylogenetically annotated orthology resource based on 5090 organisms and 2502 viruses. *Nucleic Acids Res.* **2018**, *47*, D309–D314. [CrossRef] [PubMed]

62. Overbeek, R.A.; Olson, R.; Pusch, G.D.; Olsen, G.J.; Davis, J.J.; Disz, T.; Edwards, R.; Gerdes, S.; Parrello, B.D.; Shukla, M.; et al. The SEED and the Rapid Annotation of microbial genomes using Subsystems Technology (RAST). *Nucleic Acids Res.* **2013**, *42*, D206–D214. [CrossRef] [PubMed]

63. Ogata, H.; Goto, S.; Sato, K.; Fujibuchi, W.; Bono, H.; Kanehisa, M. KEGG: Kyoto Encyclopedia of Genes and Genomes. *Nucleic Acids Res.* **1999**, *27*, 29–34. [CrossRef] [PubMed]

64. Tatusov, R.L.; Natale, D.A.; Garkavtsev, I.V.; Tatusova, T.A.; Shankavaram, U.T.; Rao, B.S.; Kiryutin, B.; Galperin, M.Y.; Fedorova, N.D.; Koonin, E.V. The COG database. *Nucleic Acids Res.* **2000**, *28*, 33–36. [CrossRef]

65. Finn, R.D.; Bateman, A.; Clements, J.; Coggill, P.; Eberhardt, R.Y.; Eddy, S.R.; Heger, A.; Hetherington, K.; Holm, L.; Mistry, J.; et al. Pfam: The protein families database. *Nucleic Acids Res.* **2013**, *42*, D222–D230. [CrossRef]

66. Que, Y.; Yue, X.; Yang, N.; Xu, Z.; Tang, S.; Wang, C.; Lv, W.; Xu, L.; Talbot, N.J.; Wang, Z. Leucine biosynthesis is required for infection-related morphogenesis and pathogenicity in the rice blast fungus Magnaporthe oryzae. *Curr. Genet.* **2019**, *66*, 155–171. [CrossRef]

67. Sun, R.-L.; Jing, Y.-L.; de Boer, W.; Guo, R.-J.; Li, S.-D. Dominant hyphae-associated bacteria of *Fusarium oxysporum* f. sp. *cucumerinum* in different cropping systems and insight into their functions. *Appl. Soil Ecol.* **2021**, *165*, 103977. [CrossRef]

68. Vijay, K.; Devi, T.S.; Sree, K.K.; Elgorban, A.; Kumar, P.; Govarthanan, M.; Kavitha, T. In vitro screening and in silico prediction of antifungal metabolites from rhizobacterium *Achromobacter kerstersii* JKP9. *Arch. Microbiol.* **2020**, *202*, 1–10. [CrossRef] [PubMed]

69. Blackman, L.M.; Cullerne, D.P.; Hardham, A.R. Bioinformatic characterisation of genes encoding cell wall degrading enzymes in the Phytophthora parasitica genome. *BMC Genom.* **2014**, *15*, 785. [CrossRef] [PubMed]

70. Gibson, D.M.; King, B.; Hayes, M.L.; Bergstrom, G. Plant pathogens as a source of diverse enzymes for lignocellulose digestion. *Curr. Opin. Microbiol.* **2011**, *14*, 264–270. [CrossRef] [PubMed]

71. Jorge, I.; Navas-Cortes, J.; Jimenezdiaz, R.M.J.M.; Tena, M. Cell wall degrading enzymes in fusarium wilt of chickpea: Correlation between pectinase and xylanase activities and disease development in plants infected with two pathogenic races of *Fusarium oxysporum* f. sp. *ciceris*. *Can. J. Bot.* **2006**, *84*, 1395–1404. [CrossRef]

Perspective

Metrology, Agriculture and Food: Literature Quantitative Analysis

Alessandra Durazzo [1,*], Eliana B. Souto [2,3], Ginevra Lombardi-Boccia [1], Antonello Santini [4] and Massimo Lucarini [1,*]

1　CREA-Research Centre for Food and Nutrition, Via Ardeatina 546, 00178 Roma, Italy; g.lombardiboccia@crea.gov.it
2　Department of Pharmaceutical Technology, Faculty of Pharmacy, University of Coimbra, Pólo das Ciências da Saúde, Azinhaga de Santa Comba, 3000-548 Coimbra, Portugal; souto.eliana@gmail.com
3　CEB-Centre of Biological Engineering, Campus de Gualtar, University of Minho, 4710-057 Braga, Portugal
4　Department of Pharmacy, University of Napoli Federico II, Via D. Montesano 49, 80131 Napoli, Italy; asantini@unina.it
*　Correspondence: alessandra.durazzo@crea.gov.it (A.D.); massimo.lucarini@crea.gov.it (M.L.)

Abstract: Great attention has been given in recent years to the relationships between metrology, agriculture, and food. This study aims at providing an analysis of the literature regarding the relationships between metrology, agriculture, and food. The Scopus online database has been used to extract bibliometric data throughout the search string: TITLE-ABS-KEY (Metrology* AND Agriculture* OR Food*), and the VOSviewer bibliometric software was used to visualize results as bubble maps. The novelty character of this perspective paper is to indicate and point out the main research themes/lines addressing the relationships between metrology, agriculture, and food by analyzing: (i) the authors of the published papers; (ii) the type of paper; (iii) the countries and institutions where the research is developed. Bibliometrics allows one to holistically examine entire scientific areas or sub-fields to get new qualitative and quantitative insights. These results represent a useful tool for identifying emerging research directions, collaboration networks, and suggestions for more in-depth literature searches.

Keywords: metrology; agriculture; food; biodiversity; literature quantitative analysis

Citation: Durazzo, A.; Souto, E.B.; Lombardi-Boccia, G.; Santini, A.; Lucarini, M. Metrology, Agriculture and Food: Literature Quantitative Analysis. *Agriculture* **2021**, *11*, 889. https://doi.org/10.3390/agriculture11090889

Academic Editor: Isabel Lara

Received: 14 July 2021
Accepted: 10 September 2021
Published: 16 September 2021

Publisher's Note: MDPI stays neutral with regard to jurisdictional claims in published maps and institutional affiliations.

1. Introduction

Nowadays, an integrated, multidisciplinary, and interoperable approach can be seen as a modern way to research food and an innovative challenge to analyze and model agro-food systems following a holistic approach [1]. A great challenge is to identify a unique dimension of food-agricultural aspects in terms of quality and safety [2]. The recent work of Brown [3], remarks how metrology remains a unique important effort, and outlines the importance of updating the concept of metrology: it proposes a new feature—'measuring measurement'—, emphasizing the characteristic meta-thought associated with the discipline, which distinguishes it from any routine measurement.

Metrology is the science of measurements, that is the discipline that deals with defining the procedures for performing correct measurements. The International Bureau of Weight and Measures in 2004 defined metrology as "the science of measurement, embracing both experimental and theoretical determinations at any level of uncertainty in any field of science and technology" [4].

Nonetheless, metrology is not only meters, kilograms, and atomic clocks; the science of measurement wants to be closer to everyday life, for example taking into account the "farm to fork model", reaching the consumer, certifying the origin of food. Metrology has always supported the needs of the technological world, and today it is accompanying the spread of completely new technologies. The application of metrology to the environment and food is the last frontier in this research area.

Today, the agri-food sector represents a strategic asset for all countries in Europe and not only there, being one of the most important socio-economic activities, and it is crucial for providing employment, the supply of healthy and high-quality food, and facilitating the integration of small and medium-sized enterprises (SMEs), representing 99% of all businesses in the European Union in the food chain [5]. The focus of consumer needs is represented by food quality and authenticity, and in this direction food traceability and safety become critical factors in ensuring the quality and protection of food consumer interests. Furthermore, providing healthy and sustainable diets is a challenge of the agri-food sector [6].

Metrology is also a tool, throughout the use of advanced data analysis methods, at the service of the so-called "precision agriculture", which combines satellite and drone images with those of sensors and actuators in order to identify, for example, the most efficient interventions in relation to the real cultivation needs and the biochemical and physical characteristics of the soil. In fact, in recent years the need to enhance the resilience of communities and territories, the reduction of the consumption of natural resources and chemical and phytosanitary fertilizers, and the optimization of agricultural production has forced the development of innovative techniques for crop management. Thus, metrology may allow, thanks to the experience in the field of meteorological forecasts, the knowledge of historical and actual agronomic data, the coordination of the activities related to the reduction of emissions and the development of information technologies for precision agriculture, such as networks of sensors, geolocation systems and agrometeorological models.

In order to achieve climate-neutrality by 2050, while also considering pandemic situations such as the COVID-19 pandemic, the challenge is to promote economic recovery and return to facing global goals, e.g., clean growth and climate change, for example by means of innovative technologies as well as the management of big data within the perspective of sharing and dissemination at the e-cloud interface [7,8]. The novelty character of the proposed perspective work is to give a current snapshot on the relationships existing between metrology, agriculture, and food, and to indicate related current directions and collaboration networks by analyzing research themes and major contributors with reference to country/regions, institutions and types of published papers. In fact, the bibliometric analysis reveals the applications of metrology principles in the agricultural and food fields.

2. Materials and Methods

The overall landscape of the literature in the research field of metrology, agriculture and food relationships has been investigated through a bibliometric analysis. The literature quantitative research analysis consisted in the following main steps: (i) literature search based on Scopus online database; (ii) data extraction and analysis.

2.1. Literature Search

In July 2021, a search for metrology, agriculture, food relationship publications was carried out based on the Scopus database content. The Scopus online database (https://www.scopus.com/home.uri, accessed on 3 July 2021) was used to extract bibliometric data using the search string: TITLE-ABS-KEY ("Metrology*" AND "Agriculture*" OR "Food*"). Publications mentioning the relevant words or their derivatives in the title, abstract, or keywords were identified throughout the applied search strategy. The evaluated parameters were: trends of publications and citations, document type, authorship, country/region and institution.

2.2. Data Extraction and Analysis

Bibliographic data, e.g., publication year, publication count, citation count, document type, authorship, countries/regions and institutions, were assessed. For the basic analyses the functions of the Scopus web online platform named "Analyze" and as "Create Citation Report" were used. The "full records and cited references" were exported to VOSviewer software (version 1.6.16, www.vosviewer.com, accessed on 3 July 2021) for additional processing procedures/operations.

The terms utilized in the title abstracts of publications and keywords of publications were analyzed by the above mentioned VOSviewer software (v.1.6.16, 2020) [9–11]: the paragraphs were broken down into words and phrases, and linked to the publications' citation data, in order to visualize the results as a bubble map. This approach has been previously used in different fields of study, including food, nutraceutical, and chemical areas [12–15]. In a term map, the bubble size indicates how frequently a term is mentioned in the articles. Two bubbles that are positioned more closely to each other reflect that the terms coappeared more often in the selected publications. The average citations per publication (CPP) are given by the color of a bubble.

Five was set as the minimum number of occurrences of a keyword. Out of the 3166 keywords, 108 met the selected threshold, and 3 of them were manually excluded.

3. Results

Three hundred and twelve publications were returned by the search: they covered the time range from 1970 to 2021 and were cited collectively 2183 times. The publication and citation trends for the relationships between metrology, agriculture and food research are reported in Figure 1. The first publication is the Proceedings of the Instrument Society of America (ISA) Silver Jubilee Conference and Exhibit on advances in instrumentation, located in Philadelphia on 26–29 October 1970 and also including subjects such as metrology, analysis instrumentation and food industry [16].

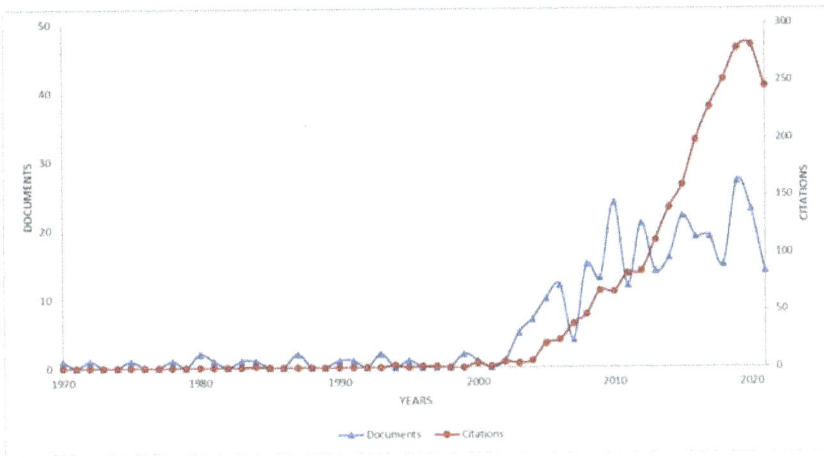

Figure 1. Publication and citation trends for the relationships between metrology, agriculture, and food research. (Bibliometric data were extracted from the Scopus online database).

The most recent "Review" is focused on the global situation of reference materials in assuring the quality and safety of the most consumed beverages in the world, i.e., coffee, cocoa, and tea, by discussing aspects related to reference material preparation processes, as well as the results of homogeneity and stability tests and their application, together with an overview of the patents developed for food. The authors remarked in the conclusion a clear need to develop certified reference materials and reference materials for these beverages for other analytes of interest, such as chlorogenic acids and other phenolic compounds [17]. The most recent "Article" showed the application of a smart pansharpening approach using kernel-based image filtering, as an example of numerous applications of remote sensing image fusion in monitoring, metrology, and agriculture [18]. Another application of innovative technologies is given by Fiorani et al. [19] on the use of a photoacoustic laser system for food fraud detection as a reliable technique for the rapid screening of counterfeited ingredients in the supply chain, which needs further development.

Additionally, the most cited paper (300 times) is a paper by De Chiffre et al. [20] which reported industrial applications of computed tomography in different fields including the food industry.

The distribution of the types of documents relative to the 312 publications retrieved are shown in Figure 2. "Article" accounts for 50.3%, followed by "Conference paper" (26.3%), "Review" (8.0%), and "Conference Review" (6.4%). Among "Book", it is worth mentioning the one published in 2007 by Meinrath and Schneider [21] entitled "Quality assurance for chemistry and environmental science: metrology from pH measurement to nuclear waste disposal": the basic metrological concepts for measurements in chemistry and geochemistry like traceability, ISO uncertainties or cause-and-effect diagrams were discussed, and applications of metrological techniques in highly complex situations, i.e., in thermodynamics, geochemical modeling, hydrology and radioactive waste disposal, were given. Another book focuses on the description of voluntary standards, mandatory technical regulations, conformity assessment (testing and measurement of products), certification, quality and quality management systems as well as other management systems such as environmental, social responsibility and food safety management systems [22].

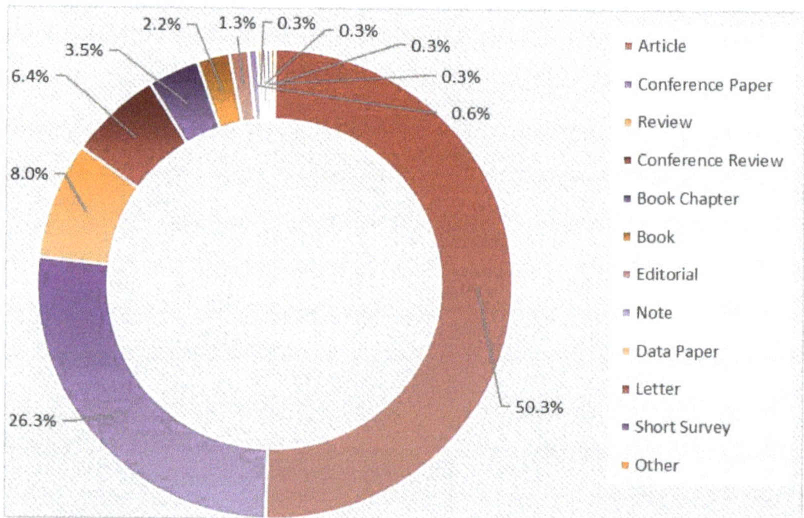

Figure 2. Distribution of the type of document. (Bibliometric data were extracted from the Scopus online database).

A book published in 2010 entitled: "Metrology in Industry: The Key for Quality" described an analysis of the metrological requirements needed to ensure quality, along with the organization of metrology, mastering the measurement process approach, the bank of measuring instruments, the traceability to national standards, measurements and uncertainties, the measuring environment, and others [23].

Among the "Editorial" category, the editorial published by Chirico and Bonavolontà [24] entitled: "Metrology for agriculture and forestry 2019" addressed recent advances in integrated monitoring and modelling technologies for agriculture and forestry.

Figure 3 reports the most productive authors. It should be noted that 'Anon' as Anonymous was originally ranked second by the Scopus 'Analyse Results' function, and it is not listed in Figure 3. Additionally, we underline that some of the most productive authors participate in the same collaborative papers.

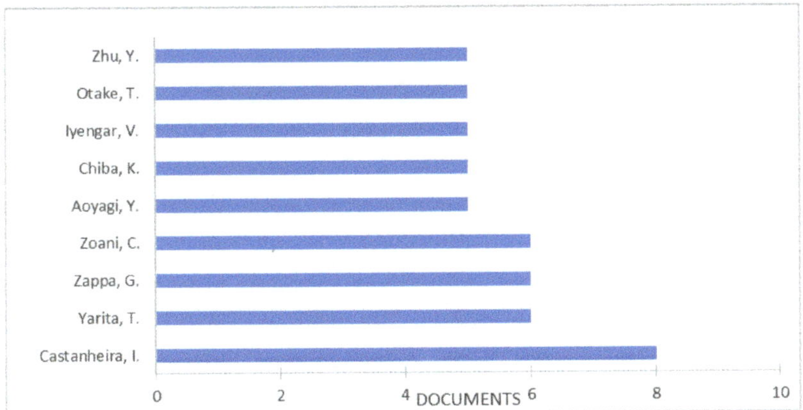

Figure 3. Most productive authors. (Bibliometric data were extracted from the Scopus online database).

Castanheira, I. (n = 8) resulted as the most productive author. The oldest work of this author regarding this matter pointed out the need to have reference materials in order to monitor the intake of nitrates [25]. Her most recent papers are represented by two conference proceedings presented at the 22nd World Congress of the International Measurement Confederation, IMEKO 2018, held from 3 to 6 September 2018 [26,27]; in one paper [26], the authors presented new Research Infrastructure METROFOOD-RI within the framework of the European Strategy Forum on Research Infrastructures (ESFRI) with the aim of promoting metrology in food and nutrition fields through the constitution of a well-organized and structured network of physical and electronic facilities: cross-cutting research activities and deliver advanced services were performed, covering several areas at the interface of different typologies of users: food business operators, research/academy, food control agencies, food policy makers, consumers/citizens. The core-services of METROFOOD-RI are the development and production of new (certified) Reference Materials (RMs). In the second one, published by Zoani et al. [27], feasibility studies for the development of new food matrix-Reference Materials (RMs) were presented. The feasibility studies included the following procedure Reference Materials preparation; procedures and guidelines definition for collecting characterization results and for processing the obtained data; Reference Materials characterization; homogeneity and stability studies; data processing and result evaluation, and three candidates for new Matrix-Reference Materials were tested: rice grains, rice flour, and lyophilized oyster tissue.

Castanheira' most highly cited paper (cited 11 times) focuses on ensuring food integrity by means of the integrated use of metrology and the application of FAIR (acronym for: findable, accessible, interoperable, and re-usable) data principles throughout the experience of the pan-European project METROFOOD-RI [28]. Among her papers, it is worth mentioning, also, the Conference Proceeding concerning the European Strategy on metrology in food composition databanks presented at the 20th IMEKO World Congress held from 9 to 14 September 2012 in Busan. In that document, the use of the International System of Units, modes of expressions and Reference Materials were indicated as being among the most important metrological tools for improving the quality of data in national Food Composition Databanks [29].

It is worth mentioning that METROFOOD-RI is composed of a Physical Infrastructure (P-RI) and an electronic infrastructure (e-RI). Among the relevant papers reported on the topic, like the ones by Zoani, C. and Zappa, G., it is worth mentioning the work of Alexandre [30], which describes the facilities which have been inventoried and classified in a database, providing an organized overview of the capacities of the distributed P-RI. These data were presented at the 3rd IMEKOFOODS Conference: "Metrology Promoting Harmonization and Standardization in Food and Nutrition", held in Thessaloniki from 1 to 4 October 2017. On the other hand, another work, presented at the same Conference by Presser et al. [31], and coauthored also by Zappa, G., described the development of a pilot e-RI where several datasets from different countries were used and interrelated to integrate national e-resources into a European-wide e-RI providing new functionalities. Additionally, a further position paper on METROFOOD-RI and its e-component were presented at the IEEE International Conference on Big Data, 2019, held in Los Angeles from 9 to 12 December 2019 [32].

The most cited paper of Iyengar, V. is a note focusing on metrological concepts for enhancing the reliability of food and nutritional measurements, including: high-quality reference standards, validated methods, robust sampling practices, proven calibration approaches, natural matrix reference materials, speciation chemistry, the assessment of measurement uncertainty and establishment of traceability links, certified reference materials to facilitate one aspect of traceability, and proficiency testing [33].

The most cited paper of Otake, T. is a research addressing the development of certified reference material—NMIJ CRM 7504—for the quantification of two pesticides in brown rice [34]. Another research from the same author was published in Food Chemistry in 2013 on the development of apple certified reference material for the quantification of organophosphorus and pyrethroid pesticides [35].

Figures 4 and 5 show the most productive countries/territories and institutions, respectively. The most productive country is represented by the United States with 48 documents. Among the documents reported for the United States, a book chapter was published on opportunities and limitations for metrology, represented by testing for foods derived from modern biotechnology [36]. An "Editorial" published by Iyengar in 2007 addressed differing perceptions of metrology in physics, chemistry, and biology [37]. Interesting results are also reported in a "Note" of Iyengar [33], previously described, and a "Note" by Koch et al. [38] on measurement science for food and drug monographs in the perspective of a global system. Among the "Article" category, the paper of Koch and Ma [39] on the approach of interfacing chemical metrology with pharmaceutical and compendial science adopted by United States Pharmacopeia indicates another relevant area of interest for the metrological approach.

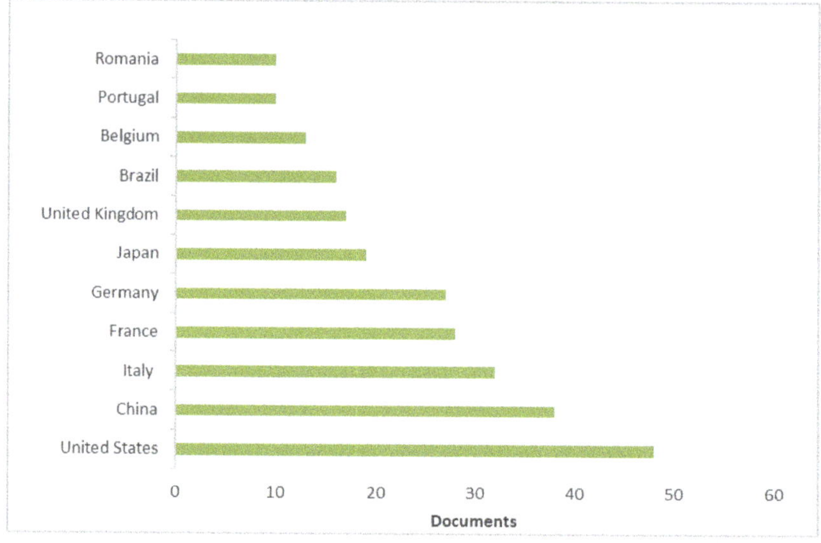

Figure 4. Most productive countries/territories. (Bibliometric data were extracted from the Scopus online database).

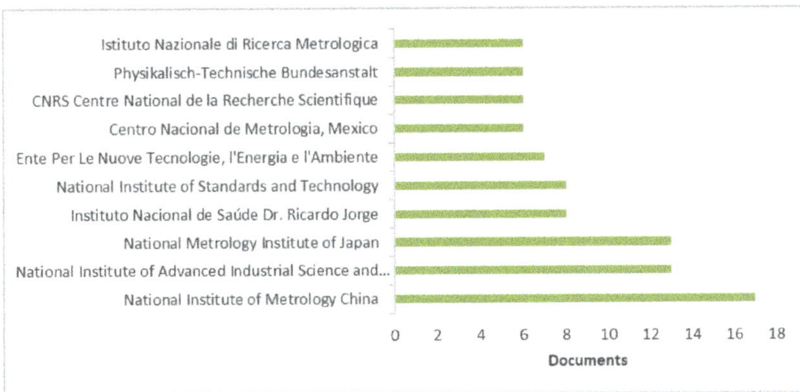

Figure 5. Most productive institutions. (Bibliometric data were extracted from the Scopus online database).

For China, the most cited "Article" (60 times) was addressed on the application of near-infrared spectroscopy to agriculture and food analysis; moreover, the authors marked the information sharing mode between the central database and end-user by using network technology and concentrating valuable resources [40]. One "Review" is reported and is focused on how a transfer program on metrology for safe food and feed in developing economies was started at the International Bureau of Weights and Measures to allow national metrology institutes or designated institutes to work together to strengthen their national mycotoxin metrology infrastructure, through a description of an application of an accurate characterization of a pure aflatoxin B1 material to avoid calibration errors. It is worth noting that mycotoxins, secondary metabolites produced from microfungi in some conditions, may represent a health threat for crops, food and feed and hence for humans through the carry-over process [41].

For Italy, among 32 documents, two reviews were relevant in the approach. One opens the door to the role of incurred materials in method development and validation in order to account for food processing effects in food allergen analysis [42]. The second is a stakeholders' guidance document for consumer analytical devices with a focus on gluten and food allergens [43]. In particular, the recommendations are based on the current known technologies, analytical expertise, and standardized AOAC INTERNATIONAL allergen community guidance and best practices for the analysis of food allergens and gluten [43].

The reported institutions have produced at least six documents. The most productive Institution is the National Institute of Metrology China with 17 documents. The most cited "Article" is a study describing an approach for the identification and determination of arsenic species in kelp extract [44]. Other important studies among the reported articles are the study of Guo et al. [45] on certified reference materials and metrological traceability for mycotoxin analysis and the study of Xue et al. [46] on reference material for the quantitative detection of *Escherichia coli*. The work of Sun et al. [47] on the comparison of maximum residue levels and the standard analytical method for pesticides in tea is another relevant example. It is also worth mentioning the recent work of Joseph et al. [48], a key comparison study on organic solvent calibration solution-gravimetric preparation and the value assignment of trans-zearalenone in acetonitrile.

For the National Institute of Advanced Industrial Science and Technology, the most cited paper is the paper of Zhu and Chiba [49] on the determination of cadmium in food samples by ID-ICP-MS with solid phase extraction for eliminating spectral interferences. Two reviews are also reported: one on a proficiency test in Japan for the elements in tea-leaf powder [50], the other on the assessment of technical problems in the analysis of inorganic elements in squid through proficiency testing [51].

One hundred and five terms in total were obtained from the literature quantitative analysis on publications, and they are visualized as a term map in Figure 6. Terms such as quality control, units of measurement, calibration, measurement/s, standard/s, reference standards, certified reference materials, and reference material appear as the top-recurring keywords: this shows the integrated research in the food and agriculture area, which is based on the control quality procedure and metrology principle (Table 1).

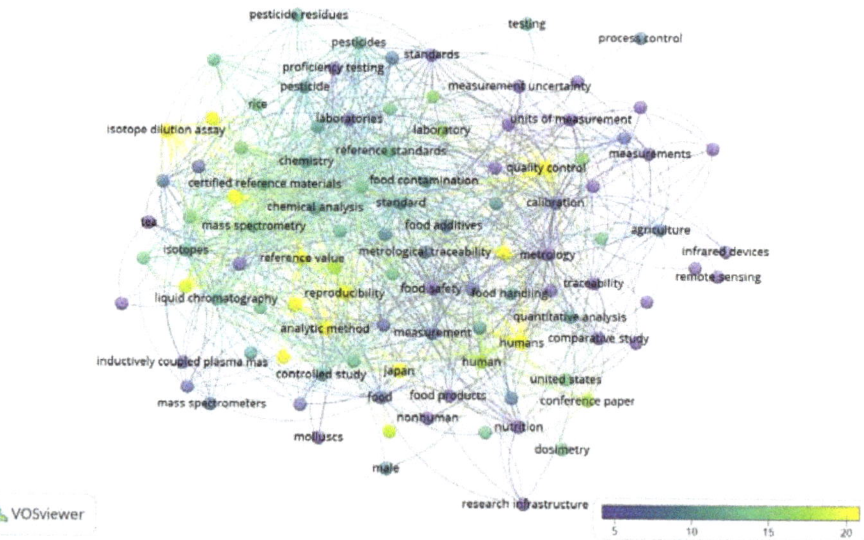

Figure 6. Term map for relationships between metrology, agriculture, and food research. The number of publications was represented by bubble size. The citations per publication (CPP) were given by bubble color. Two bubbles that are closer to each other reflect that the terms coappeared more frequently. (Bibliometric data were extracted from the Scopus online database and elaborated by VOSviewer software).

Table 1. The top-recurring terms on the relationships between metrology, agriculture and food research. (Bibliometric data were extracted from the Scopus online database and elaborated by VOSviewer software).

Term	Occurrence	Total Link Strenght
metrology	59	221
quality control	30	182
calibration	30	172
units of measurement	27	91
mass spectromentry	23	262
human	23	161
food safety	23	136
agriculture	22	69
measurement	21	167
standard	20	210
measurements	20	75
food analysis	19	176
reference standards	18	209
humans	18	139
chemical analysis	17	154
uncertainty analysis	17	90
chemistry	16	169
certified reference materials	16	167
reference material	16	67
standards	15	118

Analytical methods, reference materials, reference standards, calibration, and proficiency testing represent the key elements for quality control and ensuring the accuracy of results, as indicators of metrology traceability. The traceability of routine analysis is critical for accurate measurements.

4. Conclusions

The proposed search methodology based on a quantitative literature analysis represents a useful and potent tool to identify emerging research directions, collaboration networks, research infrastructures, and authors that are more active in the selected area of research. This may provide suggestions for more in-depth literature searches. It can be concluded that the application of metrology could provide an important contribution to the overall frontier research in agriculture and food areas worldwide, aligning investigation, research, and innovation with society's values, needs, and expectations. The proposed perspective could represent a starting point for indicating the importance of metrology in the explored areas of research, which also impact health, traceability, sustainable economy, and safety in the agro-food system, among other things.

Author Contributions: Conceptualization, A.D. and M.L.; methodology, A.D. and M.L.; investigation, A.D. and M.L.; writing—original draft preparation, A.D. and M.L.; writing—review and editing, A.D., E.B.S., G.L.-B., A.S. and M.L. All authors have read and agreed to the published version of the manuscript.

Funding: This research received no external funding.

Institutional Review Board Statement: Not applicable.

Data Availability Statement: Not applicable.

Conflicts of Interest: The authors declare no conflict of interest.

References

1. Durazzo, A. The close linkage between nutrition and environment through biodiversity and sustainability: Local foods, traditional recipes, and sustainable diets. *Sustainability* **2019**, *11*, 2876. [CrossRef]
2. Nicholson, C.F.; Stephens, E.C.; Kopainsky, B.; Jones, A.D.; Parsons, D.; Garrett, J. Food security outcomes in agricultural systems models: Current status and recommended improvements. *Agric. Syst.* **2021**, *188*, 103028. [CrossRef]
3. Brown, R.J.C. Measuring measurement—What is metrology and why does it matter? *Measurement* **2021**, *168*, 108408. [CrossRef] [PubMed]
4. The International Bureau of Weights and Measures. Worldwide Metrology: What Is Metrology? Available online: https://www.bipm.org/en/worldwide-metrology/#metrology (accessed on 3 July 2021).
5. European Commission. Internal Market, Industry, Entrepreneurship and SMEs. Available online: https://ec.europa.eu/growth/smes/sme-definition_en (accessed on 3 July 2021).
6. United Nations. Sustainable Development Goals. Available online: https://www.un.org/sustainabledevelopment/development-agenda/ (accessed on 3 July 2021).
7. Grochau, H.; Ten Caten, C.S.; de Camargo Forte, M.M. Motivations, benefits and challenges on ISO/IEC 17025 accreditation of higher education institution laboratories. *Accred. Qual. Assur.* **2018**, *23*, 183–188. [CrossRef]
8. Kunzmann, H.; Pfeifer, T.; Schmitt, R.; Schwenke, H.; Wecken-mann, A. Productive metrology–adding value to manufacture. *CIRP Ann.* **2005**, *54*, 155–168. [CrossRef]
9. Waltman, L.; van Eck, N.J.; Noyons, E.C. A unified approach to mapping and clustering of bibliometric networks. *J. Informetr.* **2010**, *4*, 629–635. [CrossRef]
10. Van Eck, N.J.; Waltman, L. Software survey: VOSviewer, a computer program for bibliometric mapping. *Scientometrics* **2010**, *84*, 523–538. [CrossRef]
11. Van Eck, N.J.; Waltman, L. Text mining and visualization using VOSviewer. *ISSI Newslett.* **2011**, *7*, 50–54.
12. Yeung, A.W.K.; Tzvetkov, N.T.; El-Tawil, O.S.; Bungău, S.G.; Abdel-Daim, M.M.; Atanasov, A.G. Antioxidants: Scientific literature landscape analysis. *Oxid. Med. Cell. Longev.* **2019**, *2019*, 8278454. [CrossRef]
13. Atanasov, A.G.; Zotchev, S.B.; Dirsch, V.M.; Supuran, C.T.; The International Natural Product Sciences Taskforce. Natural products in drug discovery: Advances and opportunities. *Nat. Rev. Drug Discov.* **2021**, *20*, 200–216. [CrossRef]
14. Lucarini, M.; Durazzo, A.; Lombardi-Boccia, G.; Souto, E.B.; Cecchini, F.; Santini, A. Wine polyphenols and health: Quantitative research literature analysis. *Appl. Sci.* **2021**, *11*, 4762. [CrossRef]
15. Yeung, A.W.K.; Souto, E.B.; Durazzo, A.; Lucarini, M.; Novellino, E.; Tewari, D.; Wang, D.; Atanasov, A.G.; Santini, A. Big impact of nanoparticles: Analysis of the most cited nanopharmaceuticals and nanonutraceuticals research. *Curr. Res. Biotechnol.* **2020**, *2*, 53–63. [CrossRef]

16. Miller, K.S. Advances in instrumentation, v 25. In Proceedings of the ISA Silver Jubilee Conference and Exhibit, Philadelphia, PA, USA, 26–29 October 1970.

17. de Carvalho Couto, C.; Gomes dos Santos, D.; Morais Oliveira, E.M.; Freitas-Silva, O. Global situation of reference materials to assure coffee, cocoa, and tea quality and safety. *Trends Anal. Chem.* **2021**, *143*, 116381. [CrossRef]

18. Smadi, A.A.L.; Yang, S.; Mehmood, A.; Abugabah, A.; Wang, M.; Bashir, M. Smart pansharpening approach using kernel based image filtering. *IET Image Proc.* **2021**, *15*, 2629–2642. [CrossRef]

19. Fiorani, L.; Artuso, F.; Giardina, I.; Lai, A.; Mannori, S.; Puiu, A. Photoacoustic laser system for food fraud detection. *Sensors* **2021**, *21*, 4178. [CrossRef] [PubMed]

20. De Chiffre, L.; Carmignato, S.; Kruth, J.-P.; Schmitt, R.; Weckenmann, A. Industrial applications of computed tomography. *CIRP Ann. Manuf. Technol.* **2014**, *63*, 655–677. [CrossRef]

21. Meinrath, G.; Schneider, P. *Quality Assurance for Chemistry and Environmental Science: Metrology from pH Measurement to Nuclear Waste Disposal*; Springer: Berlin/Heidelberg, Germany, 2007.

22. El-Tawil, A. Standards and quality. In *Standards and Quality*; World Trade Organization: Geneva, Switzerland, 2014; pp. 1–179.

23. Placko, D. Metrology in industry: The key for quality. *Invest. Manag. Financ. Innov.* **2021**, *18*, 1–270.

24. Chirico, G.B.; Bonavolontà, F. Metrology for agriculture and forestry 2019. *Sensors* **2020**, *20*, 3498. [CrossRef] [PubMed]

25. Castanheira, I.; Oliveira, L.; Valente, A.; Alvito, P.; Costa, H.S.; Alink, A. The need for reference materials when monitoring nitrate intake. *Anal. Bioanal. Chem.* **2004**, *378*, 1232–1238. [CrossRef]

26. Zappa, G.; Anorga, L.; Belc, N.; Castanheira, I.; Donard, O.F.X.; Kourimska, L.; Kukovecz, A.; Iatco, I.; Najdenkoska, A.; Nieminen, J.; et al. METROFOOD-RI: A new reality to develop reference materials for the agrifood sector. *J. Phys. Conf. Ser.* **2018**, *1065*, 232006. [CrossRef]

27. Zoani, C.; Anorga, L.; Belc, N.; Castanheira, I.; Donard, O.F.X.; Kourimska, L.; Kukovecz, A.; Iatco, I.; Najdenkoska, A.; Ogrinc, N.; et al. Feasibility studies for new food matrix-reference materials. *J. Phys. Conf. Ser.* **2018**, *1065*, 232005. [CrossRef]

28. Rychlik, M.; Zappa, G.; Añorga, L.; Belc, N.; Castanheira, I.; Donard, O.F.X.; Kouřimská, L.; Ogrinc, N.; Ocké, M.C.; Presser, K.; et al. Ensuring food integrity by metrology and FAIR data principles. *Front. Chem.* **2018**, *6*, 49. [CrossRef]

29. Castanheira, I.; Roe, M.; Bell, S.; Colomban, P.; Calhau, M.A.; Finglas, P. Metrology in Food Composition Databanks: The European Strategy. In Proceedings of the 20th IMEKO World Congress 2012, Busan, Korea, 9–14 September 2012; Volume 2, pp. 765–767.

30. Alexandre, J.; Tangni, E.K.; Zoani, C.; Donard, O.; Zappa, G.; Van Loco, J. Metrofood-RI: Inventory of the Facilities and Organization of the Physical Infrastructure. In Proceedings of the 3rd IMEKOFOODS Conference: Metrology Promoting Harmonization and Standardization in Food and Nutrition, Thessaloniki, Greece, 1–4 October 2017; pp. 193–196.

31. Presser, K.; Zoom, C.; Szymanek, J.; Ocke, M.; Zappa, G. Development of a Pilot Service for the Electronic Infrastructure of METROFOOD-RI. In Proceedings of the 3rd IMEKOFOODS Conference: Metrology Promoting Harmonization and Standardization in Food and Nutrition, Thessaloniki, Greece, 1–4 October 2017; pp. 95–99.

32. Ogrinc, N.; Seljak, B.K.; Presser, K.; Ocke, M.; Iatco, I.; Zoani, C. Promoting Metrology in Food and Nutrition: A Position Paper on METROFOOD-RI and Its e-Component. In Proceedings of the 2019 IEEE International Conference on Big Data, Los Angeles, CA, USA, 9–12 December 2019; pp. 5162–5164.

33. Iyengar, V. Metrological concepts for enhancing the reliability of food and nutritional measurements. *J. Food Compos. Anal.* **2007**, *20*, 449–450. [CrossRef]

34. Otake, T.; Itoh, N.; Aoyagi, Y.; Matsuo, M.; Hanari, N.; Otsuka, S.; Yarita, T. Development of certified reference material for quantification of two pesticides in brown rice. *J. Agric. Food Chem.* **2009**, *57*, 8208–8212. [CrossRef]

35. Otake, T.; Yarita, T.; Aoyagi, Y.; Kuroda, Y.; Numata, M.; Iwata, H.; Watai, M.; Mitsuda, H.; Fujikawa, T.; Ota, H. Development of apple certified reference material for quantification of organophosphorus and pyrethroid pesticides. *Food Chem.* **2013**, *138*, 1243–1249. [CrossRef]

36. Lipp, M. Testing for foods derived from modern biotechnology: Opportunities and limitations for metrology. In *Traceability in Chemical Measurement*; Springer: Berlin/Heidelberg, Germany, 2005; pp. 134–140.

37. Iyengar, V. Metrology in physics, chemistry, and biology: Differing perceptions. *Biol. Trace Elem. Res.* **2007**, *116*, 1–4. [CrossRef]

38. Koch, W.F.; Hauck, W.W.; De Mars, S.S.; Williams, R.L. Measurement science for food and drug monographs: Toward a global system. *Pharm. Res.* **2010**, *27*, 1203–1207. [CrossRef] [PubMed]

39. Koch, W.F.; Ma, B. The US Pharmacopeia: Interfacing chemical metrology with pharmaceutical and compendial science. *Accred. Qual. Assur.* **2011**, *16*, 43–51. [CrossRef]

40. Wang, D.J.; Zhou, X.Y.; Jin, T.M.; Hu, X.N.; Zhong, J.E.; Wu, Q.T. Application of near-infrared spectroscopy to agriculture and food analysis. *Guang Pu Xue Yu Guang Pu Fen Xi* **2004**, *24*, 447–450. [PubMed]

41. Josephs, R.D.; Li, X.; Li, H.; Wielgosz, R.I. The BIPM mycotoxin metrology capacity building and knowledge transfer program: Accurate characterization of a pure aflatoxin B1 material to avoid calibration errors. *J. AOAC Int.* **2019**, *10*, 1740–1748. [CrossRef] [PubMed]

42. Mattarozzi, M.; Careri, M. The role of incurred materials in method development and validation to account for food processing effects in food allergen analysis. *Anal. Bioanal. Chem.* **2019**, *411*, 4465–4480. [CrossRef]

43. Popping, B.; Allred, L.; Bourdichon, F.; Thompson, T.; Yeung, J. Stakeholders' guidance document for consumer analytical devices with a focus on gluten and food allergens. *J. AOAC Int.* **2018**, *101*, 185–189. [CrossRef]

44. Yu, L.L.; Wei, C.; Zeisler, R.; Tong, J.; Oflaz, R.; Bao, H.; Wang, J. An approach for identification and determination of arsenic species in the extract of kelp. *Anal. Bioanal. Chem.* **2015**, *407*, 3517–3524. [CrossRef]
45. Guo, Z.; Li, X.; Li, H. Certified reference materials and metrological traceability for mycotoxin analysis. *J. AOAC Int.* **2019**, *102*, 1695–1707. [CrossRef] [PubMed]
46. Xue, L.; Lin, J.; Sui, Z.-W.; Wang, J.; Liu, X.-X.; Fu, B.-Q.; Zhang, L.; Hao, L. Preparation of reference material for quantitative detection of *Escherichia coli. Acta Metrol. Sin.* **2015**, *36*, 652–656.
47. Sun, Z.-J.; Tang, H.; Chen, D.-Z.; Feng, J. The comparison of maximum residue levels and standard analytical method of pesticide in tea. *Acta Metrol. Sin.* **2010**, *31*, 121–124.
48. Josephs, R.D.; Daireaux, A.; Bedu, M.; Binici, B.; Gokcen, T. Key comparison study-organic solvent calibration solution-gravimetric preparation and value assignment of trans-zearalenone (trans-ZEN) in acetonitrile (ACN). *Metrologia* **2021**, *57*, 08019. [CrossRef]
49. Zhu, Y.; Chiba, K. Determination of cadmium in food samples by ID-ICP-MS with solid phase extraction for eliminating spectral-interferences. *Talanta* **2012**, *90*, 57–62. [CrossRef]
50. Zhu, Y.; Kuroiwa, T.; Narukawa, T.; Inagaki, K.; Chiba, K. Proficiency test in Japan for the elements in tea-leaf powder. *TrAC* **2012**, *34*, 152–160. [CrossRef]
51. Narukawa, T.; Inagaki, K.; Naito, S.; Fujimoto, T.; Chiba, K. Assessment of technical problems in the analysis of inorganic elements in squid through proficiency testing. *TrAC* **2016**, *76*, 216–226. [CrossRef]

agriculture

Perspective

Antioxidant Properties of Bee Products Derived from Medicinal Plants as Beekeeping Sources

Alessandra Durazzo [1,*], Massimo Lucarini [1], Manuela Plutino [2], Giuseppe Pignatti [3], Ioannis K. Karabagias [4,5], Erika Martinelli [6], Eliana B. Souto [7], Antonello Santini [8] and Luigi Lucini [6]

1   CREA-Research Centre for Food and Nutrition, Via Ardeatina 546, 00178 Rome, Italy; massimo.lucarini@crea.gov.it

2   CREA-Research Centre for Forestry and Wood, 52100 Arezzo, Italy; manuela.plutino@crea.gov.it

3   CREA-Research Centre for Forestry and Wood, 00166 Roma, Italy; giuseppe.pignatti@crea.gov.it

4   Laboratory of Food Chemistry, Department of Chemistry, University of Ioannina, 45110 Ioannina, Greece; ikaraba@uoi.gr

5   Department of Food Science and Technology, School of Agricultural Sciences, University of Patras, Charilaou Trikoupi 2, 30100 Agrinio, Greece

6   Department for Sustainable Food Process, Università Cattolica del Sacro Cuore, Via Emilia Parmense, 84, 29122 Piacenza, Italy; erika.martinelli1@unicatt.it (E.M.); luigi.lucini@unicatt.it (L.L.)

7   CEB—Centre of Biological Engineering, University of Minho, Campus de Gualtar, 4710-057 Braga, Portugal; eliana.souto@ceb.uminho.pt

8   Department of Pharmacy, University of Napoli Federico II, Via D. Montesano 49, 80131 Napoli, Italy; asantini@unina.it

*   Correspondence: alessandra.durazzo@crea.gov.it

Citation: Durazzo, A.; Lucarini, M.; Plutino, M.; Pignatti, G.; Karabagias, I.K.; Martinelli, E.; Souto, E.B.; Santini, A.; Lucini, L. Antioxidant Properties of Bee Products Derived from Medicinal Plants as Beekeeping Sources. *Agriculture* **2021**, *11*, 1136. https://doi.org/10.3390/agriculture11111136

Academic Editors: Ilaria Marotti and Ângela Fernandes

Received: 22 July 2021
Accepted: 9 November 2021
Published: 13 November 2021

Publisher's Note: MDPI stays neutral with regard to jurisdictional claims in published maps and institutional affiliations.

Abstract: Plant species are fundamental source of nectar in beekeeping since bees access nectar and pollen from flowers. Consequently, bee products are strongly linked to the bee foraging flora source, and, depending on this, they acquire defined features, including their health and medicinal properties. Medicinal plants contribute greatly to increase the beneficial properties of bee products, such as honey, pollen, royal jelly, and propolis. Bee products represent a potential source of natural antioxidants that can counteract the effects of oxidative stress underlying the pathogenesis of many diseases. The antioxidant properties of bee products have been widely studied and there is an abundance of information available in the literature. Notwithstanding, the uniqueness of the presented perspective is to provide an updated overview of the antioxidant properties of bee products derived from medicinal plants as beekeeping sources. This topic is divided and discussed in the text in different sections as follows: (i) beekeeping and the impacts of environmental factors; (ii) an overview of the role of medicinal plants for bee products; (iii) definition and categorization of the main medicinal bee plants and related bee products; (iv) the study approach of the antioxidant properties; (v) the conventional and innovative assays used for the measurement of the antioxidant activity; and (vi) the antioxidant properties of bee products from medicinal plants.

Keywords: antioxidant properties; bee products; honey; propolis; plant sources; medicinal plants

1. Introduction

Bee products represent a potential source of natural antioxidants, including phenolic acids, flavonoids, and terpenoids as well as numerous other phytochemicals, which are capable of counteracting the oxidative stress effects underlying the pathogenesis of many diseases [1]. The main action of the antioxidants is based on the capability to inhibit oxidation processes, thus reducing the production of free radicals, which result in triggering a chain reaction, which may cause harmful cellular alterations [2]. Reactive oxygen species (ROS) are produced by living organisms as a result of the normal cellular metabolism and environmental factors. The ROS are highly reactive molecules involved in many cellular signaling pathways, and can damage cell structures, such as carbohydrates, lipids, nucleic acids, and proteins, and consequently alter their functions [3,4]. Oxidative stress is defined as a condition resulting when the critical balance between free radical generation and antioxidant defenses is unfavorable [5–7]. The state of oxidative stress could be related to various degenerative diseases, such as atherosclerosis, cancer, neurological disorders, diabetes, and cardiovascular disease [8,9].

The antioxidant compounds contained in bee products have different mechanisms of action causing the decrease of the adverse consequences of reactive oxygen and nitrogen species, which lead to oxidative stress. They can inhibit the enzymes responsible for producing superoxide anions and metal chelation, break the radical chain reactions, and play a preventive role inhibiting the formation of the reactive oxidants species [10]. The antioxidant properties of bee products have been widely studied for their relevant interest [11,12]. The current trend of interest in this topic is evident by the substantial amount and typology of the existing published research papers on bee products and antioxidants. For example, a search on honey throughout the Scopus online database was carried out by means of the string TITLE-ABS-KEY (honey* AND "antioxidant property*" OR "antioxidant capacity*" OR "antioxidant assay*"). The "full records and cited references" were exported and processed using the VOSviewer software (version 1.6.16, 2020; www.vosviewer.com, accessed on 6 June 2021) [13–15]. The search returned 713 publications covering the time range from 1996 to 2021, and a total of 559 terms were identified and visualized as a term map in Figure 1. Figure 1 allows for the identification of the main terms to be correlated to research on the relationship between antioxidant properties and honey, and also identifies the main existing research lines focused on this topic. It is interesting to observe that among the top-recurring keywords, compounds such as phenols, flavonoids, phenol derivative, and polyphenols appear.

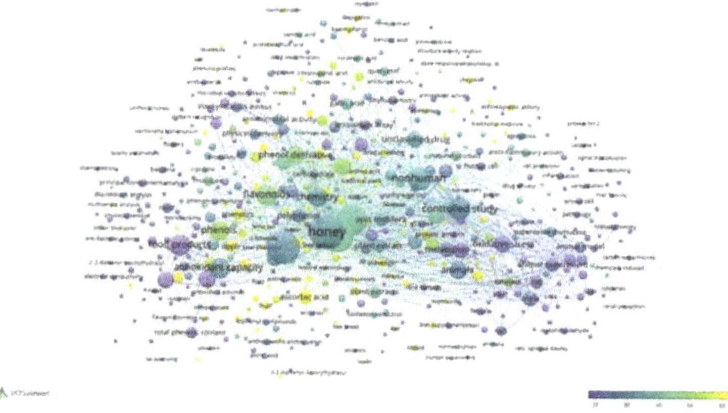

Figure 1. Term map for the relationship of honey and its antioxidant properties research. Bubble size represents the number of publications. Bubble color represents the citations per publication (CPP). Two bubbles are closer to each other if the terms co-appeared more frequently (bibliometric data were extracted from the Scopus online database and elaborated by the VOSviewer software).

Medicinal plants contribute greatly to increase the beneficial properties of bee products e.g., honey, pollen, royal jelly, and propolis, and have the potential to produce bee products with higher bioactivity. The value of honeybee products is strictly related to the plants that attract honeybees.

The uniqueness of this perspective is to provide an updated overview of the antioxidant properties of bee products derived from medicinal plants as beekeeping sources. The topic is discussed as follows: (i) beekeeping and the impacts of environmental factors; (ii) overview of the role of medicinal plants for bee products; (iii) definition and categorization of main medicinal bee plants and related bee products; (iv) study approach of the antioxidant properties; (v) conventional and innovative assays used for the measurement of the antioxidant activity; and (vi) antioxidant properties of bee products from medicinal plants. To our knowledge, these features have not been studied together in previous perspectives in the literature.

2. Bee Products, Medicinal Plant, and Environment: An Overview

An overview of the linkage of beekeeping, bee products, medicinal plants, and environment is delineated in the following sub-sections by underlining the key role of medicinal plants for bee products. The following topics are explored: (i) beekeeping and the impacts of environmental factors; (ii) overview of the role of medicinal plants for bee products; and (iii) definition and categorization of the main medicinal bee plants and the related bee products. The sizable contribution of medicinal plants to health properties of bee products is notable.

2.1. Beekeeping and the Impacts of Environmental Factors

Bees are considered significant pollinators due to their effectiveness and wide diffusion worldwide. Bee pollination provides excellent value to crop quality and quantity, improving global economic and dietary outcomes [16].

Honeybees (*Apis mellifera*) are social species and represent one of the most important pollinators for agricultural systems [17,18]. Humans have managed honeybees for thousands of years, and developed bee breeding for all continents, mostly the United States and Europe [19,20].

Bees have been used to produce honey and play a mandatory role in pollination since the time of Ancient Egypt populations [21,22]. Many centuries later, in the seventeenth and eighteenth centuries, the improvement of beekeeping techniques made it possible to maintain large bee colonies giving rise to modern apiculture [23].

In the United States and Europe, Bruckner et al. [24] showed that beekeepers faced significant seasonal problems, namely high mortality occurred over the winter. In the United States, beekeepers contain 30% colony losses each winter. Other studies showed that seasonal floral resources, insecticide use, and the availability of natural environmental habitat are major drivers of bee health and abundance [25].

Honeybees are constituted by over 20,000 species and each colony contains thousands of individuals [26,27]. In Central Europe, the number of managed honeybee colonies have decreased since the 1960s [28,29].

This decrease in number could be attributed to bee activities, their survival and antropic impact on the eco systems are closely connected; humans depend on bees for ecosystem services and the bees depend on the antropic activities for their survival [30,31]. On the other hand, antropic impact and environmental pollution are main issues regarding honeybee survival, and in this regard, it should be considered that the life of many plant species depends on bee pollination. The reduction of bee colonies poses a serious risk to many plant species survival, and can also be considered a biomarker for human health [32,33].

In an ecosystem global vision, recognizing several ecosystem services provided by bees is necessary, and so are the large variety of ecosystem services to humans, such as pollination, provision, regulation, and equilibrium [34]. If only one ecosystem service is considered, sharp declines in the provision of other ecosystem services may occur [35].

The honeybee is one of the most studied among all animals. This research has been almost entirely developed on the European honeybee *Apis mellifera*. According to the Food and Agriculture Organization (FAO) [36], only 11 new honeybee species have been recorded and identified in the past 15 years (Table 1).

Table 1. Eleven honeybees species as recorded worldwide [36–38].

Genus	Species of Honeybees
Apis	*Apis andreniformis*
	Apis binghami
	Apis breviligula
	Apis cerana
	Apis dorsata
	Apis florea
	Apis koschevnikovi
	Apis laboriosa
	Apis mellifera
	Apis nigrocincta
	Apis nuluensis

2.2. The Role of Medicinal Plants for Bee Products: An Overview

Plant species are fundamental in beekeeping as a source of nectar. The species visited by bees provide the honey with particular features: pleasant taste, typical color, and pharmaceutical beneficial properties. Honey is an exclusive vegetable product. About 80% of the world's population choose plant-based extracts for basic health care. The local population in many areas of the world uses medicinal plants holistically, and considers them an important source for the prevention and treatment of diseases, especially in low income countries where a conventional pharmacological approach may be difficult [39–42]. Honey is considered a therapeutic food, which possesses pharmacological activity [43].

The flora existing in the bees' environment is important for beekeeping since the bees collect nectar and pollen from flowers. The importance of flora in beekeeping has been observed by various authors around the world [44–46]. Plant species differ from place to place, in their flowering duration, due to climate, topography and agricultural practices. Knowing the type, density, and quality of bee flora are among the crucial factors for the success and productivity of beekeeping [47].

Forests, grasslands, agrophytocenoses (e.g., orchards, vineyards, flower crops), medicinal plant plantations, and aromatic herbs are frequently visited by bees to search for melliferous plants [48].

According to Bakour et al. [49], most beekeepers prefer plants characterized by several and numerous beneficial properties for the well-being of humans as antioxidant, anti-inflammatory, antifungal, antidiabetic, diuretic, having effects in the cure of different types of cancer, and also in the neurodegenerative, cardiovascular, and gastrointestinal tract diseases. The medicinal properties of bee products are also dependent on their botanical sources, as previously reported [50]; the several botanical origins provide bee products with numerous medicinal properties and products with therapeutic features for consumers.

Some of the most popular bee products for positive human health characteristics have been reported [51–53]:

1. Bee pollen: is a product rich in B vitamins, minerals, and unsaturated fatty acids. Many metabolic difficulties may be overcome with bee pollen supplementation and also it counteracts harmful bacteria.
2. Propolis: is effective against bacteria and purifies or disinfects. Its use is recommended for the treatment of colds, wounds, or ulcers, and diseases affecting the joints.
3. Bee bread: is a bee pollen-derived product, which acts as an activator of beneficial properties for blood circulation, is capable of healing and strengthening the immune and nervous system, and enriches polyunsaturated fatty acids intake.

The aforementioned products are strongly linked to the vegetable source from which the bees acquire their properties, including health and medicinal properties.

The importance of medicinal plants for beekeeping not only refers to the possibility of obtaining bee products from these species. In beekeeping, for some time, medicinal plant species have also been used as an alternative to pesticides [54–56]. According to

Khan et al. [57], many honey bee pathogens are contrasted with medicinal plants by beekeepers. Several medicinal plants are effective against fungi, mites as *Varroa* spp., and bacteria. Nguyen et al. [58] investigated physicochemical and viscoelastic properties of honey from medicinal plants, i.e., Tulsi, Alfalfa, and two varieties of Manuka honey derived from medicinal plants.

In the next section, bee products derived from medicinal plant sources are described and explored in detail.

2.3. Definition and Categorization of Main Medicinal Bee Plants and Related Bee Products

Many plants considered mainly as food or raw material sources have some special beneficial health effect [59–64]. Sage (*Salvia* spp.) leaves, for example, are the basis for an herbal tea, which can be used for medicinal purposes, but these are also used in food preparation as spices and seasonings (as an aromatic plant), while the essential oil is used in cosmetics (e.g., soaps, and toothpaste). In this sense, the term 'medicinal' is often understood in a wide sense, and includes several overlapping uses as herbal teas, spices, food, raw material, dietary supplements, and cosmetics containing extracts of derived compounds from plants [65].

The World Checklist of Useful Plant Species contains more than 40,000 taxon names from more than 400 families and 6000 genera with a documented human use [66]. Medicinal plants, both for human and veterinary use, account for 26,662 species, and this number constantly increases with research on uses in folk or traditional medicine systems and with the addition of new plant species. The exact number of species used as medicinal material in Europe is difficult to ascertain because of the limited amount of used material (which escape from trade catalogues), the origin from several (often undetermined) plant species and lack of documentation for local uses. About 2000 taxa commercially available are sources of medicinal and aromatic plant material, and two-third of them are represented by species native to Europe [65]. In particular, Germany (1500 taxa, 600 native), France (900 taxa, 450 native), and Spain (800, 600 native) are the leading countries in the trade of medicinal and aromatic plants in Europe.

The demand for wild-collected plants species is increasing worldwide, and has become a risk for the conservation and preservation of the natural resources [67]. Using data filtering from global checklists of medicinal plants, about 400 medicinal plant species native to Europe have been included in the European red list of medicinal plants [68]. For the species in which sufficient data are available, the 2.4% were assessed as threatened and the 4.5% near threatened, following International Union for Conservation of Nature (IUCN) Red List Categories. The highest number of species was found in the Mediterranean area and in mountain areas (e.g., Alps, Pyrenees, Massif Central, Balkan Peninsula), with a similar pattern for endemic plant species.

In the same study, the major threats have been identified as wild plant collection and loss of habitat (respectively, 26% and 30% of all species), connected to human impact, the so called anthropic effect (e.g., livestock farming, recreational activities, tourism, use of chemicals, pollution, and urban development).

Bee foraging activity on plants is dedicated to the search of nectar and honeydew as carbohydrate sources, pollen (as a protein source), and resins to produce propolis (for antimicrobial and defense purposes) [69]. Pollen is collected by the bees during their visits on the plant flowers and it is stored in pellets as a protein source into the hive. The botanical origin of both honey and pollen depends on the flora surrounding the area of foraging and influences physico-chemical, functional, and sensory properties of bee products [70–72]. More than half of the medicinal plant species of the European checklist might be considered as relevant nutrition sources for bees (Table 2).

Table 2. Main plant taxon (family, genus) of the European medicinal plant red list in relation to bee foraging sources. N = number of species belonging to the shown genus or group (W = attractive plants for wild bees). Plant data from Allen et al. [68], bee preferences for pollen and nectar from MLR, [73], MAA, [74].

Plant Family	N.	Nectar	N.	Pollen	Honeydew	N.	Wild Bees
AMARYLLIDACEAE	2	*Allium*	2	*Allium*		2	W
ANACARDIACEAE	1	*Cotinus*	1	*Pistacia*			
BETULACEAE			2	*Betula*	*Betula*		
BORAGINACEAE	3	*Borago, Symphytum, Pulmonaria*	3	*Borago, Symphytum, Pulmonaria*		3	W
CAPRIFOLIACEAE	4	*Valeriana, Viburnum, Sambucus*	4	*Valeriana, Viburnum, Sambucus*		2	W
ASTERACEAE	13	*Arctium, Aster, Cichorium, Taraxacum, Achillea, Inula, Matricaria, Silybum, Solidago, Tanacetum, Tussilago*	15	*Arctium, Aster, Cichorium, Taraxacum, Achillea, Inula, Matricaria, Silybum, Solidago, Tanacetum, Tussilago, Helichrysum*		8	W
CRUCIFERAE	1	*Brassica*	3	*Brassica, Capsella, Lepidium*		3	W
CUPRESSACEAE			4	*Juniperus*			
ERICACEAE	7	*Calluna, Erica, Vaccinium, Arbutus, Rhododendron*	6	*Calluna, Erica, Vaccinium, Rhododendron*		1	W
FAGACEAE	1	*Castanea*	4	*Castanea, Quercus*	*Castanea, Quercus*	3	W
IRIDACEAE	7	*Iris*				7	W
LAMIACEAE	34	*Hyssopus, Lavandula, Thymus, Glechoma, Nepeta, Origanum, Prunella, Rosmarinus, Salvia, Teucrium, Galeopsis, Lamium, Leonurus, Melissa, Satureja, Stachys, Mentha*	34	*Hyssopus, Lavandula, Thymus, Glechoma, Nepeta, Origanum, Prunella, Rosmarinus, Salvia, Teucrium, Galeopsis, Lamium, Leonurus, Melissa, Satureja, Stachys, Mentha, Clinopodium, Ballota*	*Thymus*	34	W
LEGUMINOSAE	7	*Melilotus, Trifolium, Medicago, Astragalus, Ononis, Pisum*	6	*Melilotus, Trifolium, Medicago, Ononis, Pisum*		7	W
MALVACEAE	3	*Malva, Althaea*	3	*Malva, Althaea*		1	W
OLEACEAE			3	*Fraxinus, Olea*	*Fraxinus*		
ORCHIDACEAE						30	W
PAEONIACEAE	3	*Paeonia*	3	*Paeonia*			
PAPAVERACEAE			3	*Papaver, Chelidonium*		3	W
PINACEAE			4	*Pinus, Abies, Larix*	*Pinus, Abies*		
PLANTAGINACEAE			6	*Plantago*		5	W
POLYGONACEAE	1	*Polygonum*	4	*Polygonum, Rumex*			
PRIMULACEAE	3	*Primula*	3	*Primula*		2	W
RANUNCULACEAE	7	*Aquilegia, Helleborus, Aconitum, Ficaria, Pulsatilla*	8	*Aquilegia, Helleborus, Aconitum, Ficaria, Pulsatilla, Hepatica*		2	W
ROSACEAE	15	*Malus, Prunus, Rubus, Agrimonia, Crataegus, Geum*	25	*Malus, Prunus, Rubus, Agrimonia, Crataegus, Geum, Rosa, Filipendula*		7	W
RUBIACEAE	3	*Galium*	3	*Galium*			
SALICACEAE	4	*Salix*	5	*Salix, Populus*	*Salix*	4	W
SCROPHULARIACEAE	5	*Digitalis, Veronica, Verbascum*	7	*Digitalis, Veronica, Verbascum*		7	W
TILIACEAE	3	*Tilia*	3	*Tilia*	*Tilia*	3	W
APIACEAE	5	*Foeniculum, Eryngium, Daucus, Angelica*	6	*Foeniculum, Eryngium, Daucus, Angelica, Carum*		5	W
VIOLACEAE	4	*Viola*	4	*Viola*		3	W

Species belonging to the *Lamiaceae* (37 medicinal plant species), *Orchidaceae* (30), *Rosaceae* (26), and *Asteraceae* (16) plant families are the main species. *Lamiaceae* plant genera (e.g., *Lavandula, Thymus, Teucrium, Salvia, Stachys*) are visited by bees mainly as nectar sources, while *Rosaceae* (e.g., *Malus, Prunus, Rubus, Crataegus*) and *Asteraceae* (e.g., *Aster, Taraxacum, Tussilago, Tanacetum, Helichrysum*) are for both nectar and pollen foraging, although many plant species offer both nutrition resources to bees. On the opposite, genera of the *Orchidaceae* family (e.g., *Anacamptis, Dactylorhiza, Ophris, Orchis*) have a limited value for foraging, although these species attract wild bees (e.g., bumble bees of the genus *Bombus*) with their flowers [75]. Overall, many plants of these families have great relevance for oligolectic wild bees, which are adapted to collect pollen only from a small number of plant species [76].

By extending the analysis from the European medicinal plant checklist to some mainly cultivated or non-native medicinal plant species (Table 3), several plant taxa are of importance as sugar or pollen resources for bees.

Table 3. Some examples of plant taxon (family, genus) of cultivated/non-native medicinal plants in Europe, in relation to bee foraging sources (nectar, pollen). Plant data from Wichtl [77], bee preferences for pollen and nectar from MLR, [73], MAA, [74].

Plant Family	Nectar and Pollen Sources
AMARYLLIDACEAE	*Allium, Galanthus*
BORAGINACEAE	*Echium, Lithospermum, Anchusa*
ASTERACEAE	*Carlina, Centaurea, Helianthus, Calendula, Eupatorium, Hieracium, Tanacetum, Artemisia, Bellis*
CRUCIFERAE	*Iberis, Isatis, Erysimum*
CUCURBITACEAE	*Cucurbita, Bryonia*
FABACEAE	*Anthyllis, Lotus, Hedysarum, Cytisus*
IRIDACEAE	*Crocus, Iris*
LAMIACEAE	*Lavandula, Teucrium, Mentha, Satureja*
MYRTACEAE	*Eucalyptus*
ONAGRACEAE	*Epilobium*
PLANTAGINACEAE	*Linaria, Plantago*
RANUNCULACEAE	*Helleborus, Nigella, Consolida*
RHAMNACEAE	*Rhamnus*
ROSACEAE	*Prunus, Geum*
RUTACEAE	*Citrus*
APIACEAE	*Coriandrum, Pimpinella, Levisticum*

Some examples include tree species of the families *Rutaceae* (e.g., *Citrus*), Rosaceae (e.g., *Prunus*), *Myrtaceae* (e.g., *Eucalyptus*), shrubs of *Fagaceae* (e.g., *Corylus*), *Eleagnaceae* (e.g., *Hippophae*), and herbaceous species *Asteraceae* (e.g., *Helianthus*), *Apiaceae* (e.g., *Coriandrum, Levisticum, Pimpinella*), which are planted for different purposes (fruit crop, timber, vegetable) or belong to an anthropic habitat. An example of an interesting new medicinal and melliferous plant, the *Perilla frutescens*, an annual herb originating from China, Japan, India, Thailand and Korea, and belonging to the mint family (*Lamiaceae*) also grows in Italy, and is described and discussed by Barbieri and Ferrazzi [78].

In Figure 2, some medicinal plants growing in Italy are shown in relation to their foraging importance for bees (e.g., as sources of nectar, pollen, propolis or honeydew).

Apart from the nectar, bees also collect honeydew as a sugar source, if available in their foraging area. The main sources of honeydew are forests and conifer trees, which originate in the secretions from the living part of the plant (e.g., the leaves) or from sap-sucking insects [71]. In Europe, honeydew honey originates mainly from fir (*Abies alba*), spruce (*Picea abies*), and *Pinus* (e.g., *Pinus halepensis, P. brutia*) trees, but also from *Salicaceae* (*Salix, Populus*), *Fagaceae* (*Castanea, Quercus*), *Oleaceae* (*Fraxinus, Olea*), *Tiliaceae* (*Tilia*), *Betulaceae* (*Betula*) and *Sapindaceae* (*Aesculus*) [71]. Due to the variety of nectar and honeydew sources in natural or artificial habitats, a wide range of different types of honey can originate [79].

Various floral honeys are regarded as medicinal honeys with high polyphenol contents. Manuka dark-colored honey, for example, originates from *Leptospermum scoparium* and *L. polygalifolium*, shrubs native to Australia and New Zealand of the *Myrtaceae* family [80,81]. Another example are the honeys from *Acacia ehrenbergina* (*Fabaceae*) and *Ziziphus* spinachristi (*Rhamnaceae*), trees native to some areas of Africa and Asia, which show high phenolic contents [82]. Propolis originates from collected vegetable material by bees and is mixed with wax. Main sources for propolis production by bees are restricted to a small number of species, which are typical for specific geographic areas [79,83]. In Europe and North America, tree species of the genera *Populus* (e.g., *Populus tremula, P. nigra* in Europe, *P. deltoides* and *P. trichocarpa* in America) and *Betula*, are known as resin resources for bees, while in tropical and subtropical areas, *Dalbergia* and *Acacia* (*Fabaceae*), *Macaranga* (*Euphorbiaceae*), *Mangifera* and *Rhus* (*Anacardiaceae*), and *Baccharis* (*Asteraceae*), are the main sources. In the Mediterranean areas, where *Populus* species might be less frequent, the source of resins for propolis are the *Cupressus* sempervirens and the *Juniperus phoenicea* [84].

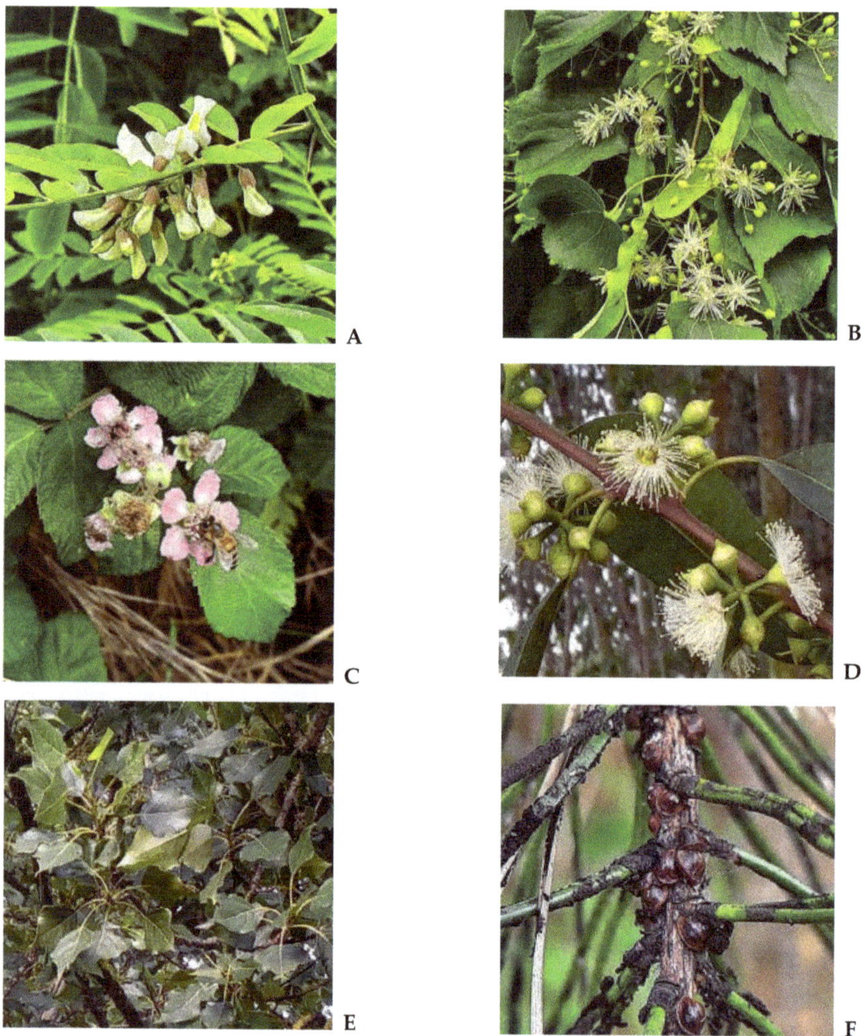

Figure 2. Examples of medicinal plant sources for bees: (**A**) *Robinia pseudoacacia* (nectar), (**B**) *Tilia cordata* (nectar, pollen, honeydew), (**C**) *Rubus ulmifolius* (nectar), (**D**) *Eucalyptus camaldulensis* (nectar, pollen, honeydew), (**E**) *Populus nigra* (propolis), (**F**) *Pinus pinaster* shoot heavily infested by *Toumeyella parvicornis* (pine tortoise scale, honeydew).

3. Bee Products from Medicinal Plants: Antioxidant Properties Measurements

Natural products, including bee products, which often contain medicinal plants containing compounds, such as honeydew secretions of *Abies*, *Betula*, *Castanea*, *Fraxinus*, *Pinus*, *Quercus*, *Rosemary*, *Thymus*, *Tilia*, and other species, are particularly appreciated by consumers for therapeutic uses as an alternative to drugs [85]. Antioxidants are sourced from the plant species. Medicinal plants generally recognized as having potential beneficial value can therefore be utilized to obtain honey with greater bioactivity and bioavailability.

Nowadays, apitherapy has gained much attention from both consumers and researchers. In principle, apitherapy is a theory of alternative medicine that utilizes bee products, such as honey, pollen royal jelly, propolis, and bee venom, for medicinal purposes. However, it remains scarce and not exactly known whether treatments with bee products are safe and how the possible health risks of using such products can be minimized [85]. Proponents of apitherapy make claims for its health benefits, which in contrast, are unsupported by traditional medicine [50,86,87].

In this context, the antioxidant properties of bee products can be considered as an expression of the *melliferous* medicinal plants' therapeutic potential. Updates and considerations on the approach towards antioxidant properties have been mentioned; nonetheless, the conventional and innovative assays for the assessment of antioxidant properties and the remarkable antioxidant properties of bee products from medicinal plant cannot be neglected.

3.1. Study Approach of the Antioxidant Properties: Updates and Considerations

The combined action of bioactive compounds, nutrients, and nutraceuticals represents the first step to study antioxidant properties and can be regarded as an indicator of the "health properties" of the food matrices [88,89].

The diversity of the chemical structure of compounds, their possible interactions, and their different mechanisms of action and biological role, make difficult the assessment of a single, adequate, and reliable procedure for the determination of antioxidant properties. Antioxidant properties are an expression of the interactions between bioactive molecules and other components in terms of both the potential health benefits of food and can be viewed as a screening method for interpretation and supporting further research. Extraction, antioxidant capacity measurements, and expression of the results can be viewed as the three key steps in the evaluation of the antioxidant properties. A study by Durazzo et al. [90] reported as the main workflow in research approach of the antioxidant properties three main steps, namely: (i) the development of a system as model study of the compounds' interactions; (ii) the investigation of extractable and non-extractable compounds; (iii) the behavior study of bioactive compounds-rich extracts.

Nowadays, the distinction between extractable and non-extractable antioxidants has been recognized as a fundamental aspect in the definition of the healthy properties in terms of the prevention of diseases [91,92]. In particular, the distinction between extractable and non-extractable antioxidants has achieved a shared consensus in the scientific community.

Indeed, new research directions point to the exploitation of new and unconventional sources for antioxidants and to the identification of new possible applications.

This research on the antioxidant properties should be based on an integrated and multidisciplinary approach, resulting from a combination of studies in several areas, such as nutrition, food chemistry, phytochemistry, and medicine. Innovative design of study research includes green procedures and sustainable technologies, and the joined up use of statistical methods, such as chemometrics.

An overall challenge is the development of dedicated databases for the antioxidant properties and the inclusion into harmonized databases; these are studies currently being carried out.

The inclusion of extractable and non-extractable compounds in current comprehensive and harmonized databases have been developed in the eBASIS BioActive Substances in Food Information System [93–95]. The development of search protocols and data collection systems have allowed to obtain new quality evaluated data on extractable and non-extractable antioxidants, used for the expansion eBASIS, leading thus, to a valuable unique data resource [96]. A total of 437 datapoints on the composition of extractable and/or non-extractable compounds were added into the database. This update of eBASIS can be viewed as the first examples of building a database dedicated to antioxidant properties. This eBASIS 'expansion provides a new and unique tool for dietitians, nutritionists, and researchers for a great range of uses, e.g., dietary assessment, epidemiological studies, and exposure studies [96].

In this context, the study of Pellegrini et al. [97], by summarizing 25 years of investigations on antioxidants research in foods and biological fluids, remarked how the availability of well-constructed Total Antioxidant Capacity databases deserves attention and must be considered. Moreover, the same authors highlighted how the appropriate use of Total Antioxidant Capacity measurement both in food and in vivo can still support interpretation of complex phenomena and can be viewed as a useful tool, for instance, for the sample screening when making a quick decision toward in-depth research investigations [97].

3.2. Antioxidant Properties Assessment: An Overview of Conventional and Innovative Assays

A variety of assays aimed at evaluating the dietary antioxidant properties have been proposed, although a reliable and commonly accepted assay has not been so far identified [98]. Overall, the methods available can be grouped into three major classes, namely in vitro, cellular, and in vivo assays. In vitro chemical assays are the most frequently used, because these are cheap and have high-throughput, but their prediction ability has been questioned in recent years [99,100]. Cell-based assays are considered a middle ground between the in vitro and in vivo tests (the latter posing ethical issues, high cost, and limited throughput) [101]. However, cellular antioxidant assays still suffer from poor standardization (e.g., differences in cell line, radical generators, fluorescent probes, etc.), making the reported results difficult to be compared across the available studies [102]. Moreover, some authors claimed that cell culturing itself may induce oxidative stress as a consequence of culture conditions, hence inducing cell acclimation and, thus, overestimating the antioxidants efficacy [103].

In this complex framework, in vitro tests still represent the most frequently used antioxidant assay methods in food science. Among others, they include the DPPH (α,α-diphenyl-β-picrylhydrazyl), the ABTS (2,2′-azinobis(3-ethylbenzothiazolin-6-sulphonate), FRAP (ferric reducing antioxidant power), the CUPRAC (cupric-ion reducing antioxidant capacity), and ORAC (oxygen radical absorbance capacity) assays, [98,104–106]. The measurement of the ability of a food or a food component to scavenge specific free radicals or to reduce a target molecule are the base for all the above mentioned methods. Differences in their principles, mechanisms, experimental conditions, and in how their end points are measured occur, and for this reason the use of several methods to estimate and/or determine the antioxidant properties is suggested. Procedures and applications for these assays should be considered for standardization [107]. Moreover, one of the main concerns on the above mentioned assays is that these are not carried out under physiological conditions, and thus, their ability to predict in vivo effects has been questioned. Notwithstanding, most of the scientific literature on bee product antioxidant capacity is referred to in vitro tests, as with other foods, with the exception of a recent paper that investigated the cellular antioxidant capacity of different Moroccan Zantaz honey samples [108].

In general, in vitro assays account for different antioxidant mechanisms that include hydrogen atom transfer (HAT), and single electron transfer (SET) rather than chelation of transition metal ions. ORAC is probably the most representative HAT-based assay, while DPPH, CUPRAC, ABTS and FRAP are SET-based methods. However, the ability to chelate Cu^{2+} and Fe^{2+}, a key initiation step in the oxidation processes, has also been considered in the context of the foods antioxidant properties [109].

Regarding the antioxidant properties of bee products, as with any other food, the limitations in these assays must be considered. Moreover, it must also be considered that in many situations, both HAT and SET occur simultaneously in vivo, and that antioxidant compounds may also act indirectly, via the regulation of antioxidant enzymes.

3.3. Antioxidant Properties of Bee Products Relates to Foraging on Medicinal Plants

The link between bee foraging on medicinal plants and antioxidant properties of bee products has received increasing attention. Nicewicz et al. [110] compared the antioxidant capacity of honey from urban areas vs. rural apiary, reporting that all antioxidant parameters were significantly higher in honey from rural than in urban areas. Such differences were not ascribed to the effect of the floral composition of honey, being rather due to the location of the honeybee colonies. In recent years, while wild pollinators are declining in many landscapes, urban areas provide high plant diversity and foraging sources for bees. Therefore, a growing interest for beekeeping in cities is observed, but the quality and safety of honey produced in urban areas impacts on the consumers' concerns. Therefore, further research in this direction is needed [75].

The following sub-sections discuss the relationship between floral diversity and bioactivity of bee products, with a specific focus on each product.

3.3.1. Honey

Honey (or honeybee honey) is a sugary foodstuff prepared by honeybees. Bees produce honey from the sugary secretions of plants (floral nectar) or from secretions of other insects of the *Aphids* family (honeydew) through their enzymatic activities and other biochemical processes, such as regurgitation and water evaporation. Honeybees store honey in wax structures called honeycombs, whereas stingless bees store honey in pots made of wax and resin [111]. Many centuries ago, in different civilizations, honey was used by ancient Greeks and Egyptians, and in Indian and Chinese traditional medicine both orally and topically to treat various illnesses. Traditional medicine reports uses towards stomach disorders, ulcers, skin wounds, and skin burns [112]. Several honey products have gained medical status and have been approved by the United States Food and Drug Administration (FDA) for their use in the treatment of wounds and burns [113]. It has been reported that honey has antioxidant, antibacterial, and antibiotic properties. The existence of numerous phytochemicals in different types of honey originating from plants known for their medicinal properties, such as *Thymus*, *Abies*, *Pinus*, *Castanea*, and *Rosemary* botanical species supports its antioxidant activity [114–118].

In more detail, Gheldof et al. [118] reported that buckwheat (*Fagopyrum esculentum*) honey increased human's serum antioxidant activity. Anand et al. [119] characterized the physico-chemical and antioxidant properties of *Agastache* honey produced from *Agastache rugosa* in comparison with commercial honeys sold in the Australia market. Their results confirmed that *Agastache* honey had a superior antioxidant capacity [119]. More recently, Adgaba et al. [120] studied, among others, the antioxidant and anti-microbial properties of some Ethiopian monofloral honeys, reporting average total antioxidant values of 320.3 $\pm$ 15.1 μM Fe(II)/100 g with a range of 225.4–465.7 μM Fe(II)/100 g. The same study reported relatively higher values (421.5 $\pm$ 23.4 and 465.7 $\pm$ 21.8 μM Fe(II)/100 g) for *Croton macrostachyus* and *Vernonia amygalina* honeys, respectively. Nonetheless, in a review article focusing, among others, on the antioxidant properties of monofloral honeys, several honey types produced in different countries, such as acacia (*Acacia* sp.), astragalus (*Astragalous microchephalus* Willd), linden (*Tilia* sp.), willow (*Salix* sp.), and others, were reported to provide antioxidant capacity using multiple antioxidant assays, such as ABTS, DPPH, FRAP, ORAC, and TEAC (Trolox equivalent antioxidant capacity) [121].

The antioxidant activity of the studied honeys was affected by the botanical source and the geographical origin [121].

3.3.2. Bee Pollen and Its Derivatives

Bee pollen is among the honeybee products that contain nourishing nutrients, which can provide energy to humans. The health-enhancing value of bee pollen is owed to plant secondary metabolites, such as tocopherols, niacin, thiamine, biotin and folic acid, polyphenols, carotenoids, phytosterols, enzymes and other co-enzymes [85,122]. However, the studies highlighting the antioxidant, anti-inflammatory, anticariogenic antibacterial, antifungicidal, hepatoprotective, anti-atherosclerotic, immune enhancing properties need to be more extensive, concerning mainly the application of cohort clinical trials. The basic hurdle in the use of bee pollen as functional component is probably related to the broad species/specific variation in its composition [122]. Such variations may differently contribute to the properties of bee-pollen and biological activity, and thus, may affect its therapeutic effects (positively or negatively). Notwithstanding, bee pollen has been recommended as a valuable dietary supplement [85]. Pollen antioxidant activity has been related to a wide range of botanical species, such as *Papaver rhoes, Chamomila recutita, Sinapis arvensis, Cistus* sp., *Trifolium* sp., *Dorycnium* sp., *Cichorium* sp., *Convolvulus* sp., *Circium* sp., *Malva sylvestris, Fumana* sp., *Eucalyptus camaldulensis, Anemone* sp., *Ononis* sp., *Asphodelus* sp., and *Quercus ilex* [122].

Moving toward bee pollen-derived products, bee bread has a similar composition to bee pollen, but with marked quantitative differences mainly related to the fermentation process, which it undergoes. For instance, bee bread delivers higher amino acids, sugar, lactic acid, and vitamin content compared to bee pollen [123,124]. In a recent study, bee bread from different regions in Greece, containing *Castanea sativa, Cistus* sp., *Hedera helix, Borago* sp., and other pollen grains belonging to the *Brassicaceae* family, showed both antibacterial and antioxidant activity [125].

3.3.3. Propolis

Propolis, commonly known as the "bee glue" is a resinous mixture that honeybees produce by mixing their saliva, which contains enzymes and beeswax, with exudate gathered from different plant materials such as leaf and flower buds, stems, and bark cracks of numerous tree species. The word propolis originates from the two Greek words "pro" and "polis", which mean "defense" and "city" or "community," respectively [50]. Propolis is typically composed of 50–60% of resins and balms (including phenolic compounds), 30–40% of wax and fatty acids, 5–10% of essential oils, 5% of pollen, and approximately 5% of other components, including amino acids, micronutrients, and vitamins (thiamin, riboflavin, pyridoxine, vitamins C, and E). More than 300 compounds belonging to polyphenols, terpenoids, steroids, sugars, amino acids, and others have been identified in propolis [126].

The antioxidant activity of propolis has been determined by the use of in vitro methods, such as DPPH, ABTS+, FRAP, and ORAC [50]. Interestingly, the antioxidant activity of propolis extracts was comparable to the synthetic antioxidant butylated hydroxytoluene (BHT) or to ascorbic acid. Moreover, studies regarding the antioxidant properties of propolis, have also been carried out on cell cultures and animals. Clinical trials investigating the antioxidant effect of propolis reported a positive modulation of cardiovascular disease markers, a mitigation of chemotherapy side effects, as well as neuroprotective effects [127–129].

Similar to other bee products, plant sources are related to the profile of bioactive compounds and antioxidant properties of propolis [130,131]. More specifically, in regions with a large diversity of trees, bees may also gather resin from flowers in the genera *Clusia* (*Clusia* L.) and *Dalechampia* (*Dalechampia* L.), which are the major plant genera that produce floral resins to attract pollinators [132]. Clusia resin contains polyprenylated benzophenones [133,134]. In some areas of Chile and Brazil, propolis contains viscidone, a terpene from *Baccharis* (*Baccharis* L.) shrubs, and prenylated acids, such as 4-hydroxy-3,5-diprenyl cinnamic acid [135,136].

3.3.4. Royal Jelly

Royal jelly is a mixture of secretions from the mandibular and hypopharyngeal glands of bees of the *Apis mellifera* species, representing the major food source for the queen honeybee [50]. Concerning its composition, royal jelly is an emulsion of proteins, sugars, and lipids in water. Moreover, it contains approximately 1.5% (*w/w*) of minerals (mainly copper, zinc, iron, calcium, manganese, potassium, and sodium) and considerable amounts of flavonoids, polyphenols, and vitamins (biotin, folic acid, inositol, niacin, pantothenic acid, riboflavin, thiamine, and vitamin E). Among the flavonoids, the flavanones (hesperetin, isosakuranetin, and naringenin), flavones (acacetin, apigenin, and its glucoside, chrysin, and luteolin glucoside), flavonols (isorhamnetin and kaempferol glucosides), and isoflavonoids (coumestrol, formononetin, and genistein) are the most abundant [137].

The antioxidant activity of royal jelly, in terms of DPPH, hydroxyl and superoxide radical scavenging, has been reported in the literature [138,139]. In this regard, both in vitro and in vivo tests have been conducted, whereas less information has been reported from clinical trials. In a recent study carried out by Pourmoradian et al. [140], the positive impact of royal jelly consumption on the parameters associated with diabetes and oxidative stress in people affected by Type 2 diabetes mellitus has been postulated. On the other hand, studies on rats and rabbits reported that royal jelly intake can be associated with antioxidant and neuroprotective effects [141,142]. This is consistent with the antioxidant activity of monophosphate nucleotides and peptides isolated from royal jelly [143].

More generally, the antioxidant activity of royal jelly may be differentiated compared to honey, bee pollen and propolis. However, the specific contribution of the botanical species available for foraging on the actual functional properties of royal jelly is still poor.

4. Conclusions and Future Directions

It is clear that medicinal plants can contribute to the antioxidant activity of bee products along with the honeybee contribution as a living organism; the antioxidant properties can be regarded as an indicator of the *melliferous* medicinal plant's potential. In this context, more research should be focused on bee products obtained from the broad range of medicinal plants and on the identification of the possible relationships between the bioactive components, which are present in plant parts and their nectars as well as the bee products.

At the same time, research and clinical trials should be conducted on humans to assess the relationship between the consumption of bee products and the aiding or treatment in health disorders. In this way, the potential use of bee products in phytomedicine (as an alternative to drugs) could be better substantiated by the scientific evidence. The complementary use of the nanotechnologies [144,145] opens new directions and new frontiers. For instance, Neupane et al. [146], by developing Himalayan honey-loaded iron oxide nanoparticles, showed that the biological activity of Himalayan honey was enhanced significantly after loading into iron oxide nanoparticles. Sarhan and Azzazy [147] developed biocompatible, antimicrobial crosslinked honey/polyvinyl alcohol/chitosan nanofibers, which hold potential as an effective wound dressing source. These aspects are relevant and trigger an additional interest for research to obtain a greater bioavailability and efficacy of bee products in the field of health, including anti-COVID-19 possible beneficial effects [148–151], also increasing the interest in studies that are carried out to assess the safety aspects of nanoformulations, which are indeed new frontiers to explore.

Author Contributions: All authors have made a substantial contribution to the writing and revision of work, and approved it for publication. All authors have read and agreed to the published version of the manuscript.

Funding: This research received no external funding.

Institutional Review Board Statement: Not applicable.

Informed Consent Statement: Not applicable.

Conflicts of Interest: The authors declare no conflict of interest.

References

1. Martinello, M.; Mutinelli, F. Antioxidant activity in bee products: A review. *Antioxidants* **2021**, *10*, 71. [CrossRef] [PubMed]
2. Birben, E.; Sahiner, U.M.; Sackesen, C.; Erzurum, S.; Kalayci, O. Oxidative stress and antioxidant defense. *World Allergy Organ. J.* **2012**, *5*, 9–19. [CrossRef]
3. Halliwell, B.; Gutteridge, J.M.C. *Free Radicals in Biology and Medicine*, 3rd ed.; Oxford University Press: New York, NY, USA, 1999.
4. Valko, M.; Leibfritz, D.; Moncol, J.; Cronin, M.T.D.; Mazur, M.; Telser, J. Free radicals and antioxidants in normal physiological functions and human disease. *Int. J. Biochem. Cell Biol.* **2007**, *39*, 44–84. [CrossRef]
5. Halliwell, B. Biochemistry of oxidative stress. *Biochem. Soc. Trans.* **2007**, *35*, 1147–1150. [CrossRef] [PubMed]
6. Genestra, M. Oxyl radicals, redox-sensitive signalling cascades and antioxidants. *Cell. Signal.* **2007**, *19*, 1807–1819. [CrossRef]

7. Pizzino, G.; Irrera, N.; Cucinotta, M.; Pallio, G.; Mannino, F.; Arcoraci, V.; Squadrito, F.; Altavilla, D.; Bitto, A. Oxidative Stress: Harms and Benefits for Human Health. *Oxid. Med. Cell. Longev.* **2017**, *2017*, 8416763. [CrossRef]

8. Uttara, B.; Singh, A.V.; Zamboni, P.; Mahajan, R.T. Oxidative Stress and Neurodegenerative Diseases: A Review of Upstream and Downstream Antioxidant Therapeutic Options. *Curr. Neuropharmacol.* **2009**, *7*, 65–74. [CrossRef]

9. Hayes, J.D.; Dinkova-Kostova, A.T.; Tew, K.D. Oxidative stress in cancer. *Cancer Cell* **2020**, *38*, 167–197. [CrossRef]

10. Huang, D.; Ou, B.; Prior, R.L. The Chemistry behind Antioxidant Capacity Assays. *J. Agric. Food Chem.* **2005**, *53*, 1841–1856. [CrossRef] [PubMed]

11. Alvarez-Suarez, J.; Giampieri, F.; Battino, M. Honey as a Source of Dietary Antioxidants: Structures, Bioavailability and Evidence of Protective Effects Against Human Chronic Diseases. *Curr. Med. Chem.* **2013**, *20*, 621–638. [CrossRef]

12. Durazzo, A.; Lucarini, M.; Cicero, N.; Gabrielli, P.; Souto, E.B.; Dugo, G.; Santini, A. The Antioxidant Properties of Honey. In *Search of Honey Authentication*; Karabagias, I., Ed.; Cambridge Scholars Publishing: Newcastle upon Tyne, UK, 2021; ISBN (10): 1-5275-6712-5, ISBN (13): 978-1-5275-6712-2.

13. Waltman, L.; van Eck, N.J.; Noyons, E.C.M. A unified approach to mapping and clustering of bibliometric networks. *J. Inf.* **2010**, *4*, 629–635. [CrossRef]

14. Van Eck, N.J.; Waltman, L. Software survey: VOSviewer, a computer program for bibliometric mapping. *Scientometrics* **2009**, *84*, 523–538. [CrossRef] [PubMed]

15. Van Eck, N.J.; Waltman, L. Text mining and visualization using VOSviewer. *ISSI Newslett.* **2011**, *7*, 50–54.

16. Khalifa, S.A.M.; Elshafiey, E.H.; Shetaia, A.A.; El-Wahed, A.A.A.; Algethami, A.F.; Musharraf, S.G.; AlAjmi, M.F.; Zhao, C.; Masry, S.H.D.; Abdel-Daim, M.M.; et al. Overview of Bee Pollination and Its Economic Value for Crop Production. *Insects* **2021**, *12*, 688. [CrossRef]

17. Robinson ACPeeler, J.L.; Prestby, T.; Goslee, S.C.; Anton, K.; Grozinger, C. Beescape: Characterizing user needs for environmental decision support in beekeeping. *Ecol. Inform.* **2021**, *64*, 101366. [CrossRef]

18. Delaplane, K.S.; Mayer, D.F. *Crop Pollination by Bees*; CABI: Wallingford, UK, 2000.

19. Dogantzis, K.A.; Zayed, A. Recent advances in population and quantitative genomics of honey bees. *Curr. Opin. Insect Sci.* **2019**, *31*, 93–98. [CrossRef]

20. Klein, A.-M.; Vaissière, B.E.; Cane, J.H.; Steffan-Dewenter, I.; Cunningham, S.A.; Kremen, C.; Tscharntke, T. Importance of pollinators in changing landscapes for world crops. *Proc. R. Soc. B Biol. Sci.* **2007**, *274*, 303–313. [CrossRef]

21. Bloch, G.; Francoy, T.M.; Wachtel, I.; Panitz-Cohen, N.; Fuchs, S.; Mazar, A. Industrial apiculture in the Jordan valley during Biblical times with Anatolian honeybees. *Proc. Natl. Acad. Sci. USA* **2010**, *107*, 11240–11244. [CrossRef]

22. Crane, E. Honeybees. In *Evolution of Domesticated Animals*; Mason, I.L., Ed.; Longman Group: London, UK, 1984; pp. 403–415.

23. Phillips, E.F. *Beekeeping*; Applewood Books: Carlisle, MA, USA, 1918.

24. Bruckner, S.; Steinhauer, N.; Engelsma, J.; Fauvel, A.M.; Kulhanek, K.; Malcom, E.; Meredith, A.; Milbrath, M.; Nino, E.; Rangel, J.; et al. *2019–2020 Honey Bee Colony Losses in the United States: Preliminary Results*. 2020, p. 5. Available online: https://beeinformed.org/wp-content/uploads/2020/06/BIP_2019_2020_Losses_Abstract.pdf (accessed on 28 October 2021).

25. Potts, S.G.; Biesmeijer, J.C.; Kremen, C.; Neumann, P.; Schweiger, O.; Kunin, W.E. Global pollinator declines: Trends, impacts and drivers. *Trends Ecol. Evol.* **2010**, *25*, 345–353. [CrossRef]

26. Tautz, J. *The Buzz about Bees: Biology of a Superorganism*; Springer: Berlin/Heidelberg, Germany, 2008.

27. Ascher, J.S.; Pickering, J. Discover Life Bee Species Guide and World Checklist (Hymenoptera: Apoidea: Anthophila). 2014. Available online: http://www.discoverlife.org/mp/20q?guide=Apoidea_species (accessed on 28 October 2021).

28. Themudo, G.E.; Rey-Iglesia, A.; Tascón, L.R.; Jensen, A.B.; Da Fonseca, R.R.; Campos, P.F. Declining genetic diversity of European honeybees along the twentieth century. *Sci. Rep.* **2020**, *10*, 1–12. [CrossRef]

29. Potts, S.G.; Roberts, S.P.; Dean, R.; Marris, G.; Brown, M.A.; Jones, R.; Neumann, P.; Settele, J. Declines of managed honey bees and beekeepers in Europe. *J. Apic. Res.* **2010**, *49*, 15–22. [CrossRef]

30. Berkes, F.; Colding, J.; Folke, C. *Navigating Social–Ecological Systems: Building Resilience for Complexity*; Cambridge University Press: Cambridge, UK, 2003; pp. 1–30.

31. Keune, H.; Dendoncker, N.; Jacobs, S. Ecosystem Service Practices. In *Ecosystem Services*; Elsevier: Amsterdam, The Netherlands, 2013; pp. 307–315.

32. Martinello, M.; Manzinello, C.; Dainese, N.; Giuliato, I.; Gallina, A.; Mutinelli, F. The Honey Bee: An Active Biosampler of Environmental Pollution and a Possible Warning Biomarker for Human Health. *Appl. Sci.* **2021**, *11*, 6481. [CrossRef]

33. Adeoye, O.T.; Pitan, O.R.; Olasupo, O.O.; Ayandokun, A.E.; Abudul-Azeez, F.I. Assessment of honeybees and bee honey as bioindicators of environmental pollution. *Aust. J. Sci. Technol.* **2021**, *5*, 460–465.

34. Horcea-Milcu, A.-I.; Hanspach, J.; Abson, D.; Fischer, J. Cultural Ecosystem Services: A Literature Review and Prospects for Future Research. *Ecol. Soc.* **2013**, *18*, 44. [CrossRef]

35. Bennett, E.M.; Peterson, G.; Gordon, L. Understanding relationships among multiple ecosystem services. *Ecol. Lett.* **2009**, *12*, 1394–1404. [CrossRef]

36. FAO. Bees and their role in forest livelihoods. In *A Guide to the Services Provided by Bees and the Sustainable Harvesting Processing and Marketing of Their Products*; FAO: Rome, Italy, 2009.

37. Michener, C.D. *The Bees of the World*; The John Hopkins University Press: Baltimore, MD, USA, 2000; p. 913.

38. Michener, C.D. *The Bees of the World*, 2nd ed.; John Hopkins University Press: Baltimore, MD, USA, 2007.

39. Chinwuba Okoye, T.; Uzor, P.F.; Onyeto, C.A.; Okereke, E.K. Chapter 18-Safe African medicinal plants for clinical studies. In *Toxicological Survey of African Medicinal Plants*; Elsevier: Amsterdam, The Netherlands, 2014; pp. 535–555.

40. Asadi, N.; Bahmani, M.; Shahsavari, S.; Asadi-Samani, M. Identification and introduction of the medicinal plants used by Honeybees in Markazi Province. *IJPPR* **2017**, *7*, 15–18.

41. Bahmani, M.; Asadi-Samani, M. Native medicinal plants of Iran effective on peptic ulcer. *J. Injury Inflam.* **2016**, *1*, e05.

42. Moradi, M.T.; Asadi-Samani, M.; Bahmani, M.; Shahrani, M. Medicinal plants used for liver disorders based on the ethnobotanical documents of Iran: A review. *Int. J. PharmTech Res.* **2016**, *9*, 407–415.

43. Molan, P. Why honey is effective as a medicine: 2. The scientific explanation of its effects. *Bee World* **2001**, *82*, 22–40. [CrossRef]

44. Verma, L.R. Beekeeping. In *Integrated Mountain Development: Economic and Scientific Perspectives*; International Centre for Integrated Mountain Development (ICIMOD): Patan, Nepal, 1990.

45. Kaur, G.; Sihag, R.C. Bee flora of India: A review. *Bee J.* **1994**, *56*, 105–126.

46. Partap, U. *Bee Flora of the Hindu Kush-Himalayas: Inventory and Management*; ICMOD: Kathmandu, Nepal, 1997.

47. Harugade, S.; Gawate, A.; Shinde, B. Bee Floral Diversity of Medicinal Plants in Vidya Pratishthan Campus, Baramati, Pune, District (M.S.) India. *Int. J. Curr. Microbiol. Appl. Sci.* **2016**, *5*, 425–431. [CrossRef]

48. Mačukanović-Jocić, M.; Jarić, S.V. The melliferous potential of apiflora of southwestern Vojvodina (Serbia). *Arch. Biol. Sci.* **2016**, *68*, 81–91. [CrossRef]

49. Bakour, M.; Laaroussi, H.; El Menyiy, N.; Elaraj, T.; El Ghouizi, A.; Lyoussi, B. The Beekeeping State and Inventory of Mellifero-Medicinal Plants in the North-Central of Morocco. *Sci. World J.* **2021**, *2021*, 1–12. [CrossRef]

50. Kocot, J.; Kiełczykowska, M.; Luchowska-Kocot, D.; Kurzepa, J.; Musik, I. Antioxidant Potential of Propolis, Bee Pollen, and Royal Jelly: Possible Medical Application. *Oxid. Med. Cell. Longev.* **2018**, *2018*, 1–29. [CrossRef]

51. Feás, X.; Vázquez-Tato, M.P.; Estevinho, L.; Seijas, J.A.; Iglesias, A. Organic bee pollen: Botanical origin, nutritional value, bioactive compounds, antioxidant activity and microbiological quality. *Molecules* **2012**, *17*, 8359–8377. [CrossRef]

52. Paulino, N.; Coutinho, L.A.; Coutinho, J.R.; Vilela, G.C.; da Silva Leandro, V.P.; Paulino, A.S. Antiulcerogenic effect of Brazilian propolis formulation in mice. *Pharmacol. Pharm.* **2015**, *6*, 580. [CrossRef]

53. Kaplan, M.; Karaoglu, O.; Eroglu, N.; Silici, S. Fatty Acid and proximate composition of bee bread. *Food Technol. Biotechnol.* **2016**, *54*, 497–504. [CrossRef]

54. Shaddel-Telli, A.; Maheri, N.; Aghajanzade, G. Using medicinal plants for controlling Varrova Mite honeybee colonies. *J. Anim. Vet. Adv.* **2008**, *7*, 328–330.

55. Bendifallah, L.; Belguendouz, R.; Hamoudi, L.; Arab, K. Biological Activity of the *Salvia officinalis* L. (Lamiaceae) Essential Oil on Varroa destructor Infested Honeybees. *Plants* **2018**, *7*, 44. [CrossRef]

56. Topal, E.; Cıpcıgan, M.; İvgin Tunca, R.; Kösoğlu, M.; Mărgăoan, R. The Use of medicinal aromatic plants against bee diseases and pests. *Bee Stud.* **2020**, *12*, 5–11. [CrossRef]

57. Khan, S.U.; Anjum, S.I.; Ansari, M.J.; Khan, M.H.U.; Kamal, S.; Rahman, K.; Shoaib, M.; Man, S.; Khan, A.J.; Khan, S.U.; et al. Antimicrobial potentials of medicinal plant's extract and their derived silver nanoparticles: A focus on honey bee pathogen. *Saudi J. Biol. Sci.* **2019**, *26*, 1815–1834. [CrossRef] [PubMed]

58. Nguyen, H.T.L.; Panyoyai, N.; Paramita, V.D.; Mantri, N.; Kasapis, S. Physicochemical and viscoelastic properties of honey from medicinal plants. *Food Chem.* **2018**, *241*, 143–149. [CrossRef] [PubMed]

59. Durazzo, A.; D'Addezio, L.; Camilli, E.; Piccinelli, R.; Turrini, A.; Marletta, L.; Marconi, S.; Lucarini, M.; Lisciani, S.; Gabrielli, P.; et al. From Plant Compounds to Botanicals and Back: A Current Snapshot. *Molecules* **2018**, *23*, 1844. [CrossRef]

60. Durazzo, A.; Lucarini, M.; Souto, E.B.; Cicala, C.; Caiazzo, E.; Izzo, A.A.; Novellino, E.; Santini, A. Polyphenols: A concise overview on the chemistry, occurrence, and human health. *Phytother. Res.* **2019**, *33*, 2221–2243. [CrossRef]

61. Durazzo, A.; Lucarini, M. Editorial: The State of Science and Innovation of Bioactive Research and Applications, Health, and Diseases. *Front. Nutr.* **2019**, *6*, 178. [CrossRef] [PubMed]

62. Durazzo, A.; Camilli, E.; D'Addezio, L.; Piccinelli, R.; Mantur-Vierendeel, A.; Marletta, L.; Finglas, P.; Turrini, A.; Sette, S. Development of Dietary Supplement Label Database in Italy: Focus of FoodEx2 Coding. *Nutrients* **2019**, *12*, 89. [CrossRef]

63. Daliu, P.; Santini, A.; Novellino, E. A decade of nutraceutical patents: Where are we now in 2018? *Expert Opin. Ther. Pat.* **2018**, *28*, 875–882. [CrossRef]

64. Santini, A.; Cammarata, S.M.; Capone, G.; Ianaro, A.; Tenore, G.C.; Pani, L.; Novellino, E. Nutraceuticals: Opening the debate for a regulatory framework. *Br. J. Clin. Pharmacol.* **2018**, *84*, 659–672. [CrossRef] [PubMed]

65. Lange, D. *Europe's Medicinal and Aromatic Plants: Their Use, Trade and Conservation*; Traffic International: Cambridge, UK, 1998.

66. Diazgranados, M.; Allkin, B.; Black, N.; Cámara-Leret, R.; Canteiro, C.; Carretero, J.; Ulian, T. World Checklist of Useful Plant Species. Produced by the Royal Botanic Gardens, Kew. Knowledge Network for Biocomplexity. 2020. Available online: https://kew.iro.bl.uk/concern/datasets/7243d727-e28d-419d-a8f7-9ebef5b9e03e?locale=en (accessed on 28 October 2021).

67. Dürbeck, K.; Hüttenhofer, T. International Trade of Medicinal and Aromatic Plants. *Med. Aromat. Plants World* **2015**, *1*, 375–382. [CrossRef]

68. Allen, D.; Bilz, M.; Leaman, D.J.; Miller, R.M.; Timoshyna, A.; Window, J. *European Red List of medicinal Plants*; Publications Office of the European Union: Luxembourg, 2014; p. 63.

69. Bankova, V.; Popova, M.; Trusheva, B. The phytochemistry of the honeybee. *Phytochemistry* **2018**, *155*, 1–11. [CrossRef]

70. Bogdanov, S.; Ruoff, K.; Oddo, L.P. Physico-chemical methods for the characterisation of unifloral honeys: A review. *Apidologie* **2004**, *35* (Suppl. 1), S4–S17. [CrossRef]

71. Pita-Calvo, C.; Vázquez, M. Honeydew Honeys: A Review on the Characterization and Authentication of Botanical and Geographical Origins. *J. Agric. Food Chem.* **2018**, *66*, 2523–2537. [CrossRef] [PubMed]

72. Thakur, M.; Nanda, V. Composition and functionality of bee pollen: A review. *Trends Food Sci. Technol.* **2020**, *98*, 82–106. [CrossRef]

73. MLR-Ministerium für ländlichen Raum und Verbraucherschutz Baden-Württemberg. Bienenweidekatalog: Verbesserung der Bienenweide und des Artenreichtums.–Stuttgart. 129 S. 2018. Available online: https://mlr.baden-wuerttemberg.de/de/unsere-themen/biodiversitaet-und-landnutzung/bienenweidekatalog/ (accessed on 28 October 2021).

74. MAA-Ministère de l'Agriculture et de l'Alimentation. Liste de plantes attractives pour les abeilles. 2017. Available online: https://www.franceagrimer.fr/content/download/51417/494444/file/290517-Plantes%20attractives-abeilles.pdf (accessed on 28 October 2021).

75. Vöth, W. *Lebensgeschichte und Bestäuber der Orchideen am Beispiel von Niederösterreich. Biologiezentrum des OÖ*; Landesmuseums: Zürich, Switzerland, 1999; Volume 65, Biologiezentrum Linz/Austria.

76. Kratochwil, A. Bees (Hymenoptera: Apoidea) as key-stone species: Specifics of resource and requisite utilisation in different habitat types. *Ber. Der Reinhold-Tüxen-Ges.* **2003**, *15*, 59–77.

77. Wichtl, M. *Teedrogen und Phytopharmaka*; Wissenschaftiche Verlag GmbH: Stuttgart, Germany, 2002; Volume 4, p. 84.

78. Barbieri, C.; Ferrazzi, P. Perilla frutescens: Interesting New Medicinal and Melliferous Plant in Italy. *Nat. Prod. Commun.* **2011**, *6*, 1461–1463. [CrossRef]

79. Oddo, L.P.; Piro, R.; Bruneau, É.; Guyot-Declerck, C.; Ivanov, T.; Piskulová, J.; Flamini, C.; Lheritier, J.; Morlot, M.; Russmann, H.; et al. Main European unifloral honeys: Descriptive sheets. *Apidologie* **2004**, *35* (Suppl. 1), S38–S81. [CrossRef]

80. Stephens, J.M.; Schlothauer, R.C.; Morris, B.D.; Yang, D.; Fearnley, L.; Greenwood, D.R.; Loomes, K.M. Phenolic compounds and methylglyoxal in some New Zealand manuka and kanuka honeys. *Food Chem.* **2010**, *120*, 78–86. [CrossRef]

81. Alzahrani, H.A.; Boukraa, L.; Bellik, Y.; Abdellah, F.; Bakhotmah, B.A.; Kolayli, S.; Sahin, H. Evaluation of the Antioxidant Activity of Three Varieties of Honey from Different Botanical and Geographical Origins. *Glob. J. Health Sci.* **2012**, *4*, 191–196. [CrossRef]

82. Al-Mamary, M.; Al-Meeri, A.; Al-Habori, M. Antioxidant activities and total phenolics of different types of honey. *Nutr. Res.* **2002**, *22*, 1041–1047. [CrossRef]

83. Bankova, V.; Popova, M.; Trusheva, B. Plant sources of propolis: An update from a chemist's point of view. *Nat. Prod. Commun.* **2006**, *1*. [CrossRef]

84. Velikova, M.; Bankova, V.; Sorkun, K.; Houcine, S.; Tsvetkova, I.; Kujumgiev, A. Propolis from the Mediterranean Region: Chemical Composition and Antimicrobial Activity. *Z. Naturforschung C* **2000**, *55*, 790–793. [CrossRef]

85. Denisow, B.; Denisow-Pietrzyk, M. Biological and therapeutic properties of bee pollen: A review. *J. Sci. Food Agric.* **2016**, *96*, 4303–4309. [CrossRef] [PubMed]

86. Russell, J.; Rovere, A. Apitherapy. In *American Cancer Society Complete Guide to Complementary and Alternative Cancer Therapies*, 2nd ed.; American Cancer Society: Atlanta, GA, USA, 2009; pp. 704–708.

87. Cassileth, B.R. Apitherapy. In *The Complete Guide to Complementary Therapies in Cancer Care: Essential Information for Patients, Survivors and Health Professionals*; World Scientific: Singapore, 2011; pp. 221–224.

88. Durazzo, A. Study Approach of Antioxidant Properties in Foods: Update and Considerations. *Foods* **2017**, *6*, 17. [CrossRef]

89. Durazzo, A.; Lucarini, M.; Santini, A. Nutraceuticals in Human Health. *Foods* **2020**, *9*, 370. [CrossRef] [PubMed]

90. Durazzo, A.; Lucarini, M. A Current Shot and Re-thinking of Antioxidant Research Strategy. *Braz. J. Anal. Chem.* **2019**, *5*, 9–11. [CrossRef]

91. Durazzo, A.; Lucarini, M. Extractable and non-extractable antioxidants. *Molecules* **2019**, *24*, 1933. [CrossRef]

92. Durazzo, A. Extractable and non-extractable polyphenols: An overview. In *Non-Extractable Polyphenols and Carotenoids: Importance in Human Nutrition and Health*; Saura-Calixto, F., Pérez-Jiménez, J., Eds.; RSC Publishing: Cambridge, UK, 2018; pp. 37–45. ISBN 978-1-78801-447-2. [CrossRef]

93. eBASIS—Bioactive Substances in Food Information System. Available online: http://ebasis.eurofir.org/Default.asp (accessed on 4 June 2021).

94. Kiely, M.; on behalf of the EuroFIR consortium; Black, L.J.; Plumb, J.; Kroon, P.A.; Hollman, P.C.; Larsen, J.C.; Speijers, G.J.; Kapsokefalou, M.; Sheehan, D.; et al. EuroFIR eBASIS: Application for health claims submissions and evaluations. *Eur. J. Clin. Nutr.* **2010**, *64*, S101–S107. [CrossRef]

95. Plumb, J.; Pigat, S.; Bompola, F.; Cushen, M.; Pinchen, H.; Nørby, E.; Astley, S.; Lyons, J.; Kiely, M.; Finglas, P. eBASIS (Bioactive Substances in Food Information Systems) and bioactive intakes: Major updates of the bioactive compound composition and beneficial bio effects database and the development of a probabilistic model to assess intakes in Europe. *Nutrients* **2017**, *9*, 320. [CrossRef]

96. Plumb, J.; Durazzo, A.; Lucarini, M.; Camilli, E.; Turrini, A.; Marletta, L.; Finglas, P. Extractable and Non-Extractable Antioxidants Composition in the eBASIS Database: A Key Tool for Dietary Assessment in Human Health and Disease Research. *Nutrients* **2020**, *12*, 3405. [CrossRef] [PubMed]

97. Pellegrini, N.; Vitaglione, P.; Granato, D.; Fogliano, V. Twenty-five years of total antioxidant capacity measurement of foods and biological fluids: Merits and limitations. *J. Sci. Food Agric.* **2020**, *100*, 5064–5078. [CrossRef]

98. Apak, R. Current Issues in Antioxidant Measurement. *J. Agric. Food Chem.* **2019**, *67*, 9187–9202. [CrossRef] [PubMed]
99. Cömert, E.D.; Gökmen, V. Evolution of food antioxidants as a core topic of food science for a century. *Food Res. Int.* **2018**, *105*, 76–93. [CrossRef]
100. Granato, D.; Shahidi, F.; Wrolstad, R.; Kilmartin, P.; Melton, L.D.; Hidalgo, F.J.; Miyashita, K.; van Camp, J.; Alasalvar, C.; Ismail, A.; et al. Antioxidant activity, total phenolics and flavonoids contents: Should we ban in vitro screening methods? *Food Chem.* **2018**, *264*, 471–475. [CrossRef]
101. Martinelli, E.; Granato, D.; Azevedo, L.; Gonçalves, J.E.; Lorenzo, J.M.; Munekata, P.E.; Simal-Gandara, J.; Barba, F.J.; Carrillo, C.; Rajoka, M.S.R.; et al. Current perspectives in cell-based approaches towards the definition of the antioxidant activity in food. *Trends Food Sci. Technol.* **2021**, *116*, 232–243. [CrossRef]
102. Kellett, M.E.; Greenspan, P.; Pegg, R.B. Modification of the cellular antioxidant activity (CAA) assay to study phenolic antioxidants in a Caco-2 cell line. *Food Chem.* **2018**, *244*, 359–363. [CrossRef] [PubMed]
103. Amorati, R.; Valgimigli, L. Advantages and limitations of common testing methods for antioxidants. *Free Radic. Res.* **2015**, *49*, 633–649. [CrossRef] [PubMed]
104. Apak, R.; Özyürek, M.; Güçlü, K.; Çapanoğlu, E. Antioxidant Activity/Capacity Measurement. 1. Classification, Physicochemical Principles, Mechanisms, and Electron Transfer (ET)-Based Assays. *J. Agric. Food Chem.* **2016**, *64*, 997–1027. [CrossRef]
105. Apak, R.; Özyürek, M.; Güçlü, K.; Çapanoğlu, E. Antioxidant Activity/Capacity Measurement. 2. Hydrogen Atom Transfer (HAT)-Based, Mixed-Mode (Electron Transfer (ET)/HAT), and Lipid Peroxidation Assays. *J. Agric. Food Chem.* **2016**, *64*, 1028–1045. [CrossRef]
106. Apak, A.; Capanoglu, E.; Shahidi, F. *Measurement of Antioxidant Activity and Capacity: Recent Trends and Applications*; Wiley: New York, NY, USA, 2018; ISBN 978-1-119-13535-7.
107. Prior, R.L.; Wu, X.; Schaich, K. Standardized Methods for the Determination of Antioxidant Capacity and Phenolics in Foods and Dietary Supplements. *J. Agric. Food Chem.* **2005**, *53*, 4290–4302. [CrossRef]
108. Elamine, Y.; Lyoussi, B.; Miguel, M.G.; Anjos, O.; Estevinho, L.; Alaiz, M.; Girón-Calle, J.; Martín, J.; Vioque, J. Physicochemical characteristics and antiproliferative and antioxidant activities of Moroccan Zantaz honey rich in methyl syringate. *Food Chem.* **2021**, *339*, 128098. [CrossRef]
109. Santos, J.S.; Deolindo, C.T.P.; Hoffmann, J.F.; Chaves, F.C.; Prado-Silva, L.D.; Sant'Ana, A.S.; Azevedo, L.; Carmo, M.A.V.D.; Granato, D. Optimized Camellia sinensis var. sinensis, Ilex paraguariensis, and Aspalathus linearis blend presents high antioxidant and antiproliferative activities in a beverage model. *Food Chem.* **2018**, *254*, 348–358. [CrossRef] [PubMed]
110. Nicewicz, A.W.; Nicewicz, Ł.; Pawłowska, P. Antioxidant capacity of honey from the urban apiary: A comparison with honey from the rural apiary. *Sci. Rep.* **2021**, *11*, 1–8. [CrossRef]
111. Crane, E. Honey from honeybees and other insects. *Ethol. Ecol. Evol.* **1991**, *3*, 100–105. [CrossRef]
112. Pećanac, M.; Janjić, Z.; Komarcević, A.; Pajić, M.; Dobanovacki, D.; Misković, S.S. Burns treatment in ancient times. *Med. Pregl.* **2013**, *66*, 263–267. [PubMed]
113. Saikaly, S.K.; Khachemoune, A. Honey and Wound Healing: An Update. *Am. J. Clin. Dermatol.* **2017**, *18*, 237–251. [CrossRef] [PubMed]
114. Majtan, J.; Klaudiny, J.; Bohova, J.; Kohutova, L.; Dzurova, M.; Sediva, M.; Bartosova, M.; Majtan, V. Methylglyoxal-induced modifications of significant honeybee proteinous components in manuka honey: Possible therapeutic implications. *Fitoterapia* **2012**, *83*, 671–677. [CrossRef]
115. Kwakman, P.H.S.; Zaat, S.A.J. Antibacterial components of honey. *IUBMB Life* **2012**, *64*, 48–55. [CrossRef]
116. Karabagias, I.K.; Dimitriou, E.; Kontakos, S.; Kontominas, M.G. Phenolic profile, colour intensity, and radical scavenging activity of Greek unifloral honeys. *Eur. Food Res. Technol.* **2016**, *242*, 1201–1210. [CrossRef]
117. Koca, I.; Tekguler, B.; Turkyilmaz, B.; Tasci, B. Some physical, chemical and antioxidant properties of chestnut (Castanea sativa Mill.) honey produced in Turkey. *Acta Hortic.* **2018**, *1220*, 227–234. [CrossRef]
118. Gheldof, N.; Wang, A.X.-H.; Engeseth, N.J. Buckwheat Honey Increases Serum Antioxidant Capacity in Humans. *J. Agric. Food Chem.* **2003**, *51*, 1500–1505. [CrossRef]
119. Anand, S.; Pang, E.; Livanos, G.; Mantri, N. Characterization of Physico-Chemical Properties and Antioxidant Capacities of Bioactive Honey Produced from Australian Grown Agastache rugosa and its Correlation with Colour and Poly-Phenol Content. *Molecules* **2018**, *23*, 108. [CrossRef] [PubMed]
120. Adgaba, N.; Al-Ghamdi, A.; Sharma, D.; Tadess, Y.; Alghanem, S.M.; Khan, K.A.; Ansari, M.J.; Mohamed, G.K.A. Physico-chemical, antioxidant and anti-microbial properties of some Ethiopian mono-floral honeys. *Saudi J. Biol. Sci.* **2020**, *27*, 2366–2372. [CrossRef] [PubMed]
121. Mărgăoan, R.; Topal, E.; Balkanska, R.; Yücel, B.; Oravecz, T.; Cornea-Cipcigan, M.; Vodnar, D. Monofloral Honeys as a Potential Source of Natural Antioxidants, Minerals and Medicine. *Antioxidants* **2021**, *10*, 1023. [CrossRef]
122. Karabagias, I.K.; Karabagias, V.K.; Gatzias, I.; Riganakos, K.A. Bio-Functional Properties of Bee Pollen: The Case of "Bee Pollen Yoghurt". *Coatings* **2018**, *8*, 423. [CrossRef]
123. Kieliszek, M.; Piwowarek, K.; Kot, A.; Błażejak, S.; Chlebowska-Śmigiel, A.; Wolska, I. Pollen and bee bread as new health-oriented products: A review. *Trends Food Sci. Technol.* **2018**, *71*, 170–180. [CrossRef]

124. Aylanc, V.; Tomás, A.; Russo-Almeida, P.; Falcão, S.; Vilas-Boas, M. Assessment of Bioactive Compounds under Simulated Gastrointestinal Digestion of Bee Pollen and Bee Bread: Bioaccessibility and Antioxidant Activity. *Antioxidants* **2021**, *10*, 651. [CrossRef] [PubMed]

125. Didaras, N.; Kafantaris, I.; Dimitriou, T.; Mitsagga, C.; Karatasou, K.; Giavasis, I.; Stagos, D.; Amoutzias, G.; Hatjina, F.; Mossialos, D. Biological Properties of Bee Bread Collected from Apiaries Located across Greece. *Antibiotics* **2021**, *10*, 555. [CrossRef]

126. Boisard, S.; LE Ray, A.-M.; Gatto, J.; Aumond, M.-C.; Blanchard, P.; Derbré, S.; Flurin, C.; Richomme, P. Chemical Composition, Antioxidant and Anti-AGEs Activities of a French Poplar Type Propolis. *J. Agric. Food Chem.* **2014**, *62*, 1344–1351. [CrossRef]

127. Mujica, V.; Orrego, R.; Pérez, J.; Romero, P.; Ovalle, P.; Zuñiga, J.; Arredondo, M.; Leiva, E. The Role of Propolis in Oxidative Stress and Lipid Metabolism: A Randomized Controlled Trial. *Evidence-Based Complement. Altern. Med.* **2017**, *2017*, 4272940. [CrossRef]

128. Salmas, R.E.; Gulhan, M.F.; Durdagi, S.; Sahna, E.; Abdullah, H.I.; Selamoglu, Z. Effects of propolis, caffeic acid phenethyl ester, and pollen on renal injury in hypertensive rat: An experimental and theoretical approach. *Cell Biochem. Funct.* **2017**, *35*, 304–314. [CrossRef]

129. Jasprica, I.; Mornar, A.; Debeljak, Ž.; Smolčić-Bubalo, A.; Medić-Šarić, M.; Mayer, L.; Romić, Z.; Bućan, K.; Balog, T.; Sobočanec, S.; et al. In vivo study of propolis supplementation effects on antioxidative status and red blood cells. *J. Ethnopharmacol.* **2007**, *110*, 548–554. [CrossRef]

130. Dezmirean, D.S.; Paşca, C.; Moise, A.R.; Bobiş, O. Plant Sources Responsible for the Chemical Composition and Main Bioactive Properties of Poplar-Type Propolis. *Plants* **2020**, *10*, 22. [CrossRef]

131. Yosri, N.; El-Wahed, A.A.A.; Ghonaim, R.; Khattab, O.M.; Sabry, A.; Ibrahim, M.A.A.; Moustafa, M.F.; Guo, Z.; Zou, X.; Algethami, A.F.M.; et al. Anti-Viral and Immunomodulatory Properties of Propolis: Chemical Diversity, Pharmacological Properties, Preclinical and Clinical Applications, and In Silico Potential against SARS-CoV-2. *Foods* **2021**, *10*, 1776. [CrossRef]

132. Mesquita, R.D.C.G.; Franciscon, C.H. Flower Visitors of Clusia nemorosa G. F. W. Meyer (Clusiaceae) in an Amazonian White-Sand Campina. *Biotropica* **1995**, *27*, 254. [CrossRef]

133. Tomás-Barberán, F.A.; García-Viguera, C.; Vit-Oliviera, P.; Ferreres, F.; Tomás-Lorente, F. Phytochemical evidence for the botanical origin of tropical propolis from Venezuela. *Phytochemistry* **1993**, *34*, 191–196. [CrossRef]

134. Bankova, V. Recent trends and important developments in propolis research. *Evid. Based Complement. Altern. Med.* **2005**, *2*, 29–32. [CrossRef] [PubMed]

135. Montenegro, G.; Mujica, A.M.; Peña, R.C.; Gómez, M.; Serey, I.; Timmermann, B.N. Similitude pattern and botanical origin of the Chilean propolis. *Phyton* **2004**, *73*, 145–154.

136. Park, Y.K.; de Alencar, S.M.; Aguiar, C.L. Botanical Origin and Chemical Composition of Brazilian Propolis. *J. Agric. Food Chem.* **2002**, *50*, 2502–2506. [CrossRef] [PubMed]

137. López-Gutiérrez, N.; del Mar Aguilera-Luiz, M.; Romero-González, R.; Vidal, J.L.M.; Frenich, A.G. Fast analysis of polyphenols in royal jelly products using automated TurboFlow™-liquid chromatography-Orbitrap high resolution mass spectrometry. *J. Chrom. B* **2014**, *973*, 17–28. [CrossRef]

138. Liu, J.-R.; Yang, Y.-C.; Shi, L.-S.; Peng, C.-C. Antioxidant Properties of Royal Jelly Associated with Larval Age and Time of Harvest. *J. Agric. Food Chem.* **2008**, *56*, 11447–11452. [CrossRef]

139. Guo, H.; Kouzuma, Y.; Yonekura, M. Structures and properties of antioxidative peptides derived from royal jelly protein. *Food Chem.* **2009**, *113*, 238–245. [CrossRef]

140. Pourmoradian, S.; Mahdavi, R.; Mobasseri, M.; Faramarzi, E.; Mobasseri, M. Effects of royal jelly supplementation on glycemic control and oxidative stress factors in type 2 diabetic female: A randomized clinical trial. *Chin. J. Integr. Med.* **2014**, *20*, 347–352. [CrossRef]

141. Mohamed, A.A.-R.; Galal, A.; Elewa, Y.H. Comparative protective effects of royal jelly and cod liver oil against neurotoxic impact of tartrazine on male rat pups brain. *Acta Histochem.* **2015**, *117*, 649–658. [CrossRef] [PubMed]

142. Aslan, A.; Cemek, M.; Buyukokuroglu, M.E.; Altunbas, K.; Bas, O.; Yurumez, Y. Royal jelly can diminish secondary neuronal damage after experimental spinal cord injury in rabbits. *Food Chem. Toxicol.* **2012**, *50*, 2554–2559. [CrossRef]

143. Xue, X.; Wu, L.; Wang, K. Chemical Composition of Royal Jelly. In *Bee Products-Chemical and Biological Properties*; Springer Science and Business Media, LLC: Berlin, Germany, 2017; pp. 181–190.

144. Souto, E.B.; Silva, G.F.; Dias-Ferreira, J.; Zielinska, A.; Ventura, F.; Durazzo, A.; Lucarini, M.; Novellino, E.; Santini, A. Nanopharmaceutics: Part I—Clinical Trials Legislation and Good Manufacturing Practices (GMP) of Nanotherapeutics in the EU. *Pharmaceutics* **2020**, *12*, 146. [CrossRef] [PubMed]

145. Souto, E.B.; Silva, G.F.; Dias-Ferreira, J.; Zielinska, A.; Ventura, F.; Durazzo, A.; Lucarini, M.; Novellino, E.; Santini, A. Nanopharmaceutics: Part II—Production Scales and Clinically Compliant Production Methods. *Nanomaterials* **2020**, *10*, 455. [CrossRef]

146. Neupane, B.P.; Chaudhary, D.; Paudel, S.; Timsina, S.; Chapagain, B.; Jamarkattel, N.; Tiwari, B.R. Himalayan honey loaded iron oxide nanoparticles: Synthesis, characterization and study of antioxidant and antimicrobial activities. *Int. J. Nanomed.* **2019**, *14*, 3533–3541. [CrossRef]

147. Sarhan, W.A.; Azzazy, H.M. High concentration honey chitosan electrospun nanofibers: Biocompatibility and antibacterial effects. *Carbohydr. Polym.* **2015**, *122*, 135–143. [CrossRef] [PubMed]

148. Durazzo, A.; Lucarini, M.; Plutino, M.; Lucini, L.; Aromolo, R.; Martinelli, E.; Souto, E.B.; Santini, A.; Pignatti, G. Bee Products: A Representation of Biodiversity, Sustainability, and Health. *Life* **2021**, *11*, 970. [CrossRef]

149. De Carvalho, F.M.d.A.; Schneider, J.K.; de Jesus, C.V.F.; de Andrade, L.N.; Amaral, R.G.; David, J.M.; Krause, L.C.; Severino, P.; Soares, C.M.F.; Caramão Bastos, E.; et al. Brazilian Red Propolis: Extracts Production, Physicochemical Characterization, and Cytotoxicity Profile for Antitumor Activity. *Biomolecules* **2020**, *10*, 726. [CrossRef]
150. De Mendonça, M.A.A.; Ribeiro, A.R.S.; De Lima, A.K.; Bezerra, G.B.; Pinheiro, M.S.; De Albuquerque-Júnior, R.L.C.; Gomes, M.Z.; Padilha, F.F.; Thomazzi, S.M.; Novellino, E.; et al. Red Propolis and Its Dyslipidemic Regulator Formononetin: Evaluation of Antioxidant Activity and Gastroprotective Effects in Rat Model of Gastric Ulcer. *Nutrients* **2020**, *12*, 2951. [CrossRef] [PubMed]
151. Ali, A.M.; Kunugi, H. Propolis, Bee Honey, and Their Components Protect against Coronavirus Disease 2019 (COVID-19): A Review of In Silico, In Vitro, and Clinical Studies. *Molecules* **2021**, *26*, 1232. [CrossRef] [PubMed]

MDPI
St. Alban-Anlage 66
4052 Basel
Switzerland
Tel. +41 61 683 77 34
Fax +41 61 302 89 18
www.mdpi.com

Agriculture Editorial Office
E-mail: agriculture@mdpi.com
www.mdpi.com/journal/agriculture